Cognitive Justice in a Global World

Graven Images

Series Editor
Leonard V. Kaplan
University of Wisconsin, Madison

The Graven Images Series will publish scholarship from across the fields of law, theology, and the humanities. The Series will analyze these disparate fields to provide a foundation for justice, both within and without state borders. In addition, it will examine those conditions that foster both spiritual and material well being for the individual. The Series aims both to engage cutting-edge thought as well as to revive relevant scholarship from the past in order to add to our contemporary understanding of law, culture, and politics.

Cognitive Justice in a Global World

Prudent Knowledges for a Decent Life

Edited by
Boaventura de Sousa Santos

LEXINGTON BOOKS
A division of
ROWMAN & LITTLEFIELD PUBLISHERS, INC.
Lanham • Boulder • New York • Toronto • Plymouth, UK

LEXINGTON BOOKS

A division of Rowman & Littlefield Publishers, Inc.
A wholly owned subsidiary of The Rowman & Littlefield Publishing Group, Inc.
4501 Forbes Boulevard, Suite 200
Lanham, MD 20706

Estover Road
Plymouth PL6 7PY
United Kingdom

British Library Cataloguing in Publication Information Available

Library of Congress Cataloging-in-Publication Data

Cognitive justice in a global world : prudent knowledges for a decent life / edited by Boaventura de Sousa Santos.
p. cm. — (Graven images)
ISBN-13: 978-0-7391-2194-8 (cloth : alk. paper)
ISBN-10: 0-7391-2194-4 (cloth : alk. paper)
ISBN-13: 978-0-7391-2195-5 (pbk. : alk. paper)
ISBN-10: 0-7391-2195-2 (pbk. : alk. paper)
1. Science—Philosophy. 2. Social sciences—Philosophy. 3. Knowledge, Theory of. 4. Conduct of life. 5. Justice (Philosophy) I. Santos, Boaventura de Sousa.
Q175.C6128 2007
121—dc22 2007017719

Printed in the United States of America

∞™ The paper used in this publication meets the minimum requirements of American National Standard for Information Sciences—Permanence of Paper for Printed Library Materials, ANSI/NISO Z39.48-1992.

Contents

Preface: How Reason Lost its Balance

Stephen Toulmin

The speculative pursuit of knowledge has played a central part in human culture for 2,500 years and more. From early on, the word "philosophy" referred to the systematic and methodical treatment of any subject. In this sense, it covered the whole range of inquiries that lent themselves to systematic investigation and debate, regardless of whether the twentieth century would classify them as science and technology or not. This spectrum reached from geometry and astronomy at one pole to autobiography and historical narrative at the other. Along the way, it embraced the study of bodies in motion; organic functions and development; law and ethics, whether in individual relations or political institutions; the patterns of change in general; aesthetics, rhetoric, and oral or written literature; the choices involved in piecing together a well-lived life; even the place of those lives in the workings of nature, not least in cosmology. From the start, speculation about these matters ranged all the way from mathematicians such as Euclid, Galileo, Riemann, and Gödel, to writers of literary essays and personal reflections such as Marcus Aurelius and Boethius, Montaigne and Santayana.

This collection of chapters continues this significant tradition. Santos has brought together a wide variety of authors dealing with a wide variety of topics, all coherently brought together by the fact that all its chapters engage in dialogue, either directly or indirectly, with the first chapter, that he authored. An indication of the solidity and cogency of the ideas is the fact that two or even more authors present them simultaneously. It is, indeed, important to note that, as early as the 1980s and 1990s, both Boaventura de Sousa Santos and I were concerned with avoiding the limitations of the social and natural sciences, which adopted a purely "theoretical" approach.

Actually, in some respects, Santos was ahead of me. He published his "A Discourse on the Sciences" (chapter 1) for the first time in 1987, when I was still planning my *Cosmopolis*.

In all these human activities "reasons" play *a* central part. They may be occasioned by particular events, the specific aims of individual actions, the goals of social policy, the factors responsible for successes or failures, the biological and physical causes of effects or phenomena, the striking features of an art object, the style or delivery of a speech, and a dozen other things. And, for more than two thousand years, all such activities were given equal consideration. No field of investigation or speculation was dismissed as intrinsically unphilosophical. A few, like astrology, might prove to be ineffective, but that was another matter.

From the mid-seventeenth century on, however, an imbalance began to develop. Certain methods of inquiry and subjects were seen as philosophically serious or "rational" in a way that others were not. As a result, authority came to attach particularly to scientific and technical inquiries that put those methods to use. Instead of a free-for-all of ideas and speculations—a competition for attention across all realms of inquiry—there was a hierarchy of prestige, so that investigations and activities were ordered with an eye to certain intellectual demands. Beside the *rationality* of astronomy and geometry, the *reasonableness* of narratives came to seem a soft-centered notion, lacking a solid basis in philosophical theory, let alone substantive scientific support. Issues of formal consistency and deductive proof thus came to have a special prestige, and achieved a kind *of certainty* that other kinds of opinions could never claim. So, as time went on, academic philosophers came to view literary authors like Michel de Montaigne—an essayist who had little use for "disciplines" and put equally little reliance on formal logic—as not being philosophers at all, let alone scientists.

It had not always been so. In mapping the reach of philosophy and human reason, the contrast between the reasonable and the rational is only one of half a dozen differences in our methods of inquiry. The contrast between the reasonableness of narratives and the rigor of formal proofs, between autobiography and geometry, is the contrast between the "soundness" of substantive *argumentation,* which has the body and force needed to carry conviction, and the "validity" of formal *arguments,* whose conclusions are determined by the starting points from which they are deduced.

Substantive arguments occur in specific situations. I am using the words *situation* and *situational* in place of the more familiar words *context* and *contextual.* It is misleading to suggest that the situation in which an argument takes place is a larger "text" within which the argument is a "subtext." Of course, a text can derive its meaning from the larger text from which it is taken. A couplet may be part of a complete poem, a chapter may describe a single episode within a novel, and so on. But situations can influence actions even before being described in human language, so the action and the situation are not related in the way that shorter text is related to the longer text from which it comes.

In analyzing arguments, the relation between acting rationally or irrationally, on the one hand, and reasonably or unreasonably, on the other, must be our primary concern. If we concentrate our attention exclusively on the propositions that figure in an argument, while ignoring the situation in which it is presented, we can be described as viewing the argument from the strict standpoint of rationality. If we focus exclusively on the devices that make an argument persuasive, by contrast, the best that can be said about a case is that we present it as reasonably as we can: only if we can balance concern for the substance of an argument with a style that is convincing but not too pressing can we be credited with a reasonableness that combines intellectual force in content with a moderation of manner. Rationality goes with focusing narrowly on matters of content, reasonableness feeling for the dozen ways in which a situation may modify both the content and the style of arguments.

Widening our attention, a capacity for rationality also tends to be inborn rather than acquired; and the same is true of the tendency to act "irrationally." A phobia toward snakes, for instance, means that a glimpse of a snake—even the picture of a snake—may provoke an instant and uncontrollable reaction: of freezing, fainting, or screaming. Such a reaction we typically describe as *irrational*, certainly not as *unreasonable*. Conversely, a habit of acting in ways that needlessly offend your friends, or regularly arguing in sloppy and unfounded terms, typically results from not having learned to pay reasonable attention to your own conduct: this is an acquired failing, not a mark of inborn irrationality. Yet a rigid contrast can rarely be insisted on; and it is a measure of the difficulty of preserving the balance of reason that the war of logic against rhetoric continued for so long after 1600.

The switch from philosophical egalitarianism to the scientific hierarchy that began in the seventeenth century did not at first exclude entirely from philosophy those essayists who set out to illuminate key features of our lives rather than trying to prove theoretical truths, and who invited us to share reflections rather than checking deductions. This change took place only gradually over three or four centuries. Before 1620, indeed, one of the most widely admired philosophers was the pioneer essayist, Michel de Montaigne. *Though* his views were quite unlike those of Bertrand Russell, Montaigne had—especially in the France of the 1580s and 1590s—the kind of reputation in educated circles that Russell had in early twentieth-century Britain. He set out, above all, to write about aspects of daily experience that all human beings can recognize. Like his successor, Francis Bacon, Montaigne wrote everyday, colloquial essays about friendship, books, tragedy, suicide, sexuality, and—finally—experience in general. His position was not a theoretical one, and he was deeply suspicious of writers who relied on abstract theories to undermine the truth of our common experience.

Montaigne himself was more a lawyer than a scientist, and he was disinclined to put much trust even in physicians: like a late twentieth-century believer in alternative medicine, he relied, first of all, on the wisdom of his own body to take care of accidental injuries and medical misadventures, and restore him to health. Like Socrates, he did not trust any single system of natural philosophy, knowing too well how the pre-Socratic philosophers told alternative stories about the world of nature, without ever arriving at a well-grounded choice among them. This skepticism about theory attracted a following in the late sixteenth century not just in France, but in other countries as well; and it was a challenge to younger thinkers and writers, who viewed the world differently, but recognized the need to meet Montaigne's challenge. The two most distinguished mathematical philosophers of early seventeenth-century France—René Descartes and Blaise Pascal—were both deeply familiar with the *Essais*, and looked for formal arguments to counter the things they considered most corrosive in Montaigne's position. Still, their writings did not display any sign of doubt that Montaigne was a true philosopher, even if a misguided one; rather, they shaped their own arguments in terms designed to undermine the skeptical, or even cynical, risks and doubts to which, in their view, Montaigne exposed his readers.

Among modern-day English-speaking academics, Montaigne is regarded as a figure in literature rather than in serious philosophy. But that is not at all how he was seen in his own time. It is anachronistic to think of European thought as having developed quite discontinuously, with the seventeenth century's exact science rapidly displacing sixteenth-century humanism; nor was the reputation of these humanists as philosophers entirely forgotten. In the late nineteenth century, for instance, Montaigne was admired by Emerson in America and by Nietzsche in Germany. But in the twentieth century, notably in America, academic philosophy became a narrowly technical subject concerned with abstract theories, and the concrete concerns of autobiographical authors like Montaigne were generally ignored as being intrinsically unphilosophical.

Theory acquired the dominant role, and practice came to be spoken of as merely a way of applying theory. Even in clinical medicine, where a fresh concern with practical, concrete, and particular issues has arisen beginning in the 1960s, prestige still tends to attach to academic studies that focus on abstract, universal theories. Thus we see such a distinguished writer as George Santayana following Montaigne into the realm of philosophical nonpersons.

If philosophers had so little to say after the mid-seventeenth century about the contrast between rationality and reasonableness, historians of philosophy have not done much better. One topic above all captures the core difference between the rival views of reason. The analysis of theoretical arguments in terms of abstract concepts, and the insistence on explanations in terms of universal laws—with formal, general, timeless, context-free, and value-neutral arguments—is nowadays the business of logic; the study of factual narratives about particular objects or situations, in the form of substantive, timely, local, situation-dependent, and ethically loaded argumentation, is at its best a matter for rhetoric, academic philosophers and serious-minded theorists in any field are concerned only with the first: the contrast between convincing and unconvincing, neatly phrased and clumsily argued argumentation is left to literary students of elocution or style. For much of the last three hundred years, and most of the twentieth century, scholars treated these investigations not merely as *distinct*, but as *separate*.

If there is anything solid to this distinction between logical analysis and rhetorical power, genuine reasoning and mere persuasion, however, how realistically or universally can it be applied? Consider all the "theories" we appeal to in one situation or another: how far are these framed in ways that are indifferent to who is presenting them to whom, where and when they are invoked, how they are presented, and so on—in a word, how they are "embodied" in human lives? Given the varied kinds of facts we observe and report in one situation or another, again, how far can these be described in terms of context-free and timeless concepts?

Presumably, a world might exist in which the relationship between language and reality let anything we say about the-world-as-we-find-it be articulated in "disembedded" terms; equally, a world might exist in which all our theoretical language could be read in strictly ahistorical and context-neutral ways. In that world, the contrast between rhetoric and logic—between the substantive appraisal of argumentation and the formal analysis of arguments—could indeed be treated as absolute, and the two resulting kinds of knowledge could be kept separate. Still, any assumption that the world-as-we-know-it is wholly or even mainly of that kind proves on examination implausible. In a word, all kinds of speech and language are more or less *situated* or *embedded* in their occasion of use, and, if abstracted from that occasion, they can in theory be more or less "desituated" or "desimbedded." Some familiar uses of language are more, some less situated or desituated; none of them are written or spoken alone.

Rather than logic and rhetoric being rivals, offering competing recipes for judging the merits or defects in our reasoning, the considerations they focus on are complementary. Formal validity is one thing. If the statements making up our *arguments* are intellectually inconsistent or claim "necessity" for inferences that later prove self-contradictory, we must ask ourselves whether they were meaningful in the first place. Substantive soundness is something quite different. Once we understand what is at issue in any piece of *argumentation*, we can ask what data its claims rest on, how solidly those data support it, and how far the resulting claims carry conviction; but only a consistent, meaningfully stated argument can have substantive strengths or weaknesses. Conversely, a conclusion presented with substantive argumentation may be beyond doubt or may be wholly ungrounded. But the strength or weakness of an argument takes its intelligibility for granted: it in no way guarantees its formal rigor or consistency. As it stands, the world of formal validity, meaningfulness, and consistency barely overlaps the world of substantive evidence, testimony, or conviction, and the statements we advance as "logical reasons" for our conclusions are only the abstract skeletons of "reasoning" that may or may not, here and now, succeed in convincing its audience.

In our time, as in Classical Greece, the tasks of reason thus fall along a spectrum. The general concepts in which we articulate our ideas and beliefs have formal implications, and it is the task of theoretical analysis to sort out and elucidate them. But, by itself, such a theoretical analysis does not tell us in what situations—how, where, or when—everyday life and practice exemplify those ideas. By contrast, the objects or situations we have occasion to notice and investigate are exposed to cultural variations and historical changes, and it is the business of empirical inquiry to explore and throw light on those vicissitudes. But, by themselves, inquiries on an empirical level do nothing to indicate what theories we can best appeal to in order to explain those vicissitudes.

So matters stood in classical Greece two and a half millennia ago, and so they still stood four hundred years ago in the intellectual and practical world of sixteenth-century humanism. Our first question here must therefore be, "Why did our ideas change after the year 1600, and what led our modes of thought to move off in such a different new direction?"

In addition, however, the seventeenth century saw the birth of the modern dream of a purely formal language, which did all the things that everyday, colloquial language could do, but "perfectly"; and the complementary dream, of a formal theory whose intellectual merits outweighed all the products of everyday experience. What lay behind this shift in our views of language and rationality, and lent a pervasive charm to the dream of formal theories and exact languages?

These questions are at the core of the epistemological, sociological and philosophical inquiries contained in this book. The answers proposed coincide largely with the orientations I suggested above. A rigorous analysis of the scientific method and its underlying presuppositions clearly shows to what extent rationality and reasonableness are closely linked. Herein lies the possibility, so insistently demanded in our time, of not separating the pursuit of truth from the pursuit of good. Hence the key to the book's subtitle: *Prudent Knowledges for a Decent Life*. In this book, however, this problematics is raised to a new, higher level, one which will probably define the new epistemological and philosophical concerns of the twenty-first century. I mean the question of cognitive justice in a globalized world, so eloquently emphasized in the title of this book. This question has to do with the coexistence of many knowledges in the world and the relation between the abstract hierarchies which constitute them and the unequal economic and political power relations which produce and reproduce increasingly more severe social injustice. For all these reasons, this book must be read not only by specialists of the disciplines it deals with, but for all those for whom the struggle for a better world is the struggle as well for a different kind of philosophy, epistemology and science.

Introduction

Boaventura de Sousa Santos

In the United States, at least, all seems to be quiet on the science wars front. We might even say that the most recent bout is gone and the science wars are now over. They came to a mysterious end, without truce or surrender treaty. Such mystery is only comparable to the break of the wars itself. But perhaps the explanation for this mystery is simpler than appears at first sight. Although they were fought among scientists, these wars were not scientific; they were political. They lasted for as long as did the enthusiasm for political debate that mobilized the scientists. However, since the scientists are professional scientists, rather than professional politicians, the mobilization was short-lived, and the scientists soon were back at work. Probably nobody convinced anybody else, and they all went about their business as usual.

This is not to say that the science wars had no impact. Of course they did, and, indeed, way beyond their obvious impact, whether positive or negative, on the prestige and professional careers of their main protagonists. The impact I would like to stress here, however, is of a different order. It concerns a deepening of the individual and collective reflexivity in scientific practice and the opportunities emerging thereby. The science wars allowed their protagonists and their readers to make explicit and claim their own assumptions and beliefs concerning scientific knowledge: truth or discourse of truth; realism or constructivism; universalism or relativism; rationality or reasonableness; general or contextual validity; mono- or multiculturalism; global or local knowledge; and adjective-less representation of reality, on the one side, and Eurocentrism, male chauvinism or racism, on the other. In this process, it became clear that the diversity of perspectives did not exist only between the community of natural scientists, on the one hand, and the community of social scientists, on the other, but also in the very midst of either community. It became clear, furthermore, that epistemological differences were articulated in a complex way with cultural and political differences, and presupposed different kinds of relations between scientific knowledge and other forms of knowledge. This new transparency made possible the opportunity, if not for more mutual tolerance or even mutual openness, at least for a greater awareness of the existent plurality of perspectives. It also created the opportunity for strengthening the relations between the natural and the social sciences with more epistemological affinities, and for opening up new transnational dialogues between scientists and other producers of knowledge.

This book starts from the idea that these opportunities have not been duly explored so far, and aims to contribute to this goal. It gathers together social scientists, natural scientists, and philosophers from different countries, all of them sharing John Dewey's idea that modern science becomes more useful to the extent that its limits are acknowledged and we accept that it is not the only form of knowledge. They also share Stephen Toulmin's idea that the rationality of modern science does not reside only, or indeed fundamentally, in its mathematical formalization, but rather in its reasonableness, that is to say, in its capacity to generate reasonable arguments about the world and the possibilities of its transformation. Moreover, the scientists who contribute to this book tend to privilege, each one in his or her own way, the perspective of science in action, its practice, and the consequences of its practice. They do not cling, therefore, to rigid disciplinary distinctions, either within the natural sciences or the social sciences, or as regards the relation between the two. They rather believe that the possibilities of mutual enrichment should be fully explored, regardless of orthodoxies that often are no more than the sediment of routine inertia. Finally, these scientists are committed to the world and to its transformation into a better, more just world, and they are aware of the fact that the social injustice that is rampant in the world today is in part the result of the cognitive injustice that a narrow and imperial conception of modern science has produced and legitimized.

In a word, the authors of this collection of chapters recognize that the idea of a reasonable and prudent knowledge (Aristotle's *phronesis*) must be present in the scientists' practical work, a work that is done in the world and must therefore be accomplished in solidarity with the world. As its title indicates, the main purpose of this book is to contribute to the construction of a prudent knowledge for a decent life.

The coherence of this book lies thus in the convergences that bring together the different scientific perspectives gathered here from such diverse disciplines as philosophy, physics, sociology, anthropology, biology, ecology, history of science, cultural studies, and economics. But one other factor gives coherence to this book as well. All the authors were invited to comment on the chapter signed by myself that opens this collection. They were urged to articulate critically their own theoretical, methodological, and epistemological reflections with the positions maintained in my chapter. My chapter, entitled, after Jean Jacques Rousseau, "A Discourse on the Sciences," was written before the most recent science wars but ended up getting involved in them against my will. It offers an epistemological critique of the dominant paradigm of modern science and broadly draws the outline of a possible new emergent paradigm. I approached a multidisciplinary and multinational group of scientists and asked them to reflect on the themes dealt with in my chapter, as well as on others that subsequently came to dominate the epistemological and sociological debates concerning science and other forms of knowledge. Contributors are scientists of different generations and from various countries, EUA, Portugal, Brazil, Germany, Belgium, India, Mozambique, Colombia. They are all internationally recognized in their respective areas of specialization.

The book opens with my chapter, which serves as catalyzer or attractor for the other chapters, and is divided into three parts.

The first part—titled *Neither Truce nor Surrender: Are the Science Wars Over?*—offers an evaluation of the science wars. In chapter 2, João Arriscado Nunes analyzes the sociological and epistemological context that gave rise to *A Discourse on the Sciences,* chapter 1 in this volume. He then goes over its arguments in the light of the scientific development and debates of the following two decades, taking into consideration as well my subsequent epistemological reflection, which became more and more concerned with issues of diversity of forms of knowledge and the social conditions of its production, appropriation and circulation. Arriscado Nunes specifically deals with the four prospective theses presented in *A Discourse*: all natural-scientific knowledge is social-scientific; all knowledge is self-knowledge; all knowledge is local and total; all scientific knowledge aims to turn itself into common sense. He brings in a series of new arguments that confirm the theses, and gives acount of the evolution of epistemological debates following the increasing protoganism of the life sciences. In particular, he highlights the way in which reductionism was overcome by complexity and the related processes of interdisciplinary crossing.

In chapter 3, Richard Lee engages in a detailed analysis of the most recent debates on science. He shows how our understanding of the science wars, particularly as they occurred in the United States, requires their contextualization within the broader realm of the cultural and political wars. What was at stake in the controversies was, on the one hand, the dominant cultural canon in the American society and the threat it faced in the emergence of cultural studies and the claims of multiculturalism; and, on the other hand, the struggle for the control of educational policies, namely university reform. At a deeper level, what is and will ever be at stake is the disappearance of the gap between the two cultures, that is to say, the separation between scientific and humanistic culture, between the study of facts and the study of values, and the challenge posed by such disappearance to the structures of production and distribution of knowledge.

In chapter 4, Peter Wagner begins by wondering about the contexts that gave rise to the science wars. Based on his sound knowledge of the history, philosophy, and politics of scientific knowledge, Wagner offers a new reading of the current debates on science, viewing them as the unfolding of a tension that is inherent to the relation between knowledge and the world. In the case of science, he argues, such a tension results from an internal contradiction between its alleged neutrality and detachment from the world, and its desired legitimacy by reason of its usefulness and efficaciousness as regards the transformation of the world. He shrewdly stresses John Dewey's philosophy of science as one of the sources of my own epistemological thinking, taking into account some of my more recent work.

Chapter 5, penned by Joan Fujimura, aims to contextualize the most recent episode in the science wars in a field of theoretical and epistemological struggles that is temporally and thematically broad, made up of a number of instances of confrontation around proposals deemed to be transgressive. Fujimura tries to enlighten the most recent debates by resorting to the history of science. The "case" analyzed by Fujimura concerns the controversy about the emergence of non-Euclidean geometry in the middle of the nineteenth century. Fujimura's intention is to show that the science wars are neither fought around the opposition between science and anti-science, nor around the opposition between objectivity and subjectivity. They are rather fought around the scientific authority to decide which kind of science must be undertaken. After a succinct, yet illuminating analysis of the *Sokal affair*, Fujimura lingers on Sokal's parody of the historicity of pi (π) in his *Social Text* hoax, and goes on to demonstrate that the value of pi (π), contrary to Sokal's thinking, does indeed have a history, a history that is recorded in the history of mathematics. To show this much is the purpose of her analysis of the controversy around non-Euclidean geometry and the fierce fights for the mathematical canon that occurred inside it. Her analysis highlights the importance of the historical perspective in contemporary methodological and epistemological debates.

In chapter 6, Immanuel Wallerstein offers to analyze the science wars in the light of the *longue durée* of the modern world system, of which he is the major theorizer. He shows how science gradually divorced philosophy to gain a dominant position in the hierarchy of knowledge, and questions, in the light of contemporary events, the separation that in this process occurred between the search for the truth and for the good. Invoking the work undergone by the Gulbenkian Commission on the Social Sciences, which he chaired, and the book where the main results were published (*To Open the Social Sciences. Gulbenkian Foundation Report on the Restructuring the Social Sciences* 1996), Wallerstein mentions the emergence of the three cultures that have shaped our knowledge for the past one hundred years: the natural sciences, the social sciences, and the humanities. He then lingers on the challenges brought to this tripartite division by the complexity studies, on the one hand, and the cultural studies, on the other.

In chapter 7, Isabelle Stengers starts out by discussing my assertion that there are no scientific reasons for the scientific explanation to have priority over other, alternative explanations (*A Discourse on the Sciences*, p. 13 in this volume) and goes on to question what she designates as "the great divide"—between science and philosophy, between facts and values—that has been the hallmark of Western knowledge. According to Stengers, the "knowledge that counts" in a given community is a hybrid of facts and values. Her example is the claim of the autonomy of science as condition of public usefulness. She resorts to Bruno Latour and his distinction between humans and nonhumans—to replace the distinction between subject and object—to justify the role of the social sciences in showing how the "knowledge that counts" is produced, the importance of entrepreneurs, and the networks of scientists and nonscientists. She concludes that the decision on the "knowledge that counts" is a highly political question, hence hostile to totalizing thought and *a priori* privileged perspectives. Following Bruno Latour, Stengers advances the concept of political ecology to signify a radical demand of democratization in this respect. While not naming it a "cultural revolution," Stengers shows nonetheless that such a demand breaks with the conceptions based on the "great divide."

In chapter 8 Frank Summers and Leonard Kaplan analyze the epistemological debates around psychoanalysis. The scientific character of psychoanalysis and its clinical application has been questioned all along and the last part of the twentieth century has even seen psychoanalysis dismissed as a brilliant fraud on the public discourse. The critiques of psychoanalysis as a science come from two different camps. On one side, the positivists and the objectivists approach psychoanalysis from a Popperian perspective. They take the view that the clinical data are too diffuse and speculative to allow for the kind of falsifiability that Popper holds necessary for science. On the other side, relativists follow the ideographic clinical model which values the particularity of each clinical situation and claims that no one story emergent from the therapeutic alliance is necessarily valid. Summers and Kaplan dialogue with both these stances, show their inherent contradictions and argue that a third way offers a greater opening for understanding the nature of the self in contemporary history, politics and culture. They plead for a different model of science, a human science model, a model not predicated in the science of the physical, and more adequate to grasp the specificity of the theoretical object implied in humanity's efforts to understand itself. Their interrogation invokes the best of phenomenology with reference to the hermeneutic enterprise of Dilthey, Husserl, Heidegger, Ricouer and Gadamer.

The second part—titled *Complex, Creative and Situated Interrogations: Science in Action*—focuses on science in action. Drawing on illustrations from various disciplines, this part tries to show the complexity and creativity of the scientific work as it confronts the complexity and creativity of the world.

In chapter 9, Hugh Lacey refers to the Rousseau question mentioned in *A Discourse on the Sciences* (are science and virtue in any way related?) in order to answer a similar question: How can science promote the welfare of human beings? Lacey begins by interrogating the relation between science and values, by highlighting the impartiality, autonomy, and neutrality of science. He then identifies three decisive moments of scientific activity: definition of priorities and research appropriate methodologies; assessment of theories or hypotheses; application of scientific knowledge. He presents a model of scientific activity based on the concept of strategy, through which the possibilities that can be explored in a given research project are made specific. He distinguishes between materialist and alternative strategies. The former are overwhelmingly dominant by converging with specifically modern values, privileging the control of natural objects. In his terminology, impartiality does not imply neutrality. In his conclusion, Lacey provides a critical assessment of *A Discourse on the Sciences*, presenting as well a set of theses on the relations between the social and the natural sciences.

In chapter 10, Olival Freire Júnior deals with Eugene Wigner's ideas on the measurement procedure in quantum physics, mentioned in chapter 1. Based on a profound knowledge of Wigner's positions and the controversy they generated, Olival Freire aims to contribute to a realist image of science, an image that may bring about a better engagement between science and society and counter irrationalist tendencies. The author lingers on the controversy that opposed Wigner to the interpretation of complementarity held by the School of Copenhagen, and describes how Wigner and his followers broke away from a consensus that seemed indestructible. Olival Freire's major thesis is that, although Wigner's conjectures on the role of consciousness in physical phenomena were not fruitful and were discarded—being today part of the history of physics and not of physics—, by formulating them against the "Copenhagen monocracy" on complementarity, Wigner gave origin to an important theoretical, experimental and philosophical problematic with significant repercussions in the last decades. Neglecting this and judging Wigner by what we know today but was not known at the time in which he formulated his ideas is to incur into the error of anachronism. The author concludes by stressing the importance of the history of science to minimize anachronism and provide a broad understanding of science, in which the scientificity of history itself may play a role.

Anna Carolina Regner is the author of chapter 11. The author calls for a broader conception of rationality, based on a new outlook on the relations between rhetoric and science, in which the distinction between scientific and rhetorical reasoning collapses. Anna Regner begins by wondering if the traditional view of scientific rationality—based on the strict distinction between demonstrations that lead to necessary and universal truths and persuasive and contextual arguments that lead to verisimilar conclusions—is prepared to account for the issues raised by the most recent findings of science. Her answer is negative and grounded on the analysis of an "exemplary case" of scientific reasoning, Charles Darwin's *Origin of the Species*. In the light of Aristotle's *Rhetoric*, she reconstructs Darwin's reasoning strategies systematically. In Darwin's narrative, observation and experimentation, submission of facts to rules, and the study of exemplary cases go hand in hand with the use of the imagination, the recourse to metaphor and analogy, the invocation of the weight of reasons, the appeal to ignorance and to mental habits, and, in general, to the sociological and psychological values and conditions of scientific research. The systematic juxtaposition of Darwin's reasoning with Aristotle's *Rhetoric* leads Regner to conclude that Darwin's reasoning strategies are "rhetorical." Far from weakening scientific discourse, rhetorical tools constitute it as well. Recognizing this much widens the conception of rationality and opens new possibilities for rational discourse.

Chapter 12, by João Ramalho-Santos, explicitly adopts the point of view of science as and when it is done in practice, in order to capture the complex processes of scientific creation and the controversies it generates. The problems discussed in this chapter show clearly why, in epistemological and sociological debates on science, biology (particularly reproduction and development biology) has since the 1990s taken over the place that in previous decades was occupied by physics. The author begins with a narrative of reproduction, of its incidents and difficulties, and the many techniques of assisted reproduction developed to overcome them. The polemic caused by these techniques is then analyzed alongside other polemics in other scientific areas, on the basis of the author's central thesis that the reconstruction of scientific transformations and the controversies they provoke must concern the moment in which they occurred and in the light of the data available at the time. To analyze them in the light of the results to which they led, or the way in which, in the long run, the mechanisms of scientific validation worked, eliminates the complexity of the processes of scientific creation and is of little use to scientists. Scientists live and work today, with the data and techniques available today, always on the frontier of what is today known and unknown. Like Olival Freire, João Ramalho-Santos warns against the error of anachronism. If anachronism is avoided, he suggests, it may be possible to have controversies beyond the science wars and rather truly rewarding for the scientists and the citizens in general.

In the third part—titled *Global Cognitive Justice: Reconstructing Knowledges and World Making*—modern science is seen as part of a cultural, political, and economic project that tried to consolidate the hegemony of the Eurocentric and Northcentric model of society on a global scale. Largely on the basis of a pragmatist perspective, the relationship between science and social transformation is considered in various thematic areas: nonprofessional or nonscientific knowledge and its articulation with science in social movements and in the management of complex ecological systems; the colonial character of modern capitalist societies and the geopolitics of scientific knowledge; the hegemony of Northcentric scientific research priorities and the proletarianization of scientists in the South; the ideological limitations imposed upon the theoretical possibilities of economic science, mainly by strictly separating between the question of justice and the question of efficiency and by focusing exclusively on the latter; the production of contextual scientific alternatives in the global South, combining scientific knowledge and technology with other knowledges and alternative technologies.

Arturo Escobar is the author of chapter 13. He begins by assessing the tendencies analyzed and the theses proposed in chapter 1 in the light of the epistemological and political developments of the two last decades, maintaining that they have been largely confirmed. He makes exception for the case of the relations between the social sciences and the natural sciences, in which he sees today the withdrawal of the former vis-à-vis the advance of the latter, contrary to what was foreseen in *A Discourse on the Sciences*. He focus on the first and fourth theses that characterize the emergent paradigm ("all natural-scientific knowledge is social-scientific" and "all scientific knowledge aims to constitute itself as common sense") in order to analyze, in their light, the production of knowledge carried out by the social movements that for the past decade have been contesting neoliberal globalization, and confronting it with the idea of an alternative form of globalization, more attentive to social justice. He resorts to the complexity and self-organization theories, developed in biology, to characterize these movements, which he conceives of as horizontal organizations structured like meshwork, whose mesh is made of heterogeneous elements and often goes in unforeseen directions, according to the interactions that occur and the catalysers that become involved. According to Escobar, the movements are producers of knowledge and, indeed, of a new common sense, whose epistemological impact in the paradigmatic transition must be acknowledged. He illustrates his thesis with the black communities of the Colombian Pacific. Since the 1990s, these communities have been producing an autonomous epistemological and cultural knowledge on the basis of the idea of the territory viewed as a political and cultural construction, capable of sustaining an adequate defense of the natural resources and biodiversity.

Peter Taylor is the author of chapter 14. He narrates his intellectual trajectory since the 1970s, when he decided to devote himself to ecology as a science. Throughout his chapter, Taylor explains how he developed a critical consciousness and a reflective stance concerning the concepts and models used in his scientific discipline to define ecological systems and their changes. Crucial for this development was the idea of the complexity of such systems and the difficulty or even impossibility of establishing clear borders between the factors considered relevant for the definition of systems and the underlying processes. At the same time, by reflecting on the developments occurring in ecology and on his own scientific practice, Taylor reached the conclusion that there is a tension (not necessarily a negative one) between science's claim of self-referentiality and the impact of the research conditions and the scientists' social concerns about the content of scientific knowledge (what he calls the sociality of science). Gradually, Taylor realized that sociality may become a resource to win new forms of knowledge. This process he designates as heterogeneous construction. He thus grasped the intersecting, complex processes present in environmental dynamics, which are hard to research (let alone communicate) and not susceptible of being accounted for by a general theory. In his conclusion, Taylor explains how he reached his idea of the "participatory planning" and "flexible engagement" of the scientist, on the basis of the kind of reflection that leads the scientist to change (reflectively) the ecological systems on which he/she produces knowledge.

Chapter 15 is penned by Francisco Louçã. The author proposes the reinvention of critical theory in economics. His aim is to give back to the science of economics the commitment to progressive social change that it used to have, but which meanwhile was "mysteriously" lost. Louçã's is an eloquent analysis of the evolution of economics during the past one hundred years. He shows that amelioration of society (by combating unemployment, recession, and inequality), as an objective shared by many of the economists that played major roles in the most important theoretical and methodological changes, ended up being boycotted or neutralized by the economists' own presuppositions, by their giving way to the dominant orthodoxy in order to maximize their capacity to influence political decisions. Social objectives led to the adoption of physics' principles by economics, hence to the latter's naturalization and formalization. Such naturalization, however, did not go along with the changes meanwhile occurred in the natural sciences. Louçã analyzes the successive generations of mathematical economics and the trajectories that led to the relinquishing of the critical stance, even in spite of the authors' original intentions. According to Louçã, since the crisis of the paradigm is obvious, economics must retrieve its call as a critical social science and prepare itself to get to know the reality, and not the abstractions into which, for the neoclassical economists, the agents and the market were transformed.

Chapter 16, by Shiv Visvanathan, deals specifically with the topic of global cognitive injustice. The author analyzes the impact of hegemonic science and the decontextualized knowledge on the countries of the South, in the case in point, in India. According to Visvanathan, the science wars in the North are of a ludic nature and mustn't be taken too seriously, after all, since they occur between scientists and philosophers (even if they present themselves as scientists). There is, however, another and far more destructive war, the war concerning the relationship between science and development. In this other war, the epistemological debates determine the life opportunities offered or refused to vast populations of the world. Giving India as example, Visvanathan shows the destruction that may result from conceptions of science that do not respect the different forms of knowledge of the populations, their memories and aspirations, their spaces and times, and, above all, their right to a voice and to democratic participation. He then goes on to analyze some of the movements that emerged to promote alternative conceptions of science, contextualized epistemologies based on a deep understanding of the idea of democratization of scientific knowledge. According to Visvanathan, the experience of these movements shows that acknowledging nonscientific forms of knowledge (i.e., peasant knowledge, so useful today for biotechnology) must not be interpreted as an antiscience stance. It contributes, rather, to render plural and diverse our conceptions of knowledge and science.

Maria Paula Meneses is the author of chapter 17. The core topic of this chapter is, once again, the global cognitive injustice derived from the geopolitical hierarchy that, on one side, opposes modern science to local forms of knowledge and, on the other, grounds the options in science policy. Other and related hierarchies are also duly taken into account: the hierarchy between the North and the South, the developed and the underdeveloped, the donor and the recipient of international philanthropy. Like Mignolo (chapter 18), Meneses considers that the origin of such hierarchies lies in the coloniality of power, here analyzed in the African, specifically Mozambican context. Using as examples medicine and biology, Meneses shows how modern science imposes Eurocentric patterns (of health, for instance), from which the equivalence of the distinctions traditional/modern and savage/civilized is derived. The coloniality of scientific knowledge consists in conceiving of the North as having knowledge and solutions, the South as having information and problems. This coloniality of knowledge, she argues, ends up affecting the Mozambican elites, as witness the bill on science and technology presented by the Mozambican government. To illustrate her argument, the author resorts to consultancy projects, which she considers to be an extreme case of decontextualized knowledge, since it ignores or renders exotic the local practices and forms of knowledge. As the author identifies the problems she herself encountered as a consultant, the chapter gains a self-reflective character. She concludes by defending the idea of the plurality of knowledges based on the multisituationality of its creation.

Chapter 18, by Walter Mignolo, deals with the global cognitive injustice produced since the seventeenth century by modern science, conceived of as an epistemic and political constellation constituted of modernity and coloniality of power. Mignolo proposes a new conception of modernity as the visible side of an historical process that cannot be understood without its other side, which until recently has been invisible. This invisible side is the colonial difference, a difference in power as well as an epistemic and ontological difference. Such a conception leads Mignolo to the conclusion that the seventeenth-century scientific revolution is less a rupture than a continuity of the theology of the previous century, since the ignorance or marginalization of other forms of knowledge worldwide is common to both paradigms. According to him, modernity must be criticized and sublated from the point of view of colonial difference, not the other way around. In colonial difference resides the double discourse of modernity, proclaiming at once the advance and the subjugation of humanity. While agreeing with my characterization of the emergent paradigm as a "prudent knowledge for a decent life," Mignolo thinks that its horizon is not postmodernity but rather transmodernity (Enrique Dussel), not universality but rather multiversity of knowledge and understanding. In his conclusion, he takes account of the contribution of "feminist epistemologies" and "ethno-racial epistemologies" to this paradigm since the 1970s.

Chapter 19 is by Orlando Fals Borda and Luis E. Mora-Osejo. Among the chapters in this book, this one is unique for being coauthored by a social scientist and a natural scientist, and for taking the form of a manifesto. Fals Borda is internationally known for the methodology he developed—participatory action-research—to overcome the discriminatory dichotomies that, to his mind, characterize Eurocentric knowledge: the distinction between scientific and grassroots knowledge, and between the scientist and the engaged citizen. In this manifesto, the authors take for granted that all knowledge is contextual and, therefore, that the modern science—both social and natural—that has been developed in the European and North Atlantic countries reflects the realities of those regions and the way in which they relate to all the other regions of the world. Eurocentric science has thus a doubly negative effect on the latter regions: it reinforces the hierarchy between developed and less developed countries, as well as the relations of internal colonialism, and prevents the construction of a form of scientific knowledge embedded in the realities of those regions. To characterize the geographical, historical and cultural differences among the various regions of the world, the authors resort to the geo-climatic distinction between tropical and temperate zones. Modern science, created in the latter zones, ignores the complexity and fragility of tropical regions, their ecosystems, their rich biodiversity, their multiethnic and multicultural communities, their peculiar forms of articulation between nature and culture. Without denying the importance of modern science, Fals Borda and Mora-Osejo propose the development of a contextualized science, an endogenous scientific paradigm, a sum of knowledges capable of valorizing popular knowledge and grounding the sustainable development of tropical regions. In their opinion, there is no other way to overcome global cognitive injustice and found new and balanced alliances between scientists of the North and the South.

Chapter 20 is a concluding chapter concerned with the limits of representation and the resulting misrepresentation of consequences in modern science. I approach such limits using as guides intellectual domains most concerned with the representation of limits, from cartography, astronomy and archeology to photography and painting. I implicitly acknowledge what Escobar refers to in his chapter: that the hopes expressed in chapter 1 of the increasingly epistemological role of the antipositivist social sciences in attracting the natural sciences did not come true, at least with the vigor that I had anticipated. I focus, thus, on the discipline of the social sciences in which positivist epistemology is most deeply entrenched and which has most radically converted Newtonian physics into the only valid scientific model. I am referring to conventional economics, approaching it from a different perspective from but convergent with Louçã's. The combination of specially strict and arbitrary limits of representation with an arrogant denial of them leads to what I call an "epistemology of blindness," whose practice results in massive misrepresentations of consequences. Such an epistemology is at the core of a form of knowing that reaches knowledge through a trajectory leading from a point of ignorance conceived as chaos to a point of knowledge conceived as order; this is what I call knowledge-as-regulation. Herein lies the epistemological foundation of the imperial order that the developed North imposes on the global South. To overcome this global social injustice and the global cognitive injustice that grounds and reinforces it, it is necessary to strive for an alternative epistemology inductive of nondestructive relations among different kinds of knowledges, that is, of ecologies of knowledges. Only through ecologies of knowledges will it be possible to think of epistemic trajectories that go from points of ignorance conceived as colonialism to points of knowledge conceived as solidarity; this is what I call knowledge-as-emancipation. Herein lies the epistemological foundation of the struggle for cognitive justice as the other side of the struggle for social justice, that is to say, the epistemological conception of a prudent knowledge for a decent life.

JAPANESE AMERICAN NATIONAL MUSEUM

TRANSPACIFIC BORDERLANDS

THE ART OF JAPANESE DIASPORA IN LIMA, LOS ANGELES, MEXICO CITY, AND SÃO PAULO

September 17, 2017 – February 25, 2018

Shinpei Takeda, *Ghost Magnet Roach Motel*, 2017. Photo by Luis Garcia.

janm.org/transpacific-borderlands

100 North Central Avenue, Los Angeles, CA 90012
213.625.0414 • janm.org • janmstore.com • DiscoverNikkei.org

Japanese American National Museum

Eduardo Tokeshi, *Bandera Uno*, 1985, latex on canvas.

Transpacific Borderlands: The Art of Japanese Diaspora in Lima, Los Angeles, Mexico City, and São Paulo will examine the experiences of artists of Japanese ancestry born, raised, or living in either Latin America or predominantly Latin American neighborhoods of Southern California. Their methods of making art are diverse, from traditional to experimental, and the work itself illustrates perspectives of the Japanese Latin American experience directly, metaphorically, and/or abstractly.

The exhibition will show how ethnic communities, racial mixing, and the concepts of homeland and cosmopolitanism inform the creativity and aesthetics of this hybrid culture. It will also provide a visual record of contemporary Japanese Latin American art and contribute to the understanding of identity in a world where the meaning of race and ethnicity is constantly evolving.

Visit **janm.org/transpacific-borderlands** for more information about the exhibition and related programs.

Major support for *Transpacific Borderlands* is provided through grants from the Getty Foundation.

Additional support for this exhibition provided by National Endowment for the Arts, Kosasa Foundation, and Pasadena Art Alliance. Media sponsor is *The Rafu Shimpo*.

Pacific Standard Time: LA/LA
Latin American & Latino Art in LA

Presenting Sponsors
The Getty
Bank of America

Transpacific Borderlands: The Art of Japanese Diaspora in Lima, Los Angeles, Mexico City, and São Paulo is part of Pacific Standard Time: LA/LA, a far-reaching and ambitious exploration of Latin American and Latino art in dialogue with Los Angeles, taking place from September 2017 through January 2018 at more than 70 cultural institutions across Southern California. Pacific Standard Time is an initiative of the Getty.

100 North Central Avenue, Los Angeles, CA 90012
213.625.0414 • janm.org • janmstore.com • DiscoverNikkei.org

1

A Discourse on the Sciences

Boaventura de Sousa Santos

When we look back into the past in an attempt to analyze the present state of the sciences, the first image that comes to mind is that of scientific progress. For the past fifty years the development of science has been so dramatic that all the preceding centuries—from the sixteenth century that gave birth to us all as modern scientists, to the nineteenth—strike us as ancient history. But a mere blink of an eye will make us realize with wonder that all the great scientists that established and charted our theoretical fields lived and worked between the eighteenth century and the first three decades of the twentieth: Adam Smith, Ricardo, Lavoisier, Humboldt, Darwin, Marx, Durkheim, Max Weber, Pareto, Planck, Poincaré, Einstein, Bohr, Heisenberg. Indeed, we might even say that in scientific terms the short twentieth century never existed. And were we to look ahead into the future, we would likewise be presented with two contradictory images. On the one hand, the potential technological applications inherent in our accumulated knowledge lead us to believe that we are on the brink of a communication and information society, free from the needs and insecurities that are still so much part of the lives of many of us. On the other hand, our concern with the limits of scientific rigor and our awareness of such increasing dangers as the ecological disaster or the nuclear war make us fear that the twenty-first century might turn the clock back to the "dark ages."

If we think of the synergy theory of Herman Haken (1977), we may say that ours is a most unstable visual system, the least fluctuation of our visual perception causing ruptures in the symmetries of what we see. Looking at one and the same figure, we see first a white Grecian urn upon a black background, then two black Grecian profiles facing each other upon a white background. Which one is the true image? Both and neither. Such is the ambiguity and complexity of our time, a time of transition—at one with so much before and after it, but at odds with itself.

As in other transition periods, which we understand and traverse with difficulty, we must go back to simple things and ask simple questions. Einstein used to say that there are questions that only children can ask, but that, once asked, shed a new light on our perplexities. In this chapter I bring along a child who two centuries and a half ago asked a few simple questions about the sciences and the scientists. He asked them at the beginning of a cycle of scientific production which many of us believe has now come to its close. The child's name is Jean-Jacques Rousseau. In his famous *Second Discourse* (1750), Rousseau raised many questions as he replied to the question posed by the Academy of Dijon (1964, 52ff.), itself rather childlike: does the progress of the sciences and the arts contribute to the purity or to the corruption of manners? This is an elementary question, at once profound and easy to understand. In order to answer it—eloquently enough to win first prize and not a few enemies—Rousseau in turn asked the following equally elementary questions: Is there a relationship between science and virtue? Is there any serious reason to replace the common sense knowledge we have of nature and of life, and which we share with the other men and women of our society, with the scientific knowledge produced by a few and unavailable to the majority of the people? Does science contribute to bridging the widening gap in our society between what one is and what one seems to be, between knowing how to say and knowing how to do, between theory and practice? To these simple questions Rousseau gives an equally simple reply: a resounding no.

It was then the middle of the eighteenth century. Modern science, which had emerged in the sixteenth and seventeenth centuries out of the scientific revolutions of Copernicus, Galileo, and Newton, was abandoning the esoteric speculations of its founders to become the ferment of an unprecedented social and technological transformation. It was an amazing period of transition, moreover, that perplexed the more alert minds and made them reflect on the very foundations of their society and on the impact wrought upon it by the transformations of the emerging scientific order. Today, two hundred and fifty years later, all of us are the products of that new order, the protagonists and living witnesses of the changes it brought about. However, in the early 2000s, things are no longer what they were thirty years ago. For reasons I shall try to sort out, we are once again perplexed, we have once more lost our epistemological confidence. We are overwhelmed by a sense of irremediable loss, all the stranger for our uncertainty about what it is that we are losing. We may even wonder at times if this sense of loss is perhaps just a fear that obscures the latest gains made in our individual and collective lives. And then again, there is always the confusion about what exactly it is that we have gained.

Hence the ambiguity and the complexity of our present time. Hence, also, the idea snared by many, that we live in a period of transition. Hence, finally, the urgent need to give answers to simple, elementary, intelligible questions. An elementary question is a question that reaches, with the technical clarity of a sling, the deepest magma of our individual and collective perplexity. Such were the questions asked by Rousseau; such must ours be. As a matter of fact, two hundred years later, our questions are still the same as Rousseau's: we are once again faced with the need to ask about the relationship between science and virtue.

Indeed, we must once again ask about the value of so-called common sense knowledge, the knowledge that we, as individual or collective subjects, create and use to give meaning to our practices, but which science insists on considering irrelevant, illusory, and false. And, finally, we must ask about the contribution of all this accumulated scientific knowledge to enriching or impoverishing our lives. In other words, we must ask if science has contributed positively or negatively to our happiness. Our difference from Rousseau is that, though our questions are equally simple, our answers will be far less so. A cycle of the hegemony of a certain scientific order has come to an end. The epistemic conditions of our questions are inscribed in the converse of the concepts we use to give them answers. We have to try to remove blinkers, walking a tightrope between being lucid and being unintelligible in our assertions. At the same time, the social and psychological bases of our queries have become different and far more complex. It is one thing to ask whether an automobile can be useful to me and bring me happiness if the question is posed when no one among my neighbors has an automobile, and another when everyone except me has one, or when I alone have one that is more than twenty years old.

Obviously, we have to be far more Rousseaunian in our queries than in our responses. Let me begin by briefly characterizing the hegemonic scientific order. I shall then analyze the signs that this hegemony is in crisis, distinguishing between its theoretical and sociological conditions. Finally, I shall speculate on the profile of the new emergent scientific order, again distinguishing between the theoretical and the sociological conditions of its emergence.

My analysis will be based on the following working hypotheses: 1) the distinction between natural and social sciences is beginning to seem meaningless; 2) the social sciences will be the catalyst of the necessary synthesis between them; 3) to achieve this synthesis the social sciences must reject all forms of empirical or logical positivism or of mechanistic idealism or materialism, thus bringing back to the center of knowledge what is conventionally called the humanities; 4) the synthesis I have in mind does not aim at a unified science or even at a general theory, but merely at a set of theoretical aqueducts into which can converge various currents which have hitherto been considered theoretically separate; 5) to the degree that such a synthesis is achieved, the hierarchical distinction between scientific and common sense knowledge will gradually disappear.

THE DOMINANT PARADIGM

The prevailing model of rationality of modern science came out of the scientific revolution of the sixteenth century and was developed primarily in the domain of the natural sciences during the following centuries. Although there were a few preliminary attempts in the eighteenth century, the model would be adopted by the emerging social sciences only in the nineteenth century. From then on, we may speak of a single global model of scientific rationality, with some internal variation to be sure, but one which distinguished itself from and defended itself quite conspicuously and consistently against two nonscientific (and therefore irrational) forms of knowledge that were potentially disturbing: common sense, and the so-called humanities (the latter including, among other things, history, philology, law, literature, philosophy, and theology).

The new scientific rationality, being a global model, was also a totalitarian model, inasmuch as it denied rationality to all forms of knowledge that did not abide by its own epistemological principles and its own methodological rules. This was the main feature of the new paradigm, the feature that best symbolized its break with the preceding scientific paradigms. It was gradually consolidated in Copernicus's heliocentric theory of the movement of the planets, Kepler's laws on the planetary orbits, Galileo's laws on the gravity of bodies, and Newton's great cosmic synthesis, and most of all in the philosophical consciousness conferred upon these findings by Bacon and especially by Descartes. This preoccupation with being the instruments of a fundamental break that allowed for only one form of true knowledge is evident in the protagonists' attitudes. They marvel at their own findings while simultaneously displaying a serene, haughty arrogance vis-à-vis their contemporaries.

In his book on *World Harmony*, published in 1619, Kepler writes, regarding the natural harmonies he had discovered in the celestial motions: "Forgive me, but I am happy; if you are angry, I shall persevere. . . . My book may have to wait a long time for its readers. But then, even God had to wait for 6000 years to have his work beheld" (Kepler 1939, 280). On the other hand, in that marvelous spiritual autobiography, *The Discourse on Method* (to which I shall come back later), Descartes writes, concerning his new method:

> Now I always try to lean toward diffidence rather than presumptions in the judgments I make about myself; and when I cast a philosophical eye upon the

> various activities and undertakings of mankind, there are almost none which I do not consider vain and useless. Nevertheless, I have already reaped such fruit from this method that I cannot but feel extremely satisfied with the progress I think I have already made in the search for truth, and I cannot but entertain such hopes for the future as to venture the opinion that if any purely human occupation has solid worth and importance, it is the one I have chosen. (Descartes 1988, 21)

To understand this epistemological confidence we have to describe, at least briefly, the main features of the new scientific paradigm. Realizing that what separates them from the still dominant Aristotelian and medieval knowledge was not so much better observation of the facts but rather a new outlook on the world and on life, the protagonists of the new paradigm engaged in a passionate struggle against all forms of dogmatism and authority. The case of Galileo is particularly exemplary, as is Descartes' assertion of intellectual independence: "I was . . . unable to choose anyone whose opinions struck me as preferable to those of all others and I found myself as it were forced to become my own guide" (Descartes 1988, 28). This new way of looking at the world and at life led to two basic distinctions: between scientific knowledge and common sense on the one hand, and between nature and human beings on the other.

Unlike Aristotelian science, modern science systematically distrusted the evidence of our immediate experience. Such evidence which is at the root of common sense knowledge was alleged to be illusory. As Einstein emphasized in his preface to the *Dialogue Concerning the Two Chief World Systems*, Galileo boldly sought to argue that his hypothesis about the rotation of the earth both around the sun and on its axis were not refuted by our inability to observe the mechanical effects of those movements, that is to say, by the fact that the earth seems to be motionless (1970, xvii). On the other hand, in modern science the separation between nature and human beings is total. Nature is mere extension and movement. It is passive, eternal, and reversible. It is a mechanism whose elements can be disassembled and then put back together again in the form of laws. It possesses no quality or dignity, which impedes us from unveiling its mysteries. Furthermore, such unveiling is not contemplative, but quite active, since it aims at knowing nature in order to dominate and control it. In Bacon's words, science will make of humanity "the master and the owner of nature" (Bacon 1933).[1]

On the basis of these assumptions, scientific knowledge advances by observing natural phenomena in a free, disinterested, and systematic way, and with as much rigor as possible. Bacon's *Novum Organum* contrasts the uncertainty of unassisted reason to the certainty of ordered experiments (Koyré 1981, 30). Contrary to what Bacon thought, experiments do not obviate the need for previous theorizing, deductive thinking, or even speculation, but require them not to omit empirical observation as part of the final demonstration. Galileo refuted Aristotle's deductions only to the degree that he found them untenable. And it was Einstein who called our attention to the fact that Galileo's experimental methods were so inadequate that, only by means of daring speculation, could he fill in the gaps in his empirical data (we need only remember that time could not be measured in that era in units smaller than seconds) (Einstein 1970, xix). As for Descartes, he proceeded unmistakably from ideas to things and not the other way around, and gave priority to metaphysics as the ultimate basis of science.

The ideas that governed observation and experimentation were those simple, clear ideas from which it was thought possible to arrive at a more profound and accurate knowledge of nature—that is to say: mathematical ideas. Mathematics provided modern science not only with its preferred analytical tool, but also with a logic of investigation, as well as a model of representing the structure of matter itself. According to Galileo, the book of nature is inscribed in geometric characters;[2] and Einstein did not think otherwise.[3]

There are two main consequences that derive from the centrality of mathematics in modern science. First, to know means to quantify. Scientific rigor is gauged by the rigor of measurements. The intrinsic qualities of the object, so to speak, do not count, and are replaced by the quantities into which they can be translated. Whatever is not quantifiable is scientifically irrelevant. Secondly, the scientific method is based on the reduction of complexity. The world is complex and the human mind cannot grasp it entirely. To know means to break down and to classify in order to establish systematic relations among these parts. Already in Descartes, one of the rules of the *Method* was precisely to "divide each of the difficulties . . . into as many parts as possible and as may be required in order to resolve them better" (Descartes 1988, 29). The primordial distinction is between the "initial conditions" and the "laws of nature." The initial conditions are the realm of complexity and contingency, where it is necessary to select those conditions that are relevant for the facts being observed. The laws of nature are the realm of simplicity and regularity, where it is possible to observe and measure with accuracy. This distinction between initial conditions and the laws of nature was, of course, far from "natural." It was, indeed, totally arbitrary, as Eugene Wigner has noted (1970, 3). Nonetheless, it was the very basis of modern science.

The theoretical nature of scientific knowledge derives from the epistemological presuppositions and from the methodological rules already mentioned. It is a causal knowledge, which aims at formulating laws in the light of observed regularities and with a view to foreseeing the future behavior of phenomena. The discovery of the laws of nature, then, rests on the assumption, on the one hand, that the relevant initial conditions can be identified (for example, in the case of falling bodies, the initial position and its velocity), and on the other hand, that the outcome will be independent of the place and time of the initial conditions. In other words, the discovery of the laws of nature is based on the principle that absolute position and absolute time are never relevant initial conditions. This is, according to Wigner (1970, 226), the most important theorem of invariance in classical physics.

The laws, insofar as they are categories of intelligibility, depend on a concept of causality, chosen (but not arbitrarily) among those of Aristotelian physics. Aristotle distinguished four kinds of causes: the material cause, the formal cause, the efficient cause, and the final cause. The laws of modern science are a kind of formal cause that gives priority to *how-it-works* as against *who-is-the-agent* or *what-is-the-purpose* of things. Thus, scientific knowledge breaks with common sense knowledge. Indeed, whereas in common sense (and hence in the practical knowledge it produces) cause and intention coexist comfortably, scientific determination and the formal cause entails ignoring intention. It is this type of formal cause that makes possible the prediction of reality, hence interference with it, and which ultimately allows modern science to answer the question about the foundation of its claims to accuracy and truth by pointing to its successes in manipulating and transforming reality.

Knowledge that is based on the formulation of laws has as its meta-theoretical presupposition the idea of order and stability in the world, the idea that the past repeats itself in the future. In Newtonian mechanics, the world of matter is a machine whose operations can be precisely determined by means of physical and mathematical laws—an eternal and static world hovering in an empty space, a world which Cartesian rationalism makes knowable by dividing it into its constituent parts. This idea of a machine-like world was so strong that it became the great universal hypothesis of the modern era. It may be surprising, even paradoxical, that such a way of knowing could become one of the pillars of the idea of progress that has pervaded European thought since the eighteenth century and that was the intellectual sign of the rise of the bourgeoisie.[4] The truth is, however, that order and stability in the world are the precondition for the technological transformation of reality.

Mechanistic determinism provides a clear horizon for a form of knowledge that was meant to be utilitarian and functional, acknowledged less for its capacity to understand reality at the deepest level than for its capacity to control and transform it. At the social level, this was also the cognitive horizon most consonant with the interests of the rising bourgeoisie, who considered the society they were beginning to control as the final stage in the evolution of mankind (Comte's positive state, Spencer's industrial society, Durkheim's organic solidarity). Hence, Newton's prestige and the simple laws to which he reduced all the complexities of the cosmic order readily turned modern science into the hegemonic model of rationality that then spilled over from the study of nature into the study of society. As it had been possible to discover the laws of nature, so would it be possible to discover the laws of society.

Bacon, Vico, and Montesquieu were the great precursors. Bacon affirmed the plasticity of human nature, hence its perfectibility given appropriate political, legal, and social conditions, which can be accurately known. Vico suggested that there are laws that govern the evolution of societies deterministically and that allow for the prediction of the outcome of collective actions. With remarkable premonition, Vico ([1725] 1968) identified and solved the contradiction between the freedom and unpredictability of individual human action and the determinism and predictability of collective action. Montesquieu ([1748] 1989) may be considered a precursor of the sociology of law, when he established a relationship between the man-made laws of the legal system and the inescapable laws of nature.

In the eighteenth century these preliminary efforts were expanded and deepened into that intellectual ferment—the Enlightenment—which would create the conditions for the emergence of the social sciences in the nineteenth century. The philosophic understanding of modern science, first formulated in Cartesian rationalism and Baconian empiricism, evolved into nineteenth-century positivism. Since, for positivism, there are only two forms of scientific knowledge—the formal disciplines of logic and mathematics and the empirical sciences following the mechanistic model of the natural sciences—the social sciences could not but be empirical. The way in which the mechanistic model was followed varied, however. It is common to distinguish two main tendencies. The dominant one consisted in applying to the study of society, to the degree possible, all the epistemological and methodological principles that had dominated the study of nature since the sixteenth century. The other one, long a marginal tendency but with an increasing number of followers, consisted in claiming for the social sciences its own distinct epistemological and methodological status, based upon the specificity of the human being radically different from nature. These two conceptions have usually been taken to be antagonistic, the former being subject to the positivist yoke, the latter free from it, both claiming the monopoly of social-scientific knowledge. I shall offer a different interpretation, once I have briefly characterized them.

The first conception—whose epistemological commitment is clearly symbolized in the name of "social physics" which the scientific study of society was initially called—takes for granted that the natural sciences are the concrete application of a model of knowledge that is universally valid, and indeed the only valid one. Therefore, no matter how large the differences between natural and social phenomena, it is always possible to study the latter as if they were the former. Admittedly, such differences work against social phenomena, or rather, they make the methodological canon harder to accomplish and the knowledge arrived at less accurate. But there are no qualitative differences between scientific procedure in the social sciences and in the natural sciences. In order to study social phenomena as if they were natural phenomena, that is, in order to conceive of social facts as things (as envisioned by Durkheim [1982], the founder of academic sociology), it is necessary to reduce social facts to their external, observable, measurable dimensions. The causes of the rise in the rate of suicide in Europe at the turn of the century are not to be looked for in the motives invoked by those committing suicide in their letters, as had been the custom, but rather by checking the regularities in such conditions as the sex and marital status of those committing suicide, whether or not they had children, their religion, and so on (Durkheim 1951).

Because such reductionism is not always easy and not always possible without grossly distorting the facts or even reducing them to near irrelevance, the social sciences have a long way to go before they can be made compatible with the criteria of scientificity of the natural sciences. The obstacles are enormous but not insurmountable. *The Structure of Science* by Ernest Nagel is a good example of the efforts made in this field to identify the obstacles and the ways of overcoming them. Some of the principal obstacles he identified are the following. There are no explanatory theories in the social sciences that would allow them to abstract from reality in such a way as to be able to search for adequate proof in that reality in a methodologically controlled way. The social sciences cannot establish universal laws because social phenomena are historically conditioned and culturally determined. The social sciences cannot make reliable predictions because human beings change their behavior according to how much is known about it. Social phenomena are naturally subjective and as such they cannot be understood as objective behavior. The social sciences are not objective because the social scientist cannot free his observations from the values that underlie his general practice, and hence also his practice as a scientist (Nagel 1961, 447ff.).

For each of these obstacles, Nagel tried to demonstrate that the opposition between the social and the natural sciences is not so linear as commonly thought and that, whatever differences there may be, they are either surmountable or negligible. He recognized, however, that overcoming obstacles is not always easy, and that this accounts for the backwardness of the social sciences vis-à-vis the natural sciences. The idea of the backwardness of the social sciences is central to this kind of methodological reasoning, as is the idea that, with time and money, this backwardness may be reduced or even eliminated.

In Thomas Kuhn's theory of scientific revolutions the backwardness of the social sciences was explained by their pre-paradigmatic nature, as opposed to the paradigmatic nature of the natural sciences (1962). The development of knowledge in the natural sciences has allowed for the formulation of a set of principles and theories about the structure of matter, which are unquestionably accepted by the entire scientific community. This acceptance is what we mean by a paradigm. But in the social sciences there is no paradigmatic consensus, which is why the debate tends to involve every kind of acquired knowledge. The strain and the waste this involves are both cause and effect of the backwardness of the social sciences.

The second conception claims an independent methodological status for the social sciences. According to this view, the obstacles identified above are insurmountable. Some reject the very notion of a science of society; others argue that it is a different kind of science. The main argument is that human action is radically subjective. Unlike natural phenomena, human behavior cannot be described, let alone explained, on the basis of its external, objectifiable characteristics, since the same external act may have multiple interpretations. The social sciences will always be a subjective science, not an objective science like the natural sciences. The social sciences must understand social phenomena in terms of the attitudes and the meanings that the agents ascribe to their actions. That requires research methods and epistemological criteria different from those used in the natural sciences, qualitative rather than quantitative methods, in order to arrive at intersubjective, descriptive, empathetic knowledge, as opposed to objective, explanatory, nonmothetic knowledge.

The latter conception of the social sciences acknowledges being anti-positivist. Its philosophical tradition is phenomenology in its many varieties, from a more moderate version, as in Max Weber (1968), to a more radical one, as in Peter Winch (1970). However, close inspection reveals that this view, as it has been elaborated, is more dependent on the model of rationality of the natural sciences than it at first seems. It shares the nature/human beings dichotomy, which amounts to a mechanistic view of nature, to which it contrasts, as one might expect, the specificity of human beings. This distinction, which was crucial for the scientific revolution of the sixteenth century, led in turn to others, such as those between nature and culture and between humans and animals, culminating in the eighteenth century in the celebration of the unique character of humanity. The line thus delineated between the study of humanity and the study of nature remained a prisoner of the cognitive priority assigned to the natural sciences, since although, on the one hand, a biological determinism of human behavior was denied, on the other hand, biological arguments were used to establish the specificity of the human being.

The inescapable conclusion then is that both these conceptions of science belong to the paradigm of modern science, even if the second conception points to a crisis in the paradigm and already contains some elements of transition toward a new scientific paradigm.

THE CRISIS IN THE DOMINANT PARADIGM

There are many clear signs today that the model of scientific rationality I have outlined earlier is going through a profound crisis. In this section I shall argue, first, that the crisis is not only profound but also irreversible; secondly, that we are living in a time of scientific revolution that began with Einstein and quantum physics, and that it is not yet clear when it will end; thirdly, that, although the signs in question allow for nothing more than mere speculation about the paradigm that will emerge out of this revolutionary time, we may already assert with certainty that the basic distinctions underlying the dominant paradigm described previously will collapse.

The crisis of the dominant paradigm is the result of a series of interacting conditions. I make a distinction between social and theoretical conditions. I shall pay more attention to theoretical conditions and begin with them. My first remark, which is not as trivial as it sounds, is that the identification of the limits and structural shortcomings of the modern scientific paradigm is the outcome of the great advance in knowledge it made possible. The deepening of knowledge revealed the fragility of the pillars on which it rested.

Einstein was responsible for the first rupture in the paradigm of modern science, indeed a wider rupture than he himself would ever have been able personally to admit. One of Einstein's profoundest insights was the relativity of simultaneity. Einstein distinguished between the simultaneity of events happening in the same place, and the simultaneity of distant events, particularly events separated by astronomical distances. As far as the latter are concerned, the logical problem is the following: how can the observer establish the temporal order of events in space? To be sure, by measuring the velocity of light, assuming, as Einstein's theory does, that in nature there is no velocity that is greater. However, upon measuring velocity going in one direction (from A to B), Einstein was confronted with a vicious circle. In order to determine the simultaneity of distant events the velocity must be known, and in order to measure velocity the simultaneity of events must be known. In a flash of genius, Einstein broke this circle, by demonstrating that the simultaneity of distant events cannot be verified, but merely defined. It is, therefore, arbitrary, and thus, as Reichenbach (1970, 60) pointed out, when we measure, the results cannot be contradictory insofar as they give us back the simultaneity that we have introduced into the measuring system by definition.

This theory has revolutionized our conceptions of time and space. Since there is no universal simultaneity, Newton's absolute time and space do not exist. Two events that are simultaneous in one system of reference are not simultaneous in another. The laws of physics and geometry are based on local measurements. "The instruments for measuring, be they clocks or yardsticks, have no independent magnitude; rather, they adjust themselves to the metric field of space, the structure of which manifests itself most clearly in the rays of light" (Reichenbach 1970, 68).

Considering the local character of measurements, and hence of the accuracy of the knowledge thus obtained, led to the second theoretical condition of the crisis of the dominant paradigm, namely quantum physics. If Einstein relativized the accuracy of Newton's law in the field of astrophysics, quantum physics did the same in the field of microphysics. Heisenberg and Bohr demonstrated that it is not possible to observe or measure an object without interfering with it, without actually changing it in such a way that, after being measured, the object is no longer the same as it was before. As Wigner pointed out, "the measurement of the curvature of space caused by a single particle could hardly be carried out without creating new fields which are many billion times greater than the field under investigation" (1970, 7). The idea that we know nothing about the real but what we ourselves bring to it, that is to say, that we know nothing of the real except our intervention in it, is well expressed in Heisenberg's uncertainty principle—we cannot simultaneously reduce the errors of the measurement of velocity and of the position of particles; whatever we do to reduce the error of the one will increase the error of the other (Heisenberg 1958, 1971). This principle, and therefore the demonstration of the subject's structural interference in the observed object, is of great consequence. On the one hand, since the rigor of our knowledge is structurally limited, we can aspire only to approximate results, which make the laws of physics merely probabilistic. On the other hand, the hypothesis of mechanistic determinism is no longer viable, since the whole of reality is not reducible to the sum of the parts into which we divide it in order to observe and measure it.

Finally, the subject/object distinction is far more complex than it may seem at first. It loses its dichotomous contours to assume the form of a continuum.

The accuracy of measurement that quantum physics put into question would be even more deeply shaken if we question the accuracy of the formal vehicle used to express the measurements, mathematical rigor. This happened with the work of Gödel, which for that reason I consider the third condition of the crisis ([1931] 1962). The theorem of incompleteness or the theorems about the impossibility, in certain circumstances, of finding within a given formal system the proof of its consistency demonstrate that, even if the rules of mathematical logic are strictly followed, it is possible to formulate undecidable propositions, which can be neither demonstrated nor refuted, one of the propositions being precisely the one that postulates the noncontradictory character of the system.[5] If the laws of nature base their rigor on the rigor of the mathematical formalizations in which they are expressed, then Gödel's findings demonstrate that the rigor of mathematics is itself in need of a foundation. After this, it is possible not only to question the accuracy of mathematics but also to redefine it as one form of accuracy that contrasts with alternative forms of rigor. In other words, it is a form of rigor whose conditions of success in modern science can no longer be taken for granted as obvious and natural. The philosophy of mathematics itself has engaged in the creative problematization of these themes, and today it is recognized that mathematical accuracy, like any other form of rigor, is based on a criterion of selectivity, thus having both a constructive and a destructive side.

The fourth theoretical condition of the crisis of the Newtonian paradigm derives from the advances of knowledge in the fields of microphysics, chemistry, and biology over the past thirty years. Let me cite, by way of example, the findings of Ilya Prigogine. His theory of dissipative structures and his principle of "order through fluctuations" established that, in open systems, that is, in systems that function far from equilibrium, evolution is explained by fluctuations of energy which, at certain not entirely predictable moments, spontaneously generate reactions, which in turn, by means of nonlinear mechanisms, pressure the system beyond its utmost limit of disequilibrium. The situation of bifurcation, that is to say, the critical point at which the slightest fluctuation may lead to a new state, represents the potentiality of the system to be attracted to a new state of lesser entropy. Thus the irreversibility of open systems means that they are the product of their history (Prigogine and Stengers 1984; Prigogine 1980; 1981, 73ff.).

The importance of this theory is that it rests on the new conception of matter and of nature, which is hard to reconcile with the one we inherited from classical physics. In place of eternity, we now have history; in place of determinism, unpredictability; in place of mechanicism, interpenetration, spontaneity, irreversibility, and evolution; in place of order, disorder; in place of necessity, creativity, and contingency. Prigogine's theory revives even such Aristotelian concepts as potentiality and virtuality, which the sixteenth-century scientific revolution appeared to have definitively cast into the dustbin of history.

But the greatest importance of this theory is that it is not an isolated phenomenon. It is rather part of a converging movement, which has gained strength mainly since the 1970s, and which traverses the various natural sciences and even the social sciences. It has indeed a transdisciplinary calling that Jantsch (1980) calls the self-organization paradigm and which is developed, among others, by Prigogine's theory, Haken's synergetics (Haken 1977; 1985, 205), Eigen's concept of the hypercycle and his theory on the origin of life (Eigen and Schuster 1979), Maturana and Varela's concept of autopoiesis (Maturana and Varela 1973; 1975),[6] Thom's catastrophe theory (Thom 1983), Jantsch's evolutionary theory (Jantsch 1981, 83ff.), David

Bohm's theory of the "implicate order" (Bohm 1984), and Geoffrey Chew's S-matrix theory and its underlying "bootstrap" philosophy (Chew 1970, 762ff.; 1970, 23ff.).[7] This scientific movement, along with the theoretical innovations that I have defined earlier as so many theoretical conditions of the crisis of the dominant paradigm, have precipitated a profound epistemological reflection on scientific knowledge. This is such a rich and varied reflection that, better than anything else, it exemplifies the intellectual situation of our time.

There are two important sociological facets to this reflection. Firstly, it is predominantly carried out by the scientists themselves, those who have mustered the necessary philosophical competence and concern to problematize their own scientific practice. We can safely state that there have never been so many philosopher-scientists as today, a trend that is not intellectually accidental. After the nineteenth-century scientistic euphoria and the concomitant aversion to philosophical reflection, epitomized by positivism, we have at the end of the twentieth century been seized by the near desperate desire to complement our knowledge of things with our knowledge of our knowledge of things—in other words, with a knowledge of ourselves. The second facet of this reflection is that it deals with questions previously left to sociologists. The analysis of the social conditions, of the cultural contexts, of the organizational models of scientific research, which previously had been the separate realm of the sociology of science, have now come to play a key role in epistemological reflection.

Let me now give a few examples of the main themes of this reflection. First, the concept of laws and the related concept of causality are put into question. The formulation of the laws of nature is based on the idea that the observed phenomena are independent of all but a fairly small number of conditions—the initial conditions—whose interference is observed and measured. This idea, it is now recognized, necessarily creates broad distinctions between phenomena, distinctions which furthermore are always provisional and precarious, since the verification of the noninterference of certain factors is always the result of imperfect knowledge, however more nearly correct it may become. The laws are thus probabilistic, approximate, and provisional, as clearly laid out in Popper's falsifiability principle (1962). But above all, the simplicity of laws constitutes an arbitrary simplification of reality that confines us to a minimal horizon beyond which lie other kinds of knowledge about nature, probably richer and of far greater human interest.

In biology, where interactions among phenomena and forms of self-organization in nonmechanical totalities are more visible, but also in the other sciences, the notion of law has been partially replaced by the notions of system, structure, pattern, and finally by the notion of process. The decline of the hegemony of laws is parallel to the decline of the hegemony of causality. The questioning of causality in modern times has a long tradition, going back at least to David Hume and logical positivism. Critical reflection has addressed both the ontological problem of causality (what are the characteristics of the causal nexus? does such a nexus exist in reality?) and the methodological problem of causality (what are the criteria of causality? how can a causal nexus be identified or a causal hypothesis tested?). Today, the relativization of the concept of cause stems mainly from the acknowledgment that the central place it has occupied in modern science has its explanation less in ontological or methodological than in pragmatic reasons. The concept of causality is well suited to a science that aims at intervening in reality and that measures its success by the scope of its intervention. After all, a cause is something that can be acted upon. Even advocates of causality, such as Mario Bunge (1979), recognize that it is merely one of the forms of determinism and that it therefore plays a limited, though irreplaceable, role in scientific knowledge.[3] The truth is that, under the aegis of biology and even of microphysics, "causalism" as a category for the intelligibility of the real is now giving way to "finalism."

The second great theme of epistemological reflection deals with the content rather than with the form of scientific knowledge. Since it is a minimal knowledge that closes the door to many other ways of knowing the world, modern scientific knowledge is a sad and disenchanted knowledge that turns nature into an automaton or, as Prigogine says, into an awfully stupid interlocutor (Prigogine and Stengers 1984). This vilification of nature ends up vilifying the scientist himself inasmuch as it reduces the alleged experimental dialogue to an act of prepotency over nature. Scientific rigor, because it is based on mathematical rigor, quantifies, and because it quantifies, it disqualifies. It is a rigor that, by objectifying phenomena, objectivizes and degrades them; in characterizing the phenomena, it caricatures them. In sum, scientific rigor is a form of rigor, which, in asserting the scientist's personality, destroys the personality of nature. In this way, knowledge gains in rigor what it loses in richness. The vaunted successes of technology obscure the limits of our understanding of the world and suppress the question of the human value of a scientific endeavor thus conceived. This question is, however, inscribed in the subject/object relation presiding over modern science, a relationship that internalizes the subject at the cost of externalizing the object, thus making both of them self-enclosed and unable to communicate with each other.

The limits of this kind of knowledge are thus qualitative and cannot be overcome by more research and more precise instruments. Indeed, the qualitative precision of knowledge is itself structurally limited. For example, as far as information theories are concerned, Brillouin's theorem demonstrates that information is not without cost (Brillouin 1956; Parain-Vial 1983, 122ff.). Any observation performed on a physical system increases the system's entropy in the laboratory. The gain of a given experiment has therefore to be defined by the relation between the information obtained and the concomitant increase of entropy. But, according to Brillouin, the gain is always less than one and only rarely even close to one. In this view, a rigorous experiment is impossible, for it would require an infinite expenditure of human activity. Finally, precision is limited because, if it is true that knowledge advances only by the progressive subdivision of the object (as is attested by increasing scientific specialization), this also proves the irreducibility of wholes, whether organic or inorganic, to their constituent parts. Thus, the knowledge gained from observing the parts is necessarily distorted. The observed facts are beginning to break out of the solitary confinement to which science has subjected them. The frontiers of objects are less and less clear. The objects themselves are like rings interlocked in such complex chains that they end up being less real than the relations between them.

At the beginning of this section, I said that the crisis of the paradigm of modern science is explained not only by theoretical conditions, which I have just partially indicated, but also by social conditions. The latter cannot be dealt with in detail here. I shall merely suggest that, no matter what the structural limits of scientific rigor, there can be no doubt that what science has gained in rigor over the past forty or fifty years, it has lost in capacity for self-regulation. The beliefs about the autonomy of science and about the disinterestedness of scientific knowledge, which has long constituted the spontaneous ideology of the scientists, have collapsed due to the global industrialization of science, especially since the 1930s and 1940s. In contemporary capitalist societies the industrialization of science has made it tainted by the centers of economic, social, and political power, which came to play a decisive role in the definition of scientific priorities.

The industrialization of science occurred both at the level of applied science and at the level of the organization of scientific research. The Hiroshima and Nagasaki bombs were a tragic symbol of applied science. Initially viewed as accidental and fortuitous, they are today, in the face of the ecological disaster and the danger of a nuclear proliferation, increasingly perceived as the manifestation of a mode of scientific production prone to transform accidents into regular events. Science and technology are showing themselves to be two sides of a historical process in which military and economic interests tend to converge to the point of becoming virtually indistinguishable. As far as the organization of scientific research is concerned, the industrialization of science has brought about two main results. On the one hand, the scientific community has become stratified, the power relations among scientists have become more unequal and authoritarian, and the great majority of scientists have undergone a process of proletarianization within laboratories and research centers. On the other hand, capital-intensive research (dependent on scarce and expensive instruments) made free access to equipment impossible, thus widening the gap, in terms of scientific and technological development, between core and peripheral countries.

Under the social and theoretical conditions just mentioned, the crisis of the paradigm of modern science does not constitute some gloomy quagmire of skepticism or irrationalism. Rather, we find ourselves observing the portrait of an intellectual family that is large and unstable, but also creative and fascinating, at the moment of its rather sad farewell to conceptual points of reference, both theoretical and epistemological, ancestral and intimate, that are no longer persuasive and reassuring, a farewell in search of a better life in the surroundings where optimism is better founded and rationality more plural and where at last knowledge will once again become an enchanted adventure. The characterization of the crisis of the dominant paradigm brings with it the profile of the emergent paradigm. It is that profile I shall now attempt to draw.

THE EMERGENT PARADIGM

We can only speculate about the precise configuration of the dawning paradigm. Such speculation is, of course, based on the signals emitted by the crisis of the present paradigm, though they do not determine the outcome. Indeed, as Rene Poirier has said, and was noted by Hegel and Heidegger before him, "we can know the global coherence of our physical and metaphysical truth only retrospectively" (Poirier 1983, 10). So, when we speak of the future that we feel we are already living, what we say about it is always the product of a personal synthesis steeped in imagination, in my own case, in the sociological imagination. No wonder then that, though they sometimes converge, the syntheses presented up to now are so different. Ilya Prigogine (1979; 1980; 1981), for example, speaks of the new alliance and the metamorphosis of science. Fritjof Capra (1983; 1984) speaks of the new physics and the Tao of physics; Eugene Wigner (1970, 215ff.), of shifts of the second type; Erich Jantsch (1980; 1981), of the self-organization paradigm.

As for myself, I shall speak of the paradigm of prudent knowledge for a decent life. By this phrasing I wish to signify that the scientific revolution we are undergoing today is structurally different from the sixteenth- and seventeenth-century revolution. Because it is a scientific revolution occurring in a society that has already undergone a scientific revolution, its emergent paradigm cannot be merely a scientific paradigm (the paradigm of prudent knowledge), but must also be a social paradigm (the paradigm of a decent life). I shall present the emergent paradigm as a set of theses along with the justification of each.

1. All Natural-Scientific Knowledge Is Social-Scientific

The dichotomy of natural and social sciences is becoming increasingly meaningless. This distinction is based on a mechanistic conception of matter and of nature with which it contrasts, it is presumed obviously, the concepts of human beings, culture, and society. The newest findings of physics and biology question the distinction between the organic and the inorganic, between living beings and inert matter, and even between the human and the nonhuman. The characteristics of self-organization, of metabolism, and of self-reproduction, which were previously thought to be specific to living beings, are nowadays ascribed as well to precellular systems of molecules. Furthermore, they are ascribed traits and behavior that were previously believed to be specific to human beings and to social relations. All the recent scientific theories I have mentioned ascribe to matter the concepts of historicity and progress, freedom and self-determination—and even consciousness, which men and women had previously held to be inalienably their own. I am alluding to Prigogine's dissipative structures, Haken's synergy, David Bohm's "implicate order," Geoffrey Chew's matrix-S and its underlying "bootstrap" theory, as well as Fritjof Capra's synthesis of contemporary physics and Eastern mysticism.

All these theories have a holistic vocation, and some are even intended to resolve the inconsistencies between quantum physics and Einstein's relativity. It is as if men and women had set out in search of the most distant and alien objects, and then, once they had gotten there, had discovered themselves as though they were being reflected in a mirror. At the beginning of the 1960s, extrapolating from quantum mechanics, Eugene Wigner maintained that the "inanimate" is not a different quality but only an extreme or limiting case that the distinction between body and soul has long ceased to have meaning, and that physics and psychology would eventually merge into one and the same science (Wigner 1970, 271). Today it is possible to go far beyond quantum mechanics. If quantum mechanics made consciousness part of the act of knowing, we now have to make it part of the object of knowledge itself, and thereby the distinction between subject and object will undergo a radical change.

Not unlike Leibniz's pan-psychism, Bateson's "wider consciousness" refers to a psychic dimension of nature, of which the human mind is but a part, a mind immanent to the global social system and to the planetary ecology that some call God (Bateson 1985). Geoffrey Chew postulates the existence of consciousness in nature as a necessary element for nature's self-consciousness, which would mean that future theories of matter will have to include the study of human consciousness. At the same time, we witness a renewed interest in Jung's "collective unconscious." Indeed, Capra claims that Jung's ideas—primarily the idea of synchronicity to explain the relationship between inner and outer reality—are confirmed in particle physics by the recent concepts of local and nonlocal interactions.[9] As in Jungian synchrony, nonlocal interactions are instantaneous and cannot be predicted in accurate mathematical terms. They are, therefore, not the outcome of local causes; we might, at the very most, speak of statistical causality. Capra considers Jung one of the theoretical alternatives to Freud's mechanistic conception, and Bateson asserts that, just as Freud expanded the concept of mind inwardly (allowing us to grasp the subconscious and the unconscious), we now need to expand it outwardly (by recognizing the existence of mental phenomena other than those of individuals and humans).

Similarly, David Bohm's theory of the implicate order, which according to its author may constitute a common base for quantum theory and relativity, considers consciousness and matter as interdependent, though not linked by any causal nexus. They are rather two related projections of a higher reality that is neither matter nor consciousness. According to the emergent paradigm, then, knowledge tends to be nondualistic. It is knowledge based on the superseding of familiar and obvious distinctions that were taken for granted until very recently—nature/culture, natural/artificial, animate/inanimate, mind/matter, observer/observed, subjective/objective, and animal/person. This relative collapsing of dichotomous distinctions has its repercussions on the scientific disciplines that derive from them. As a matter of fact, there have been sciences that never felt very comfortable with these distinctions, so much so that they had to fracture internally in order to conform to them minimally. I refer to anthropology, geography, archeology, and psychology. More than any others they have reflected the contradictions brought about by the separation between natural and social science. That is why, in this period of paradigmatic transition, it is so important, from an epistemological point of view, to observe what is going on within these sciences.

However, it is not enough to stress the tendency to supersede the distinction between the natural and the social sciences; we must understand the meaning and content of this supersession. Once again, in physical terms, the question is whether the "parameter of order" (Haken) or "attractor" (Prigogine) of this supersession will be the natural or the social sciences. Precisely, because we are living in a state of turbulence, the fluctuations of the new paradigm behave unequally in the various regions of the dominant paradigm, and so the signs of the future are ambiguous. Some interpret them as the emergence of a new naturalism revolving around the biological presuppositions of human behavior. This is the argument of Konrad Lorenz or of sociobiology. For them, the supersession of the dichotomy natural/social sciences is occurring under the aegis of the natural sciences. Against this view, it might be said that its conception of the future is the same conception with which the natural sciences have justified within the dominant paradigm their current scientific, social, and political prestige. It therefore sees in the future only that which will repeat the present. If, on the other hand, we consider more deeply the theoretical content of those sciences that are more advanced in their knowledge of matter, we shall realize that the emergent intelligibility of nature is infused with concepts, theories, metaphors, and analogies from the social sciences. We need only think of Prigogine's "dissipative structures" and Haken's "synergy." Both theories explain the behavior of particles by such concepts as social revolutions, violence, slavery, domination, nuclear democracy—all of which are borrowed from the social sciences (sociology, political science, history, etc.). The same applies to Capra's theories on the relation between physics and psychoanalysis, in which the patterns of matter and the patterns of mind are seen to reflect each other. Even though these theories blur the borderlines between the objects of physics and the objects of biology, the latter has no doubt absorbed the explanatory models of the social sciences more deeply in recent decades. The concepts of teleomorphism, autopoiesis, self-organization, organized potentiality, originality, individuality, and historicity do indeed ascribe human behavior to nature. In his book on the life sciences, Lovelock (1979) states that our bodies are made up of cell cooperatives.

That the explanatory models of the social sciences have been behind the development of the natural sciences for the past decades is further indicated by the fact that, once they have been formulated in their specific domain, the natural-physical sciences are easily applied to the social domain. Thus, for example, Peter Allen (1981), one of Prigogine's closest collaborators, has applied the theory of dissipative structures to economic processes and to the evolution of cities and regions. Haken (1985), in turn, has stressed the potentialities of synergetics to explain revolutionary situations in society. It is as if Durkheim's motto had been reversed. Rather than studying social phenomena as if they were natural phenomena, scientists now study natural phenomena as if they were social phenomena.

The fact that the supersession of the dichotomy between natural and social sciences is being carried out under the aegis of the social sciences, however, is not enough to characterize the model of knowledge in the emergent paradigm. Since, as mentioned above, the social sciences themselves were formed in the nineteenth century according to the models of rationality of the classical natural sciences, to speak of the aegis of the social sciences may turn out to be misleading. I did say, however, that the social sciences were constituted according to two different tendencies: one of them closely linked to the positivist epistemology and methodology of the natural sciences; the other, of an antipositivist vocation, molded in a complex philosophical tradition of phenomenology, interactionism, myth-symbolism, hermeneutics, existentialism, and pragmatism. The latter claim the particularity of the study of society, while at the same time assuming a mechanistic conception of nature. The power of the latter trend in the past decades indicates that it is a model of social sciences, which, in a time of scientific revolution, carries within itself the postmodern sign of the emergent paradigm. It is indeed a transitional model, for it defines the specificity of the human in opposition to a conception of nature that the natural sciences are increasingly considering superseded; but it is a model less strongly attached to the past than to the future. In sum, to the degree that the natural sciences are getting closer to the social sciences, the social sciences are getting closer to the humanities. The subject, cast into the diaspora of irrational knowledge by modern science, is returning, charged with the task of rebuilding a new scientific order.

That this is the global trend of our present scientific revolution is also suggested by the ongoing reconceptualization of the epistemological and methodological conditions of social scientific knowledge. I have enumerated above some of the obstacles to the scientificity of the social sciences, which, according to the still dominant paradigm, are supposedly responsible for the backwardness of the social sciences vis-à-vis the natural sciences. But it is also the case, as I have already indicated, that new developments in the knowledge of the natural sciences, and the epistemological reflection that they have brought about, have shown that the obstacles to the scientific knowledge of society and culture are actually conditions of knowledge in general, be it of social or of natural objects. In other words, what used to be considered the explanation of the greater backwardness of the social sciences today is seen as the greater advance of the natural sciences. Hence, Thomas Kuhn's notion of the preparadigmatic (i.e., less developed) character of the social sciences (1962) must be considerably revised, if not abandoned.

The supersession of the natural sciences/social sciences dichotomy tends, therefore, to revalue the humanities. But for this revaluation to take place, the humanities need themselves to be profoundly changed. What there is in them of the future is that they have resisted the separation between subject and object, and that they have preferred to understand the world rather than to manipulate it. Their genuine core was, however, often trapped in mystifying preoccupations (dreamy esotericism and empty erudition). The ghetto into which the humanities chose to retire was in part a strategic defense against the assault of the social sciences that triumphantly wielded the scientistic bias. But it was also the result of the void they felt, once their space had been taken over by the scientistic model. That is what happened in historical studies with quantitative history, in legal studies with the pure theory of law and legal dogmatics, and in philological, literary, and linguistic studies with structuralism. The genuine core of the humanities must be recovered and put to the service of a global reflection about the world. The text, which has always been the concern of philology, is one of the basic analogies upon which the knowledge about nature and society will be built in the emergent paradigm.

The humanistic conception of the social sciences, as the catalyst of the gradual melting together of natural and social sciences, places the person, as author and subject of the world, at the very center of knowledge, but, unlike what happened in the traditional humanities, it places what we today call nature at the center of the person. There is no human nature because all nature is human. It is therefore necessary to look for global categories of intelligibility, hot concepts capable of melting down the frontiers into which modern science has divided and enclosed reality. Postmodern science is an admittedly analogical science that knows what it knows less well through what it knows better. I have already invoked the textual analogy. Other important basic categories of the emergent paradigms are, to my mind, the analogies of play, of drama, and even of biography. Today the world is natural or social, tomorrow it will be both, and will be looked at as if it were a text or a play, a stage or an autobiography. Clifford Geertz spoke of these humanistic analogies (1983, 19ff.) but he restricted their use to the social sciences; I conceive of them as universal categories of intelligibility.

It will not be long before particle physics shall speak of particles playing, or biology of the molecular theater, or astrophysics of the heavenly text, or chemistry of the biography of chemical reactions. Each of these analogies unveils a corner of the world. Total nakedness, which will always be the nakedness of those who see themselves in what they see, will arise out of the configurations of analogies we may be capable of imagining. After all, the play presupposes a stage, the stage requires a text, the text is its author's autobiography. Play/text, stage, or biography, the world is communication. Streams of meanings and their constellations are converging into rivers, whose sources are in our local practices, which drag along with them the sands of our molecular, individual, communal, social, and planetary trajectories. This is not a jumble of meanings (which would be noise, not meaning), but rather interactions and intertextualities organized around local projects of undivided knowledge. From this arises the second characteristic of postmodern scientific knowledge.

All Knowledge Is Local and Total

In modern science knowledge advances by specialization. Knowledge is ever more rigorous as its object is restricted. Indeed, herein lies what is today recognized as the basic dilemma of modern science: its rigor increases in direct proportion to the arbitrariness with which it straitjackets reality. As a disciplinary knowledge, it is prone to be a disciplined knowledge—that is to say, it organizes knowledge by policing the borders and repressing all trespassers. It is recognized today that the extreme fragmentation and disciplinarity of scientific knowledge turns the scientist into a specialized ignorant person—a negative development whose effects are visible primarily in the applied sciences. Technology today is concerned about its destructive impact on the ecosystems. Medicine is realizing that the hyperspecialization of medical knowledge has transformed the patient into a meaningless checkerboard of parts, when in fact we are only ever ill as a whole person. Pharmacy is discovering the destructive side of drugs, all the more destructive to the degree that they are more specific, and looks for a new logic of chemical combination heedful of organic equilibrium. Law, having reduced the complexity of legal life to the dryness of dogmatics, is rediscovering the philosophical and sociological world in its search for a lost prudence. Economics, after having legitimated quantitative and technocratic reductionism claiming thereby success in economic predictions, is being forced to recognize, given the poverty of the results, that the human and sociological quality of the economic agents and processes is now sneaking in through the window after having been forced out the door[10]. In order to garner the recognition of its users (who, whether public or private, institutional or individual, have always been in a position of power vis-à-vis those being analyzed), applied psychology has favored efficient and manageable tools, such as tests, which have reduced personality's richness to the functional demands of unidimensional institutions.

The evils of this fragmentation of knowledge and the arbitrary reductionism it carried with it are now being recognized, but the measures offered to correct them usually end up doing nothing more than reproducing them in another guise. New disciplines are created to solve the problems brought about by the old, and thus the same model of scientificity is reproduced. To give one example, the general practitioner, resurrected in the hope of compensating for medical hyperspecialization, is running the risk of being transformed into simply one more specialist. This perverse effect suggests that there is no solution for this problem within the dominant paradigm, for the latter is precisely the problem, from which all the others arise.

In the emergent paradigm, knowledge is total. Its horizon is Wigner's universal totality or Bohm's undivided totality. But it is as local as it is total. It is gathered around themes adopted by concrete social groups at a given time as projects of local life, be it to recover the history of a place, to preserve a green space, to build a computer adequate to the local needs, to make the infant mortality rate decrease, to invent a new musical instrument, to wipe out an illness, etc., etc. Postmodern fragmentation is thematic rather than disciplinary. Its themes are like galleries along which the various kinds of knowledge move, meeting each other. Unlike what happens in the present paradigm, knowledge advances as its object grows larger. Like a tree, it grows by differentiation and the spreading of its roots, in search of new and more varied interfaces.

But though it is local, postmodern knowledge is also total, because it reconstitutes the local cognitive projects, stressing their exemplarity and thus turning them into enlightened local thinking. While claiming to be analogic, as I said earlier, the science of the emergent paradigm also claims to be a translator. That is to say, it encourages the emigration of concepts and theories developed locally to other cognitive spheres and their utilization outside their original context. Such a procedure, which was suppressed in the form of knowledge that conceived via operationalism and generalized via quantification and uniformity, is quite normal in the form of knowledge that conceives via the imagination and generalizes via quality and exemplarity.

Though total, postmodern knowledge is not deterministic; though local, it is not descriptive. It is knowledge about the conditions of possibility of human action projected into the world from local time-spaces. Such knowledge is relatively unmethodical, since it springs from methodological plurality. Each method is a language, and reality replies in the language of the question. Only a constellation of methods can capture the silence between each language asking questions. In the phase of scientific revolution we are traversing, this plurality of methods is possible only through methodological transgression.[11] If it is true that each method clarifies only that which is convenient for it, and when it does clarify anything, does not allow for any major surprises, scientific innovation consists in inventing persuasive contexts that allow the application of methods outside their natural habitat. Since the narrowing of the gap between the natural and the social sciences will bring the former nearer to the latter, we might wonder whether it is possible, for example, to do a philological analysis of an urban project, to interview nature, or to conduct participant observation among computers.

Methodological transgression affects the literary styles and genres that govern scientific discourse. Postmodern science does not adopt a unidimensional, easily identifiable style; it uses a configuration of styles constructed according to the scientist's criteria and personal imagination. Discursive tolerance is the other side of methodological plurality. In this transition phase there are already clear signs of this process of stylistic fusion, of interpenetration of writing canons. Clifford Geertz (1983, 20) studied this phenomenon in the social sciences and presented a few examples: philosophical investigation that sounds like literary criticism in Sartre's work on Flaubert; baroque fantasies under the guise of empirical observations in Jorge Luis Borges; parables presented as if they were ethnographic research in Carlos Castañeda; epistemological studies in the form of political texts, as in Paul Feyerabend's *Against Method*. And like Geertz, we might ask if Foucault is a historian, a philosopher, a sociologist, or a political scientist. The individualized, transdisciplinary composites indicated by these examples suggest a movement toward a greater personalization of highest valued scientific knowledge. And here we arrive at the third characteristic of scientific knowledge in the emergent paradigm.

All Knowledge Is Self-Knowledge

Modern science consecrated man as epistemic subject but expelled him, as it had expelled God, as empirical subject. Rigorous, factual, objective knowledge could not permit the interference of human or religious values. This is the ground for the dichotomous distinction between subject and object. However, this distinction was never as acceptable in the social sciences as in the natural sciences, and, as already stated, this was the supposed cause of the former's backwardness. After all, the objects of study were men and women, like those studying them. The epistemological distinction between subject and object had to be methodologically articulated with the empirical distance between subject and object. This is quite obvious if we compare the methodological strategies of cultural and social anthropology on the one hand, and of sociology on the other. In anthropology the empirical distance between subject and object was enormous. The subject was the "civilized" anthropologist; the object was the "primitive" (or "savage") people. In this case, it was acceptable, even necessary, that the distance between the subject and the object be narrowed down by means of methodologies calling for a greater intimacy with the object, namely ethnographic fieldwork and participant observation. In sociology, on the other hand, the empirical distance between subject and object was small or even nonexistent. Scientists did research on their fellow citizens. In this case, the epistemological distinction required a widening of the distance by means of methodologies of detachment: for example, the sociological questionnaire, content analysis, statistics, and the structured interview.

Anthropology, between postwar decolonization and the Vietnam war, and sociology, since the late 1960s, have come to question this methodological status quo, as well as the underlying notions of social detachment. All of a sudden, the savages were seen to be within us, within our societies, and sociology proceeded more often to utilize methods (like participant observation), which earlier had been almost the monopoly of anthropology. At the same time, in anthropology the objects had become peers, full-fledged members of the United Nations, and had to be studied according to sociological methods. The effect of these shifts in distinction between subject and object in the social sciences finally exploded in the poststructuralist period.

In the field of the natural sciences, the return of the subject had already been announced by quantum mechanics when it demonstrated that the act of knowledge and the product of knowledge were inseparable. The developments in microphysics, astrophysics, and biology during the last decades have restored to nature the properties of which they had been deprived by modern science. The deepening of knowledge conducted according to a materialist matrix finally emerged as an idealistic knowledge. The new dignity of nature was further strengthened by the realization that disordered technological development had separated us from, rather than united us with, nature, and that the exploitation of nature entailed the exploitation of man. The uneasiness that the distinction between subject and object had always provoked in the social sciences thus spread to the natural sciences. The subject was coming back donning the object's garb. Furthermore, Bateson's concepts of the "immanent mind," the "wider mind," and the "collective mind," as well as many others, represented dispersed indications that that other fugitive of modern science, God, might be about to return. He will return transfigured, with nothing divine about him except our desire for harmony and communion with all that surrounds us and which is, we now see, our innermost self. A new gnosis is in the process of gestation.

We can assert today, to paraphrase Clausewitz, that the object is the continuation of the subject by other means. Therefore, all scientific knowledge is self-knowledge. Science does not discover; rather it creates. And the creative acts performed by each scientist and by the scientific community as a whole must be understood intimately before we can use this knowledge to know reality. Metaphysical presuppositions, systems of belief, value judgments do not come before or after the scientific explanation of nature or society; they are part and parcel of it. Modern science is not the only possible explanation of reality; and there is no scientific reason whatsoever that it should even be considered better than the alternative explanations of metaphysics, astrology, religion, art, or poetry[12]. The reason why we give priority today to a form of knowledge based on the prediction and control of phenomena has nothing to do with science. It is a value judgment. The scientific explanation of phenomena is the self-justification of science as the central phenomenon of our contemporaneity. Thus, science is autobiographical.

The consecration of modern science in the course of the past 400 years has naturalized the explanation of reality, to the extent that we cannot conceive of the real except in ways that science offers us. Without such categories as space, time, matter, and number—the cardinal metaphors of modern physics according to Roger Jones (1982)—we are incapable of thinking, even if we are now capable of considering them conventional, arbitrary, metaphorical categories. The process of naturalization was slow. The protagonists of the scientific revolution had a clear sense that the innermost proof of their personal convictions preceded and gave coherence to the external proofs they developed. Descartes revealed better than anyone else the autobiographical character of science. He wrote in his *Discourse on Method*: "I shall be glad . . . to reveal in this discourse what paths I have followed, and to present my life in it as in a picture, so that everyone may judge it for himself; and thus, learning from public response the opinions held of it, I shall add a new means of self-instruction to those I am accustomed to using" (Descartes 1988, 21). Today we know, or suspect, that our personal trajectories and that of the collective scientific community, as well as the values, the beliefs, and prejudices they bring with them, are the innermost proof of our knowledge, without which our laboratory or archival research, our calculations or our field work would be no more than a tangle of meaningless efforts, from beginning to end. Nonetheless, such knowledge of our life trajectories and values, of which we may or may not be aware, flows in subterranean, clandestine ways, in the unspoken presumptions of our current scientific discourse.

In the emergent paradigm, the autobiographical and self-referential character of science is fully acknowledged. Modern science has bequeathed us a functional knowledge of the world, which has enlarged to an extraordinary degree our prospects of survival. Today, the question is not how to survive, but how to live. This requires another form of knowledge, holistic, intimate knowledge, that does not separate us from, but rather connects us personally with, whatever we study. The uncertainty of knowledge, which modern science has always viewed as a technical limitation to be gradually overcome, is transformed into the key to understanding of a world that must be contemplated rather than controlled. This has nothing to do with the medieval wonderment before a hostile reality haunted by the divine spirit. It is rather a sense of prudence before a world, which, even though it be tamed, reveals to us each day the precarious meaning of our life, however stable it may be at the level of survival. The knowledge of the emergent paradigm is more contemplative than active. The quality of knowledge is assessed not only by what it controls or operates in the external world but by the personal satisfaction it brings to whoever enjoys and partakes of it.

The esthetic dimension of science has been acknowledged by scientists and philosophers of science from Poincaré to Kuhn, from Polanyi to Popper. Roger Jones (1982, 41) thought that Newton's system was a work of art as much as a work of science. Scientific creation in the emergent paradigm is becoming ever nearer to literary or artistic creation. They share the belief that the active dimension of the transformation of reality (the sculptor chiseling the block of stone) should be subordinated to the contemplation of the result (the work of art). Scientific discourse, in its turn, is getting increasingly close to the discourse of literary criticism. In a way, literary criticism presages the subversion of the subject/object relation, which the emergent paradigm seeks to effect. In literary criticism, the object of study, as we would call it in scientific language, has always been, in effect, a supersubject (a poet, a novelist, a dramatist) in relation to whom the critic was no more than a secondary subject or secondary author. It is true that, in recent times, the critic has been tempted to outdo the writer under scrutiny, to the point that we might even speak of a struggle for supremacy[13]. But precisely because it is a struggle, it involves two subjects rather than a subject and an object. Each is the other's translator; both are creators of texts. Their texts are written in different languages, but both languages are necessary to learn how to appreciate the words and the world.

Thus resubjectified, scientific knowledge teaches us how to live and becomes everyday know-how. Hence the fourth and last characteristic of postmodern science.

All Scientific Knowledge Aims at Becoming Common Sense

We know today that modern science teaches us very little about our way of being in the world, and that the little that it teaches us will always be scanty, no matter how much we increase it, since its scantiness is inscribed in the very form of knowledge it constitutes. Modern science produces both knowledge and ignorance. If it turns the scientist into someone with specialized ignorance, it turns the ordinary citizen into someone with generalized ignorance.

On the other hand, postmodern science knows that no single form of knowledge is in itself rational; only their collective configuration can be rational. Therefore it tries to enter into a dialogue with other forms of knowledge and permits itself to be influenced by them. The most important other form is common sense knowledge, the ordinary, practical knowledge that guides our everyday behavior and gives meaning to our life. Modern science built itself against common sense, which it deemed superficial, illusory, and false. Postmodern science tries to rehabilitate common sense, for it recognizes in this form of knowledge some capacity to enrich our relationship with the world. Common sense knowledge tends to be a mystified and mystifying knowledge, but in spite of that, and in spite of its conservative quality, it does have a Utopian and liberating dimension that may be enhanced by its dialogue with scientific knowledge. This Utopian, liberating quality may be seen to flourish in many different characteristics of our common sense knowledge.

Common sense collapses cause and intention. It rests on a worldview based on action and on the principle of individual creativity and responsibility. Common sense is practical and pragmatic. It reproduces knowledge drawn from the life trajectories and experiences of a given social group, and asserts that this link to group experience renders it reliable and reassuring. Common sense is self-evident and transparent. It mistrusts the opacity of technological objectives and the esoteric nature of knowledge, arguing the principle of equal access to discourse, to cognitive and linguistic competence. Common sense is superficial, because it disdains structures that cannot be consciously apprehended, but for the same reason it is expert at capturing the horizontal complexity of conscious relations both among people and between people and things. Common sense knowledge is nondisciplinary and nonmethodical. It is not the product of a practice expressly devised to create it; it reproduces itself spontaneously in the daily happenings of life. Common sense accepts what exists as is. It favors actions that do not provoke significant ruptures in reality. Common sense is rhetorical and metaphorical; it does not teach, it persuades. Finally, common sense, in Dewey's words, fuses use with enjoyment, the emotional with the intellectual and the practical.

In the light of what was said about the emerging paradigm, these characteristics of common sense hold the virtue of foreknowledge. Left to itself, common sense is conservative and may well legitimate claims to superior knowledge. However, once articulated with scientific knowledge, it may be the source of a new rationality—a rationality comprised of multiple rationalities. For this configuration of knowledge to occur, I propose a *double epistemological break*. By it I mean that, once the first epistemological break is accomplished (thus allowing modern science to distinguish itself from common sense). There is another important epistemological break to perform, and that is to break with the first epistemological break so as to transform scientific knowledge into a new common sense. In modern science the epistemological break symbolizes the qualitative leap from common sense knowledge to scientific knowledge; in postmodern knowledge the most important leap is that from scientific knowledge to common sense knowledge. In other words, postmodern knowledge must break with the mystified and mystifying conservative common sense, not in order to create a separate, isolated form of superior knowledge, but rather to transform itself into a new, emancipatory common sense. Postmodern scientific knowledge fulfills itself only insofar as it becomes translated into common sense. Only thus will it be a clear science that fulfills Wittgenstein's dictum: "Everything that can be said, can be said clearly" (Wittgenstein 1981, 4.116). Only thus will it be a transparent science that does justice to Nietzsche's desire that "all commerce among men aims at letting each one read upon the other's soul, common language being the sound expression of that common soul" (Nietzsche 1971, 139).

By becoming common sense, postmodern science does not shun the knowledge that produces technology, but does believe that, as knowledge must translate into self-knowledge, so technological development must translate into life wisdom. The latter points out the markers of prudence for our scientific adventure, prudence being the acknowledgment and control of insecurity. Just as Descartes, at the threshold of modern science, experienced doubt rather than suffered it, we, too, at the threshold of postmodern science, should know insecurity rather than suffer it.

In this phase of scientific transition and revolution, such insecurity derives from the fact that our epistemological reflection is far more advanced and sophisticated than our scientific practice. None of us could at this moment visualize concrete research projects that might fully correspond to the emergent paradigm I have sketched here. That is precisely because we are in a period of transition. We have sufficient doubts about the past to imagine the future, but we live too much in the present to accomplish the future in it. We are divided and fragmented. We know we are on the path, but not how far along in the journey we are. The epistemological condition of science is visible in the existential condition of the scientists. In the end, if all knowledge is self-knowledge, then all ignorance is self-ignorance.

NOTES

1. According to Bacon, "the ways that lead man to power and to science are very close, indeed they are almost the same" (1960, 110). Bacon also says that, if the aim of science is to dominate nature, it is equally true that "[n]ature to be commanded must be *obeyed*" (1960, 39; my italics). The latter assertion has, however, not always been stressed as it ought to have been in interpretations of Bacon's theories of science.

2. Among many other passages of the *Dialogue Concerning the Two Chief World Systems*, see the following speech by Salviati:

> Taking man's understanding intensively, in so far as this term denotes understanding some propositions perfectly, I say that the human intellect does understand some of them perfectly, and thus in this it has as much absolute certainty as Nature itself has. Of such are the mathematical sciences alone; that is, geometry and arithmetic, in which the Divine intellect indeed knows infinitely more propositions, since it knows all. But with regard to those few which the human intellect does understand, I believe that its knowledge equals the Divine in objective certainty, for here it succeeds in understanding necessity, beyond which there can be no greater sureness. (Galileo 1970, 103)

3. Einstein's admiration for Galileo is well expressed in the "Preface" referred to above (Einstein 1970). His (instinctively) radical way of "seeing" the mathematical nature of the structure of matter partly explains his long struggle over the interpretation of quantum mechanics (especially against the Copenhagen interpretation). On this point, see Hoffmann (1973, 173ff.).

4. See, among many others, Pollard (1971, 39).

5. The impact of Gödel's theorems on philosophy of science has been assessed in different ways. See, e.g., Ladrière (1967, 312ff.), Jones (1982, 158), Parain-Vial (1983, 52ff.), Thom (1985, 36), and Briggs & Peat (1985, 22).

6. See Benseler, Heij, and Koch (1980).

7. See Capra (1979, 11ff.).

8. Bunge writes: "The causal principle is, in short, neither a panacea nor a myth; it is a general hypothesis subsumed under the universal principle of determinacy,

and has an approximate validity in its proper domain" (Bunge 1979, 353). Armando de Castro has produced remarkable theoretical work in this field (Castro 1975; 1978; 1980; 1982; 1987).

9. See Bowen (1985, 213ff.).

10. See chapter 20.

11. On the concept of methodological transgression, see Santos (1981, 275ff.).

12. This statement has stirred vivid controversy. See also chapter 7 in this volume. By it I want to stress two ideas. First, there is a plurality of knowledges circulating in society and their renditions of natural/social reality are either complementary or conflicting. No practice, not even the laboratorial practice, is informed by a single form of knowledge. Second, all the different knowledges generate their own epistemologies (their criteria for validating knowledge). Since each one of them (including science) evaluates the others in its own terms, there is no overarching epistemological reason to declare, in absolute or general terms, some evaluations more credible or consistent than others. The criteria—whose presence is crucial to prevent constructivism from being confused with relativism—must emerge out of argumentative, context-dependent, pragmatic, and democratically generated audiences.

13. A struggle exemplarily illustrated by the literary criticism of Harold Bloom, starting with his seminal book on the anxiety of influence (1973).

REFERENCES

Alien, Peter. 1981. "The Evolutionary Paradigm of Dissipative Structures," in E. Jantsch, ed., *The Evolutionary Vision*. Boulder, CO: Westview Press, 25–72.

Bacon, Francis. 1960. *The New Organon and Related Writings*. Ed. Fulton H. Anderson. New York: Liberal Arts Press.

Bateson, Gregory. 1985. *Mind and Nature*. London: Fontana.

Benseler, Frank, Peter M. Heijl, and Wolfram K. Koch, eds. 1980. *Autopoiesis, Communication, and Society: The Theory of Autopoietic Systems in the Social Sciences*. Frankfurt: Campus.

Bloom, Harold. 1973. *The Anxiety of Influence: A Theory of Poetry*. New York: Oxford University Press.

Bohm, David. 1984. *Wholeness and the Implicate Order*. London: Ark Paperbacks.

Bowen, Margarita. 1985. "The Ecology of Knowledge: Linking the Natural and the Social Sciences." *Geoforum* 26 (2):213–25.

Briggs, John, and F. David Peat. 1985. *Looking-Glass Universe: The Emerging Science of Wholeness*. London: Fontana.

Brillouin, Leon. 1956. *Science and Information Theory*. New York: Academic Press.

Bunge, Mario. 1979. *Causality and Modern Science*. New York: Dover.

Capra, Fritjof. 1979. "Quark Physics without Quarks: A Review of Recent Developments in S-Matrix Theory." *American Journal of Physics*, 47, no. 1, Jan.:11–23.

———. 1983. *The Turning Point*. New York: Bantam Books.

———. 1984. *The Tao of Physics*. New York: Bantam Books.

Castro, Armando de. 1975–1987. *Teoria do conhecimento científico*. 5 vols. Porto, Portugal: Limiar.

Chew, Geoffrey. 1970. "Harden Bootstraps: Triumphy or Frustration?" *Physics Today,* 23, no. 10, Oct.:23–28.

Descartes, René. 1988. *Selected Philosophical Writings*. New York: Cambridge University Press.

Durkheim, Emile. 1951. *Suicide: A Study in Sociology*. Glencoe, IL: Free Press.

——. 1982. *The Rules of Sociological Method*. London: Macmillan.

Eigen, M., and P. Schuster. 1979. *The Hypercycle: A Principle of Natural Self-Organization*. Heidelberg: Springer.

Einstein, Albert. 1970. "Preface." In *Dialogue Concerning the Two Chief World Systems*. C. Galileo. Berkeley: University of California Press.

Galileo, Galilei. 1970. *Dialogue Concerning the Two Chief World Systems*. Berkeley: University of California Press.

Geertz, Clifford. 1983. *Local Knowledge: Further Essays in Interpretative Anthropology*. New York: Basic Books.

Gödel, Kurt. 1931. *On Formally Undecidable Propositions of Principia Mathematica and Related Systems*. Edinburgh: Oliver and Boyd, 1962.

Haken, Herman. 1977. *Synergetics: An Introduction*. Heidelberg: Springer.

——. 1985. "Synergetics: An Interdisciplinary Approach to Phenomena of Self-Organization." *Geoforum*, 26 (2):205–11.

Heisenberg, Werner. 1958. *The Physicist's Concept of Nature*. New York: Harcourt, Brace.

——. 1971. *Physics and Beyond*. London: Alien & Unwin.

Hoffmann, Banesh. 1973. *Albert Einstein: Creator and Rebel*. New York: New American Library.

Jantsch, Erich. 1980. *The Self-Organizing Universe: Scientific and Human Implications of the Emerging Paradigm of Evolution*. Oxford: Pergamon.

——. 1981. *The Evolutionary Vision: Toward a Unifying Paradigm of Physical, Biological, and Sociological Evolution*. Boulder, CO: Westview Press.

——. 1981. "Uniting Principles of Evolution." Ed. Erich Jantsch, 83–115.

Jones, Roger. 1982. *Physics as Metaphor*. New York: New American Library.

Kepler, Johannes. 1939. *Welt-Harmonik*. Munich: Oldenbourg.

Koyré, Alexandre. 1981. *Entrtliens sur Descartes*. New York and Paris: Brentano's.

Kuhn, Thomas. 1962. *The Structure of Scientific Revolutions*. Chicago: University of Chicago Press.

Ladriére, Jean. 1967. "Les limites de la formalization." In *Logique et connaissance scientifique*, ed. Jean Piaget. Paris: Gallimard, 312–32.

Lovelock, James E. 1979. *Gaia: A New Look at Life on Earth*. Oxford: Oxford University Press.

Luckham, Robin, ed. 1981. *Law and Social Enquiry: Case Studies of Research*. Uppsala: Scandinavian Institute of African Studies.

Maturana, Humberto R., and Francisco Varela. 1973. *De máquinas y seres vivos*. Santiago de Chile: Editorial Universitaria.

——. 1975. *Autopoietic Systems*. Urbana, IL: Biological Computer Laboratory of Illinois.

Montesquieu, Charles de Secondat, baron de. 1989. *The Spirit of the Laws*. Cambridge: Cambridge University Press.

Nagel, Ernest. 1961. *The Structure of Science: Problems in the Logic of Scientific Explanation*. New York: Harcourt, Brace & World.
Nietzsche, Friedrich. 1971. "Rhetorique et langage." *Poétique*, 2 (5):99–144.
Parain-Vial, Jeanne. 1983. *Philosophie des sciences de la nature: tendances nouvelles*. Paris: Klincksieck.
Poirier, Rene. 1983. "Preface" to J. Parain-Vial.
Popper, Karl. 1962. *Conjectures and Refutations; The Growth of Scientific Knowledge*. New York: Basic Books.
Prigogine, Ilya. 1980. *From Being to Becoming*. San Francisco: Freeman.
———. 1981. "Time, Irreversibility, and Randomness." Ed. E. Jantsch, 73–82.
Prigogine, Ilya, and Isabelle Stengers. 1984. *Order out of Chaos: Man's New Dialogue with Nature*. Boulder, CO: New Science Library.
Reichenbach, Hans. 1970. *From Copernicus to Einstein*. New York: Dover Publications.
Rousseau, Jean Jacques. 1964. *The First and Second Discourses*. New York: St. Martin's Press.
Santos, Boaventura de Sousa. 1981. "Science and Politics: Doing Research in Rio's Squatter Settlements." Ed. R. Luckham, 261–89.
Thom, René. 1985. *Paraboles et catastrophes*. Paris: Flammarion.
Vico, Giambattista. 1968. *The New Science of Giambattista Vico*. Ithaca, NY: Cornell University Press.
Weber, Max. 1985. *Methodologische Schriften*. Frankfurt am Main: Fischer.
Wigner, Eugene. 1970. *Symmetries and Reflections: Scientific Essays*. Cambridge: Cambridge University Press.
Winch, Peter. 1970. *The Idea of a Social Science and its Relation to Philosophy*. London: Routledge & Kegan Paul.
Wittgenstein, Ludwig. 1981. *Tractatus Logico-Philosophicus*. Trans. C. K. Ogden. London: Routledge.

I

NEITHER TRUCE NOR SURRENDER: ARE THE SCIENCES WARS OVER?

2

Pathways to Cosmopolitan Knowledges

In the Aftermath of the Discourse on the Sciences

João Arriscado Nunes

A Discourse on the Sciences is a concise, dense and accessible treatment of three analytically differentiated though intertwined thematic orientations. The first consists of a general characterization of the history of the sciences, dealing with the latter as constituting a paradigm, but focusing as well on their diversity, and on the tensions and epistemological dilemmas that have cut across them. The second orientation identifies the signs of a crisis of that paradigm, based on a range of symptoms that have emerged throughout the twentieth century and, in particular, through some of the debates of the 1980s. The latter involved different forms of questioning, in a critical and transversal way, the general assumptions of modern science as a unifying project, as well as the specific assumptions of the different sciences. The expression "crisis" does not mean, in this context, the collapse of the modern sciences, but rather a condition opening up spaces and opportunities for transforming interventions without a guarantee of what the outcomes of those interventions will be.[1] What is at stake here, as Isabelle Stengers has put it, is the identification of the "critical spaces" where "distinctions of scale" disappear and "multiple dimensions which are usually kept apart" are made to resonate (Stengers 1996/97, vol. 3, 99). The last line of argument consists of what the author himself describes as an exercise in speculation on how the emerging forms of knowledge look like and what the possible consequences of the abovementioned crisis may be.

The text is organized as a characterization of a critical state, which, on the one hand, allows the reconstruction of a past and, on the other, questions emerging signs which, in the present, point toward possible futures. The reconstruction of the past makes it possible to create links and to bring together themes converging within a space of turbulence that cuts across the disciplinary order of the modern sciences. The identification of modern science with a paradigm, an epistemological model common to the different disciplines and areas of knowledge, but with some room for internal diversity—as stressed by the author—allows the confrontation of a range of themes, objects, theories, exemplars and conceptions which "make" dominant science with the outlines of an alternative paradigm.

The extension toward the future appears in the form of a speculative reflection on the possibilities opened up by the paradigmatic crisis. Clues to these possibilities are identifiable by whoever travels across the "disunified" territory of the sciences. The latter may be defined, in a general way, as organized modes of exploring unknown areas of the world and of putting to the test beliefs and conjectures, through procedures, which allow the latter to be confronted with the resistance or recalcitrance of the world and of the entities that exist in it. But this task may be fulfilled in different ways, on the basis of different epistemological models and drawing on a variety of research procedures, including experimenting, controlled observation, building models and theories, historical and documentary analysis, statistical analysis, simulation, different versions of fieldwork, interviewing, and so on. A consequence of this variety is that the competence and skills acquired in one area or discipline does not imply competence in others. From this point of view, a scientist is no different from the "common" citizen, who is usually an expert in one area (a profession or occupation, for instance) and a "layperson" in others. Besides, if it is true that all disciplines and scientific areas exist in society, if they are "vascularized" through their articulations with other domains of social life, the articulations themselves appear in different modes.

The confrontation between the paradigm (in crisis) of modern science and the emerging order of knowledge—itself described as an alternative paradigm—should not conceal their radical asymmetry. The author himself points at this asymmetry when he states that the "synthesis I have in mind [of the natural and social sciences and the humanities] does not aim at a unified science or even at a general theory, but merely at a set of theoretical aqueducts into which can converge various currents which have hitherto been considered theoretically separate" (Santos 1992, 12).[2] The unfolding of the reflections advanced in *A Discourse* underline the importance of what appears to some readers and critics as the most controversial statement included in the text (the opening and closing sentences of this passage are often omitted by critics):

> Metaphysical presuppositions, systems of belief, value judgments do not come before or after the scientific explanation of nature or society: they are part and parcel of it. Modern science is not the only possible explanation of reality; and there is no scientific reason whatsoever that it should even be considered better than the alternative explanations of metaphysics, astrology, religion, art, or

poetry. The reason why we give priority today to a form of knowledge based on the prediction and control of phenomena has nothing to do with science. It is a value judgment. The scientific explanation of phenomena is the self-justification of science as the central phenomenon of our contemporaneity. Thus, science is autobiographical. (Santos 1992, 41)

As is clear from the argument of *A Discourse* as a whole, what is at stake in this passage is not a disqualification of science *versus* other modes of engagement with the world, but the need to define the relevance of the various modes of knowledge (and of their articulations) as a function of the context, the situation and the aims of those who mobilize them. Science is thus denied the privilege of legislating on other forms of knowledge or experience, whose problems and questions cannot be reduced to those scientific knowledge defines as relevant. In his subsequent work, Boaventura de Sousa Santos has clarified this stance through his contrast of the two main forms of knowledge bequeathed to us by modernity: knowledge-as-regulation and knowledge-as-emancipation. The former treats the diversity of modes of knowing and of intervening in the world as symptoms of a chaos, made of irrationality and of prescientific notions openly hostile to science, which can be turned into order only through the exercise of rationality as enacted by science. For the latter, in contrast, the pretension of science and scientific rationality to legislate on other forms of knowledge and experience corresponds to a situation of colonialism, built on the marginalization, discrediting or sheer termination of that which cannot be reduced to the demands of the scientific/technical rationality. The author proposes, instead, a solidaristic conception of knowledge, based on the coexistence, dialogue and articulation of modes of knowledge and experience, avoiding mutual disqualification. This is the form of knowledge, which should be privileged during the current period of transition. Far from being a plea for an epistemological "anything goes," this stance requires that different modes of knowledge be assessed according to the contexts and situations in which they are enacted and of the aims of those who enact them, without making them dependent on global demands for rationality which ignore the situated features of the production and appropriation of all forms of knowledge and of their consequences for persons and places bearing a singularity which is the outcome of their histories (Santos, 1991, 1995, 2002a).[3]

CHANGING THE EPISTEMOLOGICAL DEBATE

The original publication of *A Discourse* in 1987 offered an interpretation of the epistemological debates carried out "by the scientists themselves, those that have mustered the necessary philosophical competence and concern to problematize their own scientific practice (...). The second facet of this reflection is that it deals with questions previously left to sociologists" (Santos 1992, 26). Since physics appears as the discipline more often called upon as a model of scientificity, it will not be surprising that some of the most important questions on the concepts of law or causality or on the issues of time and irreversibility were raised within physics, namely through the recovery and reformulation of problems which, throughout the history of the discipline, had been pushed into oblivion, marginality or irrelevance.[4] *A Discourse* was an attempt at exploring the implications of those questions both for the "transversal" changes in the conditions of scientific knowledge production and for the conception of knowledge itself.

In later work, the author expanded, deepened and developed many of the themes dealt with in *A Discourse* and, in the process, identified emerging problems. *Introdução a uma Ciência Pós-Moderna* (1989) explored themes such as the "double epistemological break," the rhetorical and hermeneutic dimensions of the production and social appropriation of scientific knowledge, the development of an epistemology inspired by philosophical pragmatism, the epistemological problems which are specific of the social sciences or the social and political conditions of scientific activity, including the discussion of the performance of the autonomy of scientific activity through "regulatory conversion." At the time of its publication, the book was the object of a lively public debate of the author with a physicist and a sociologist.

Over the following years, and through a series of contributions which had their peak in *Toward a New Common Sense* (1995), the approach outlined in *A Discourse* went through a process of "de-centering," in order to take into account the relevance of the experiences of the various forms of colonialism, of multiculturalism and of the diversity of modes of knowledge for the historical and contextual understanding of the modern sciences. This was set within a broader and more detailed approach of the global changes in the capitalist world-system over the last five centuries and of the processes of globalization underway since the later decades of the twentieth century, emphasizing the diversity of modes of knowledge and of the social conditions of their production, appropriation, circulation and situated articulation is given particular emphasis. The different spaces of social practices which structure social life are themselves spaces of production of forms of power, rights and knowledges which cannot be reduced to state power and to state law, nor to the kinds of scientific or formal knowledge sanctioned and certified by scientific and academic institutions.

This work has been further developed through a focus on North/South relations, and, in particular, on the expression of these relations through rival knowledges, and the issue of biodiversity and intellectual property, thus broadening the original concerns of *A Discourse*, in order to include a discussion of the epistemological and social conditions of the emergence of multicultural, cosmopolitan and emancipatory conceptions of knowledge (Santos 2002, 2004).

The broadening of the ground originally covered by *A Discourse* kept pace with the reflection and empirical research on the momentous changes undergone by the world, by knowledge and science and by science-society relations since the mid-1980s, such as the end of the Cold War and the ensuing trend toward the relative decline of those disciplines and research areas more directly linked to the development of weapons or whose funding was largely dependent on their contributions to technologies with potential military applications. Physics was one of the most strongly affected disciplines, and the episode of the suspension of the super-accelerator-collider in the United States, in the early 1990s, was perhaps the strongest hint at how times were changing. The parallel rise of the life sciences—namely of genetics—and their move toward a "big science" model of organization, thanks to a large extent to the Human Genome project and to its sequels and extensions, was to push biology into a central role in the debates on knowledge and science throughout the 1990s.

The rise of biology had a significant role in the formulation of new questions and of new research themes with implications which cut across different areas of knowledge: what is a gene, and what genes can and cannot do; the relationship between genetics, developmental biology and evolutionary biology; the nature-nurture link; the implications of the use of different languages to deal with biological—and, in particular, genetic—phenomena, such as the physical-chemical definitions and descriptions of life or the language of information and code; the relationships between the biotechnology industries and research in the life sciences; the issue of the property of living entities or processes related to life; the regulation of research in, and social uses of, genetics and, more generally, new tools for the manipulation of life and of reproduction. These were just some of the issues, which were to extend and broaden, within and beyond the life sciences, the controversies and debates discussed in *A Discourse*.[5] This emergent space of research, innovation and debate included not only biologists working within different subdisciplines and specialties, biochemists and physicists who "migrated" to the life sciences, but philosophers, historians, sociologists, anthropologists, political scientists, cognitive scientists, linguists, psychologists, ethicists, legal scholars, theologians, representatives of the biotechnology industry, and environmental, consumer, patient, human, and animal rights activists as well.

Throughout the 1990s, it became clear that some of the new directions of research would be easily integrated by more conventional disciplinary orientations. Others, however, were to open up new fields and contributed to reorientations, which often drew on questions and on currents, which had been marginalized or forgotten throughout the history of the stabilization of disciplinary orthodoxies.[6] In a first moment, the "new sciences," as some have called them, encompassed a broad range of themes, such as uncertainty and complexity, entropy, dynamic systems, self-organization, information, time, irreversibility and contingency, fractal geometry and chaos, among others.[7]

The term "complexity" was to become the main "attractor" of emerging approaches. But what some regarded as a new "science of complexity" (Lewin 1992) soon went through further bifurcations and differentiations which questioned the attempts at retrospectively endowing the set of themes subsumed under that word with a greater coherence than the one associated with the "disunified" territory of the traditional scientific disciplines.

In fact, a double process seems to have occurred: on the one hand, a "transversalization" of the themes associated with complexity and, on the other, a localized and differentiated appropriation of those themes within a variety of disciplinary spaces or domains of research, with different consequences and impacts, but leading in many cases to the opening up of new spaces of debate and controversy cutting across disciplines, subdisciplines and research areas. The example mentioned above of recent developments within biology is a good illustration of the tensions that arose between reductionistic (or, rather, gene- or molecule-centered) orientations and orientations toward complexity, and of how they shape the production of knowledge.[8] These developments had a significant influence on the turn of epistemological debates toward a growing acknowledgment of the relevance of epistemological orientations inspired by pragmatism and of conceptions of knowledge as the outcome of local, situated and heterogeneous processes of construction.[9]

The part of *A Discourse* which is more vulnerable to the passage of time and to confrontation with recent history is the one where the author, in a mood which is avowedly speculative but nonetheless informed by his previous reflections, provides an outline of what the new configurations of knowledge and the emerging features of the sciences may come to look like. Are there any signs, symptoms or exemplars shaping—even if in a partial and spatially confined way—the features which, according to Santos, allow us to read in the present time alternatives to the modern sciences? There are, undoubtedly, "strong" exemplars of transformations which open up spaces criss-crossed by contradictions and processes which are not always compatible, but which, nonetheless, offer opportunities for change, even if it is not clear whether they point toward a "prudent knowledge for a decent life." Studies of science, technology and society (STS) provide a privileged observatory of those spaces and exemplars. But we can find precious contributions to this task as well in both the developments of the "immanent" reflection carried out within the sciences, on the one hand, and the proliferation of critical movements and discourses promoting alternative modes of producing knowledge (which should not be confused with the ubiquitous and tentacular "antiscience movement" feeding the conspiratorial imagination of so many self-appointed defenders of science . . .).

ALL NATURAL-SCIENTIFIC KNOWLEDGE IS SOCIAL-SCIENTIFIC

Studies of science, technology and society have shown that even the modern sciences are the emergent and situated outcome of the intersection and dynamic articulation of human actors, nonhuman living entities, different kinds of materials, instruments, different skills, institutional and financial resources. The process through which these sets of heterogeneous entities are articulated in order to generate something that did not exist before and is more than the sum of previously existing elements is usually called construction. Both knowledge and technological objects are, in this sense, always constructed, within specific ecologies of practices (Stengers 1996/97). Opposing the "real" and the "constructed" thus makes little sense. The relevant distinction is to be made between "good" and "bad" constructions, as they are assessed within specific ecologies of practices. If they are "well" constructed according to the relevant criteria, these new entities may become, under certain conditions, autonomous form the contexts and situations they were constructed in, thus becoming available for further appropriations in other contexts and by other actors. In other words, both these processes of construction and the objects that are their outcomes will be assessed according to their consequences and to the difference they make. But the heterogeneous resources mobilized in construction cannot be manipulated in an arbitrary way. They resist these manipulations and are recalcitrant to them in different ways, thus turning construction into a process of mutual adjustment and accommodation over time. In this sense, construction, as the notion is used here, is always, in fact, a process *of co-construction*.

This perspective allows a different understanding, on the one hand, of the emergence of new domains and objects of research which cannot just be classified as natural or social and, on the other, of the proliferation, in contexts of daily life, of objects and situations which are the outcome of the "interweaving" of science, nature and technology (Michael 2000). Concepts enjoying a wide currency in STS, such as those of heterogeneous construction or heterogeneous engineering provide useful tools for the description and analysis of the production of these "new" objects and the questioning of the subject-object distinction (Taylor 1995; Latour 1999). Global climate change, human biodiversity, primatology, environmental science or cognitive science are instances of new domains whose objects are, at the same time, "natural" and "social," and which have led to innovative articulations of modes of knowledge and to cooperation and convergence among researchers of areas that have traditionally been kept apart by the "great divide" of the "two cultures." In some cases, the range of forms of knowledge acknowledged as relevant includes some of the "rival knowledges" of populations, social movements and communities in both the Southern and Northern hemispheres. It is not yet clear, though, to what extent the propensity to make a distinction between the "ethno-knowledges" of the "others" and "our" scientific knowledge can be counteracted in a durable way by the symmetrical recognition of the various kinds of knowledge as being, without exceptions, partial and situated and, thus, deserving of equal recognition and respect.

The case of biodiversity provides a good illustration of the implications of this approach. The definition of a biological order based on the establishment of boundaries of "populations" cannot be cut off from discourses on social and cultural differences and hierarchies, on the determination of those to whom the "biodiversity goods" belong, the conditions under which others may have legitimate access to these goods, or the uses these goods can be put to—such as trade and medical research on diseases affecting mostly Southern hemisphere populations—and the criteria and means for the redistribution of the positive outcomes—economic and others—of the uses of "biodiversity goods." Biodiversity is, in fact, a process of co-constructing an order which is inextricably political and scientific, technical and moral, biological and cultural, involving a differentiated and heterogeneous range of actors—scientists, indigenous peoples and their organizations, scientific institutions from both North and South, governments and local authorities and ecosystems (Santos 2004; Santos, Meneses and Nunes 2004).

ALL KNOWLEDGE IS SELF-KNOWLEDGE

The exploration of disciplinary histories in search of orientations and currents which, at some point in time, were defeated, marginalized or ignored, but which may provide new ways of questioning the present status of knowledge, is being articulated with critical voices arising from different quadrants—from indigenous movements defending their right to the resources of biodiversity and to the recognition of the value of their knowledges and experiences to environmentalist movements producing alternative or critical forms of knowledge and the feminist and multicultural critiques of the biases of research programs, theories and methodologies. Their aim is to influence the reorientation of scientific and technological research toward greater social and environmental responsibility.[10] Feminist perspectives are particularly relevant from that point of view, notwithstanding the caricatures often drawn by their critics. These perspectives were undoubtedly influential in opening up the access of women to scientific activity, in promoting changes in the organization of scientific careers and scientific work and in redefining the very contents of scientific knowledge in a range of disciplines, including the definition of research subjects, languages, images, research procedures and interpretations of results. A further outcome of feminist criticism was the redefinition of the boundaries of science and of the divide between science and other forms of knowledge, allowing the recognition of practices usually performed by women but generally excluded from certified knowledge, such as local economic activities or the local management of environment, health and reproduction.

These influences were certainly diverse across countries and scientific domains and disciplines. Among these, the health sciences, certain areas of biology—such as developmental biology or evolutionary biology, primatology, archaeology, psychology, sociology, anthropology, history or geography have incorporated—in diverse ways and degrees, to be sure—new questions and innovative, critical perspectives inspired by feminism. In other areas, such as physics or mathematics, the influence of feminist approaches seems to have been more limited or, at least, more difficult to assess. A particularly interesting case is that of primatology, where the majority of qualified researchers are women (80% of primatologists with a Ph.D.), the discipline often being described as an exemplary case of incorporation of feminist perspectives (Strum and Fedigan 2000).[11]

But two further consequences of feminist critique and of the debates over science and gender should be kept in mind. The first concerns the denaturalization and critical questioning of the historical processes of "masculinization" of the science worlds, sustained by institutions, practices and professional ideologies. The second has to do with the definition of the conditions associated with the constitution of the subjects of knowledge which are considered relevant to the establishment of objective knowledge—not only gender, but ethnicity, class, nationality or religion, for instance, a "strong" form of objective knowledge, linked to specific situations or "standpoints" (Harding 2003). The aim of feminist criticism is thus not the constitution of a "separate" science, but to contribute to the transformation of actually existing science, pushing forward and renewing the critical outlook that was at the origin of modern science, through the inclusion of new questions, perspectives, themes and practices, within renewed institutional and professional settings, in search of what Schiebinger (1999) describes as "sustainable science."

Another orientation which is particularly interesting as it carries forward the idea that all knowledge is self-knowledge is the one associated with recent research within cognitive science on the embodiment of knowledge and skills in human beings (Lakoff and Johnson 1999). This topic is of great relevance to areas such as public health and environmental protection and, in particular, to the intersections of both, as in dealing with "environmental illness." This type of health problem requires the use of one's own body and senses as means of producing knowledge on the environment and on environmental hazards and their effects on health. There is a need, however, to overcome the propensity to reduce embodiment to incorporation in the brain and to explore in more detail the social and cultural conditions of situated processes of embodiment.[12] As I shall argue in the next section, this approach is interesting as well for understanding the relationship between embodied capacities for dealing with, for instance, quantities, distances and relationships and the learning of formal skills in mathematics.

ALL KNOWLEDGE IS LOCAL AND TOTAL

The claim that "all knowledge is local and total" is well illustrated by the proliferation of forms of knowledge that enact it. The convergence of disciplines or knowledge domains in projects aimed at providing responses to social problems, to the development of technologies for sustainable development or to environmental or public health problems identified at the local level is expressed in a variety of experiments whose wealth is still largely unknown. The many projects within what has come to be called "mode 2" of knowledge production (Nowotny et al. 2001) are mostly carried out through cooperations among universities, the state and business and most often take private companies as the main locus of innovation.

But there are many other initiatives which try to mobilize scientific and cognitive resources of different kinds to respond to the needs of local populations, communities and citizen groups through inclusive, participatory processes and democratic decision-making involving all the parties and actors interested or affected. Experiments in this vein include the largely European-based science shops, community-based research, movements of scientists for public-interest research, collaborative work involving health authorities, researchers and activists on AIDS and other instances of joint work of research institutions, social and citizen movements and NGOs. These initiatives articulate different modes of knowing: analytical and formal knowledge (*episteme*), technical knowledge (*techné*), knowledge associated with prudent and fair decision (*phronesis*) and practical, local knowledges (*metis*), thus giving shape to the configurations of rationalities and reasonableness that make up "prudent knowledge for a decent life" (Santos 1995, 2002a; Toulmin 2001; Visvanathan and Escobar, this volume).

The criteria for assessing these forms of knowledge do not require a confrontation with universal, idealized patterns of what "good science" should look like, but rather situated assessments, questioning the "how" and the "what for" of knowledge production, the relationship between the work of constructing knowledges and their consequences or effects. The recent work of James Scott (1998), with its detailed accounts of the failures to achieve large scale, "rational" transformations of the world and of society, and how they ignored the relevance and need of the local, practical, "cunning" kinds of knowledge the Greeks referred to—often in a dismissive way—as *metis*, provides further support for one of the strongest ideas pervading *A Discourse*: "[N]o single form of knowledge is in itself rational: only their collective configuration can be rational" (Santos 1992, 44).

ALL SCIENTIFIC KNOWLEDGE AIMS AT BECOMING COMMON SENSE

The landscape described in the previous section takes us straight into the theme of the relationship between the sciences and common sense. This, too, was the subject of intense debate and research over the years since the publication of *A Discourse*. Under the label of "public understanding of science" or of the "promotion of scientific culture," we have witnessed the proliferation of initiatives allegedly conducive to "scientific citizenship" or "technical democracy," and to the growing understanding of the importance of "lay" knowledges, skills and experiences for the production of situated knowledges in contemporary societies, both North and South (Gonçalves 2000, 2002; Michael 2002; Callon et al. 2001; Irwin and Wynne 1996; Irwin and Michael 2003; Santos et al. 2004).

A crucial feature of these developments is the recognition that scientific knowledge cannot just be "inscribed" onto people's minds as if the latter were blank sheets. Throughout their life courses, all human beings acquire skills linked to a variety of situated forms of activity, which provide the starting points—as resources or as obstacles—to the acquisition of new skills and knowledges. It is easy to recognize the convergences between these themes and the approach proposed by Boaventura de Sousa Santos, in his more recent work, to the connections between the contexts of social practices, forms of power, of law and of knowledge. Achieving an adequate understanding of these forms of power, law and knowledge require that they be analyzed within the contexts of their mobilization, according to criteria, aims and values that are specific to these contexts. In many daily life situations or in situations of exception (such as natural disasters or industrial accidents, for instance), different forms of knowledge may be found to converge and confront one another or be articulated in a variety of ways. Depending on the kind of technology and the context of activity and of learning, the modes and channels of acquisition and proliferation of skills and knowledges can be very diverse. The media and information technologies are not the only competitors of school and family as contexts of learning. The importance of other contexts and processes—such as experiences at work and leisure or the links to experts who master science-based forms of knowledge, such as physicians or engineers—is increasingly acknowledged.

This question is particularly relevant for the recurrent discussions on the alleged limitations and deficits of knowledge in areas such as mathematics, including gender differences in mathematical skills. These limitations and deficits are most often identified through institutionalized forms of assessing knowledges and skills in the context of formal schooling. As suggested by recent research in cognitive science, however (see, for instance, Lakoff and Nuñez 2000), the skills associated with mathematics have their roots in a range of skills needed to deal with quantities, relationships, sets or spatial orientation, for instance, which may be represented through a set of "ground metaphors," such as "object collections," "object construction," "measurement pattern" or "movement along a trajectory." These, in turn, may undergo further metaphorical transformations into formal mathematical statements. Leaving aside, here, the detailed discussion of the arguably polemical approach proposed by Lakoff and Nuñez, the latter fits well with some of the work carried out by cognitive anthropologists on the uses of numeracy and of mathematics in daily life settings. This work has shown that the skills needed to deal with quantities, distances or relations are developed through participation in different forms of activity, and that they may display variations depending on type of activity, context and experience (Lave 1988, 1996). The metaphors drawn upon may be different, but it is possible nonetheless to assess the conditions under which they may be transformed into formal mathematical statements. The corollary of this approach is that, instead of considering mathematics as a formal and self-referential kind of knowledge, one may start from the ground metaphors and from the situated experiences of actors to understand how the formal statements of mathematics are acquired, without throwing away the experiences and skills of the actors which prove so useful in the different contexts and situations they deal with in daily life. Thus, an interesting and effective way of promoting the learning of formal mathematical skills would be to draw upon the knowledges of the actors as resources for formal learning, instead of disqualifying them as obstacles to learning.

The problems associated with what the German sociologist Ulrich Beck (1992) christened the "risk society" and, in particular, the uncertainties deriving from the social and environmental impacts of technological innovation, have contributed as well to transform the subject of the relationship between scientific and expert knowledge, on the one hand, and the "lay" knowledges and experiences of citizens, on the other, into a central and controversial aspect of the search for new forms of knowledge production and political decision-making under conditions of uncertainty associated with complex processes (Nunes 1998/99). Controversies, problems or disasters are often the "channel" through which "lay" citizens become interested in scientific and technical knowledge, which, under "normal" circumstances is encased within technological "black boxes."

There is still a long way to go until the diverse modes of knowledge engage mutually in a symmetrical and nondisqualifying way. But, as has been shown by detailed empirical research in the social studies of science and technology, this is a promising path toward broadening scientific knowledge without wasting other forms of knowledge and experience. The perspectives that frame recent collaborative projects, such as "Reinventing Social Emancipation," directed by Boaventura de Sousa Santos, are significant contributions to the broadening and redefinition of knowledge and of experience in innovative, multicultural and cosmopolitical directions, in a world characterized by the contradictory dynamics of hegemonic and counter-hegemonic globalizations.[13]

AN ANACHRONISTIC AND OUT-OF-PLACE POLEMICS

In 2002, *A Discourse on the Sciences* was the target of a violent attack by a Portuguese science popularizer (Baptista 2002). Besides the timing of the attack—fifteen years and thirteen editions after the original publication of *A Discourse* in Portugal—it is interesting to notice that the attack had as its broader target science and technology studies and the social sciences as a whole. No mention was made of the intellectual context in which *A Discourse* was published, nor of the transformations that cut across the sciences and the relationships of science and society over these fifteen years. Although the attack had some features, which are linked to the specificities of the Portuguese context, its author was explicit in his attempt to present it as a move in the "science wars." This, in fact, amounted to missing the context, the content and the aims of the discussions *A Discourse* was meant to be a contribution to, treating it as an episode in the "science wars" saga or as an extension of some of the "classical" debates in epistemology and the philosophy of science. This requires some clarification in order to assess the contribution of *A Discourse* to past and current debates on the sciences.

The "science wars" reached their peak in the 1990s, first in Britain, where the primary target of the "science warriors" included currents within the sociology of scientific knowledge and their most prominent proponents, such as the Strong Program of the Edinburgh School (David Bloor, Barry Barnes, David Edge) and the so-called Bath School (Harry Collins). In the United States, the "science wars" were to become a part of the broader struggle for cultural authority within a multicultural society known as the "culture wars." Within this struggle, physics was again promoted as the exemplary enactment of science as a true explanation of the world, and the language of physics as a meta-discourse of scientific rigor and truth. As seen by outside observers, the "science wars" thus appeared as localized phenomena, with diverse manifestations, to be sure, but hardly international in scope, and which took up again some of the time-honored topics of the debates on the "two cultures" (of the humanities and the sciences) and on epistemology (such as the confrontations of realist and antirealist currents and the polemics over relativism).[14]

The "export" of these polemics beyond the Anglo-Saxon world proved not to be an easy task. Its high point was the publication in France, in 1997, of *Impostures Intellectuelles*, by physicists Alan Sokal and Jean Bricmont. The book was based on a set of "personalized" denouncements of the references to science and of the uses of scientific language by prominent figures of French philosophy, apparently selected according to their influence within the U.S. academy and turned into tailor-made targets of an *aggiornamento* of the controversies over the two cultures and epistemology.[15] The translation of the book into several languages met with heterogeneous receptions, often appearing—as was the case in Portugal—as a strange body in contexts where the logic and the alignments of the "science wars" were only marginally relevant or where ongoing debates were focused on different issues.

It is noteworthy that, with the exception of the denouncing of the alleged misuses by philosophers and scholars in the humanities of terms such as "chaos," "complexity" and "nonlinearity"—an operation which assumes that these are terms with single meanings, fixed once and for all by their status as scientific concepts, thus disqualifying their use in other contexts and discursive configurations—the subjects dealt with by the "new sciences," their developments in different scientific areas and the questions they raise for conventional knowledge were hardly dealt with by the "science warriors"—notwithstanding some rare exceptions.[16] It is interesting to notice, too, how the self-proclaimed defenders of science tended to ignore the relevance of developments in other disciplines or areas of knowledge besides physics and mathematics, such as the life sciences, the cognitive sciences or the social sciences, or the broadening and deepening of the issues raised by the emerging constellations of knowledge which were the object of an early discussion in *A Discourse*.[17]

A detailed discussion of the internationalization of the "science wars" is beyond the scope of this chapter. But the allusion to the "science wars" is justified here, since there were attempts by the "science warriors" to enroll the author of *A Discourse* on the side of the enemies of science, ignoring, first, that this text had been written before the onset of the "science wars" and, secondly, that its purpose had not been to take a stand on the debate over the "two cultures," nor to resume past epistemological debates, but rather to provide an account of, and to intervene in, an emerging space of debate and controversy which, as Santos himself reminds us, is the outcome of reflections by scientists within disciplines such as physics, chemistry and biology, but reaching beyond disciplinary boundaries to acquire a "transversal" relevance.[18]

The interest of these debates for the social sciences is twofold. On the one hand, the discussion on the criteria of scientificity of the social sciences has always been associated with assumptions deriving from a model of science inspired by physics. Even those who argued for an alternative epistemological model for the social sciences did so on the basis of the differences between the objects of the natural and the social sciences, but they did not question the adequacy of Newtonian physics as a model for the natural sciences. On the other hand, the new space of debate focused on themes such as time, history, contingency, irreversibility or complexity, precisely some of the topics that, in the past, had fed the discussions on the differences between the natural and the social sciences and their objects.

This is not to suggest that the questions raised by "conventional" debates over epistemology or the problem of cultural authority which took such a central place in the "science wars" were irrelevant to understand the origins of the new currents of thought and work which, by the second half of the 1980s, had been subsumed under the label "new sciences" (and, later, "sciences of complexity") or that, conversely, the former were not affected by the latter.[19] Let me just mention two of the most interesting consequences of the opening up of this new space of debate. The first rests upon the proposal of a radical redefinition of the epistemological debate, shifting its focus from the opposition of realism and constructivism, the criteria for defining truth or the problem of relativism to the themes of the heterogeneity and constitutive complexity of reality, of self-organization and *autopoiesis*, of nonlinearity, of time and duration, of irreversibility and history, of language and rhetoric, of the cultural resources and the institutional forms that both constrain and make possible the production of scientific knowledge, of the participation of the researchers in the process of constructing knowledge, of the limits and validity of the statements of "classical" scientific disciplines, of the themes and concerns that are "transversal" to the natural sciences, the social sciences and the humanities. The second major consequence was the new attention given to the diversity of forms of knowledge and to the possible relationships between them based on experiences of shared authority, of mutual recognition of the partial and situated character of all knowledge and on giving up the disqualification of those forms of knowledge labeled as "nonscientific."

Throughout the 1980s and 1990s, these two types of consequences converged to stimulate the expansion and diversification of the research field known as social studies of science; social and cultural studies of science;

studies of science, technology and society; or just science studies. The diversity of designations itself is a sign of the proliferation of themes, theoretical orientations, disciplinary contributions and research programs they describe. In different countries and regions of the world, both North and South, the field of science studies has moved in different directions, sometimes identifying convergences in concerns and research questions, but always exploring problems related to specific historical paths and contexts. It will not be surprising then that within different contexts, concerns and debates are emerging which do not just recapitulate (even if under new guises and new vocabularies) past epistemological debates or import the ongoing polemics raging in some core countries. The attempt at importing the "science wars" into Portugal, for instance, failed to take into account the development of science studies in the country; the relationships between researchers in science studies and the wider scientific community; and the themes, orientations and spaces of debate that have accompanied the remarkable effort, particularly during the 1990s, to create a system of scientific and technological research to support scientific research and to train new generations of young researchers in all areas of the natural and social sciences and of the humanities.

Therefore, any critical assessment of *A Discourse on the Sciences* has to include a discussion of the intellectual context in which the text was produced, as well as of the links of that context to the changes in society and in science that have been going on since its original publication. This is a precondition of any serious evaluation of the ways in which Santos's text may help us understand the current conditions of production and social appropriation of the sciences within emerging configurations of knowledge, avoiding the confinement of this evaluation to the limits of conventional epistemological controversies or of the debate on the "two cultures," namely of its recent incarnation in the "science wars." In a world going through a period of turbulence, and given the possibility of human interventions to influence the dynamics of the world toward either the deepening and exacerbating of current forms of oppression, violence and inequality or a broadening and intensification of cosmopolitan, solidaristic and emancipatory initiatives, it makes a lot of difference, as Boaventura de Sousa Santos has shown us, what kind of knowledge we choose to construct. The aim of these reflections is to make a contribution—albeit modest to the indispensable debate and collective reflection on the options before us and on their (possible or likely) consequences.

NOTES

1. A key feature of Boaventura de Sousa Santos's work is the centrality of critical activity as a condition of possibility of interventions in the world aimed at chang-

ing it. This, in turn, rests upon the acknowledgment that whatever exists is the outcome of a historical process and, thus, neither necessary nor eternal. This is a key to understanding the interpretation of the crisis of modern science as it is depicted in *A Discourse on the Sciences*, and a condition for avoiding the frequent confusion between critical postmodern thought and "affirmative" or "celebratory" postmodernism. The failure to make this distinction often takes the form of indiscriminately labelling as "anti-science" all those who do not share authoritarian, exclusionary and ahistorical conceptions of science and knowledge. See Santos 1998a, 2000; and Nunes 1998/99, 2001.

2. The term "paradigm" was originally associated by Thomas Kuhn (1970) with specific disciplines and with "normal science." Santos uses it to characterize a "single global model of scientific rationality, with some internal variation to be sure, but one which distinguished itself from and defended itself quite conspicuously and consistently against two non-scientific (and therefore irrational) forms of knowledge that were potentially disturbing: common sense and the so-called humanities (the latter including, among other things, history, philology, law, literature, philosophy, and theology (Santos 1992, 13). The paradigm emerging from the crisis of modern science will itself be the outcome of a "scientific revolution," but one taking place in a "society, which has already undergone a scientific revolution." This paradigm will have to be both a scientific paradigm, the "paradigm of a prudent knowledge," and a social paradigm, the "paradigm of a decent life" (Santos 1992, 30–31). Whether the outcome of this transition is to be described as a new paradigm or as a "post-paradigmatic" condition is an interesting question, but which has to remain open for the time being.

3. In her recent work, Isabelle Stengers proposes the project of an "ecology of practices," defined by a "cosmopolitical" relationship of forms of knowledge which are produced, circulate, communicate, articulate and confront themselves in different spaces, without disqualifying one another (Stengers 1996/97).

4. For a detailed account of this process, see Stengers 1996/97.

5. On the debates within the life sciences and their articulations with other disciplines and research areas, see, among others, Oyama, Griffiths and Gray 2001, and Singh, Krimbas, Paul and Beatty 2001.

6. On the consequences of these developments for the social sciences, see Santos 1995, 2000, 2001; Wallerstein 1999; and the report of the Gulbenkian Commission on the Restructuring of the Social Sciences (1996). On economics, see Louçã 1997. Some more epistemologically "conservative" approaches can be found in Eve, Horsfall and Lee 1997. Hayles (1990) includes an interesting discussion of the impacts of the "new sciences" on the humanities.

7. The bibliography on the "new sciences" put together by Richard Lee in 1992 displays the heterogeneity of themes and of disciplinary and theoretical orientations. The extension of the bibliographical repertoire to the present would certainly reveal significant continuities and discontinuities and the emergence of new research themes. For a recent overview, see Dalmedico 2002. It would be interesting to compare the ways in which topics such as complexity or uncertainty were incorporated into the concerns of scientists across a broad range of disciplines and areas with the spread of systems approaches during World War II and the Cold War. On the latter, see Hughes and Hughes 2000.

8. In 1992, in one of the most influential general treatments of the emerging landscape of complexity studies, Lewin described the latter as an "emerging science," dealing with "life on the edge of chaos." He relied on statements by scientists from different areas with diverse and sometimes even opposing views on what this "science" was about and what its objects were, including the very concept of "complexity." A decade later, in another general treatment of complexity, Mark Taylor (2001) defined the latter as a phenomenon transversal to the sciences, the humanities, the arts, religion and education, associated with the "acceleration of change" in the post-Cold War world and with the emergence of a "network culture." Instead of "life at the edge of chaos," complexity would be more appropriately described as "more is different." The distance between these two conceptions is a symptom of the ways in which complexity and other associated topics define a variety of agonistic spaces, which can hardly be described as a science or a theory. On this, see Stengers 1992 and 1996/97 (especially volume 6) and Thrift 1999.

9. This is a central theme of the work by Santos published after *A Discourse*, namely, Santos 1999 and 2004. Within the social studies of science and related historical and philosophical approaches, the recent inflexion of the epistemological debate is recognizable in the contributions of Peter Galison, John Dupré, Alan Fine or David Stump included in Galison and Stump 1996, in the chapters by Fujimura, Escobar, Visvanathan and Taylor in this volume, or in the contributions to Law and Hassard 1999, to name only a few. In spite of the convergences, there is still a broad space of difference and divergence within the field.

10. The author's approach to the recovery of those marginalized, forgotten or defeated conceptions or to making visible the absences and emergences draw on procedures which, depending on topic and circumstance, are close to foucauldian archaeology and genealogy and to anamnesis as proposed by Stengers (1996/97). But it is a highly original approach, which is presented in detail in several publications, namely Santos 1995, 1998b, 2000 and 2002b and in the closing chapter of this volume.

11. For detailed discussions of feminist approaches to science and knowledge, see Schiebinger 1999, and Harding 2003; on the relevance of feminist and postcolonial studies to epistemological debates, see Harding 1998. Strum and Fedigan (2000) provide an important collection of contributions on science, society and gender focused on primatology.

12. An exemplary study articulating the different aspects and processes involved in the constitution of "sexed" bodies is Fausto-Sterling 2000.

13. The project, funded by the Macarthur and Calouste Gulbenkian Foundations, focused on experiences of popular and citizen initiatives in six countries (Brazil, Colombia, India, Mozambique, South Africa and Portugal), aimed at building an alternative, "bottom-up" globalization, including the production and mobilization of rival forms of knowledge. See Santos 2004.

14. The links between the "science wars" and more traditional epistemological controversies was more visible where they involved sociologists of scientific knowledge, as suggested by the contributions to the volume edited by Labinger and Collins (2001). In this case, political questions related to cultural authority were framed as extensions of the debate on the "two cultures" and on the need to create forms of dialogue between them. It should be noticed that one of the earliest attacks

on alleged antiscience stances within epistemology was launched by two physicists, Theocaris and Psimopoulos, who in an article published in *Nature* in 1987, chose as their targets several philosophers of science, including Popper, Kuhn, Feyerabend and Lakatos. Other approaches to the "science wars" focused on the political issue of cultural authority and on its links to the post-Cold War situation of science in the United States (Ross 1996). The frequent labelling by "science warriors" of their allegedly "antiscience" opponents as "relativists" or "postmodernists" (even when the latter are explicit in their critique of relativism and of "affirmative" postmodernism) is a very unscientific move which often replaces the effort of actually reading what they criticize.

15. The responses to Sokal and Bricmont's book and assessments of the debate itself include, among others, Jurdant 1998; Jeanneret 1998; Gyerin 1999, 336–362; Labinger and Collins 2001 (this book was organized as a dialogue involving sociologists and historians of science close to the sociology of scientific knowledge and scientists from different areas, located on both sides of the debate, including Sokal, Bricmont and Steve Weinberg); and, for responses generally sympathetic to Sokal and Bricmont's book, Koertge 1998.

16. These exceptions include Sokal and Bricmont 1997, 123–33, and Bricmont 1996, who discusses the concept of "time's arrow." But they remain within the boundaries of physics and mathematics and play down the broader significance of this discussion for epistemological debates. A different reflection on these topics, with a focus on problems of the language of physics, on the relationship of the latter to other languages and on the critical dimension of scientific practice can be found in the work of Jean-Marc Lévy-Leblond, a physicist who is an outspoken critic of Sokal and Bricmont. See, in particular, Lévy-Leblond 1996.

17. The recent public positions taken by Sokal and Bricmont, together with many of those who had been their opponents in the "science wars," against the U.S. administration's post 9/11 political orientations raise another issue, which has been discussed elsewhere (Santos 1995, 508–10, on Chomsky; Nunes 1998/99, 44–45): that of the (far from simple and linear) links between epistemological and political positions. Santos calls for rethinking the possibilities and limits of an emancipatory politics, which stops short of a "cosmopolitical" relationship between modes of knowledge (Santos 1995, 2002a, 2002b, 2004). See the symposium "Debating Knowledge" in the *European Journal of Social Theory*, on an article by Santos (Santos 2001b, 2001c; Pieterse 2001; Wagner 2001; Caraça 2001).

18. Jean-Marc Lévy-Leblond has repeatedly stressed that the interpretation of some passages of Heisenberg's work (including those quoted in *A Discourse*) has been a matter of debate among physicists themselves.

19. On this, see the chapters by Wallerstein, Lee and Wagner, in this volume.

REFERENCES

Baptista, António Manuel. 2002. *O discurso pós-moderno contra a ciência: Obscurantismo e irresponsabilidade*. Lisboa: Gradiva.

Beck, Ulrich. 1992. *Risk Society: Toward a New Modernity*. London: Sage.

Callon, Michel, Pierre Lascoumes, and Yannick Barthe. 2001. *Agir dans un monde incertain. Essai sur la démocratie technique*. Paris: Seuil.

Eve, Raymond A., Sara Horsfall, and Mary E. Lee, eds. 1997. *Chaos, Complexity, and Sociology: Myths, Models, and Theories*. Thousand Oaks, CA: Sage.

Fausto-Sterling, Anne. 2000. *Sexing the Body: Gender Politics and the Construction of Sexuality*. New York: Basic Books.

Galison, Peter, and David J. Stump, eds. 1996. *The Disunity of Science: Boundaries, Contexts, and Power*. Stanford, CA: Stanford University Press.

Gonçalves, Maria Eduarda, ed. 2000. *Cultura Científica e Participação Pública*. Oeiras, Portugal: Celta.

——, ed. 2002. *Os Portugueses e a Ciência*. Lisboa: Dom Quixote.

Gulbenkian Commission. 1996. *Open the Social Sciences: Report of the Gulbenkian Commission on the Restructuring of the Social Sciences*. Stanford, CA: Stanford University Press.

Harding, Sandra. 1998. *Is Science Multicultural? Postcolonialisms, Feminisms, and Epistemologies*. Bloomington: Indiana University Press.

——, ed. 2003. *The Feminist Standpoint Theory Reader: Intellectual and Political Controversies*. New York: Routledge.

Hayles, N. Katherine. 1990. *Chaos Bound: Orderly Disorder in Contemporary Literature and Science*. Ithaca, NY: Cornell University Press.

Hughes, Agatha C., and Thomas P. Hughes, eds. 2000. *Systems, Experts, and Computers: The Systems Approach in Management and Enginneering, World War II and After*. Cambridge, MA: MIT Press.

Irwin, Alan, and Mike Michael. 2003. *Science, Social Theory and Public Knowledge*. Maidenhead: Open University Press/McGraw-Hill Education.

Irwin, Alan, and Brian Wynne, eds. 1996. *Misunderstanding Science? The Public Reconstruction of Science and Technology*. Cambridge: Cambridge University Press.

Kuhn, Thomas. 1970. *The Structure of Scientific Revolutions*. 2nd ed. Chicago: University of Chicago Press.

Labinger, Jay L., and Harry Collins, eds. 2001. *The One Culture? A Conversation about Science*. Chicago: University of Chicago Press.

Lakoff, George, and Mark Johnson. 1999. *Philosophy in the Flesh: The Embodied Mind and Its Challenge to Western Thought*. New York: Basic Books.

Lakoff, George, and Rafael Núñez. 2000. *Where Mathematics Comes From: How the Embodied Mind Brings Mathematics into Being*. New York: Basic Books.

Latour, Bruno. 1999. *Pandora's Hope: Essays on the Reality of Science Studies*. Cambridge, MA: Harvard University Press.

Lave, Jean. 1988. *Cognition in Practice: Mind, Mathematics and Culture in Everyday Life*. Cambridge: Cambridge University Press.

——. 1996. "The Savagery of the Domestic Mind." In *Naked Science: Anthropological Inquiry into Boundaries, Power, and Knowledge*, ed. Laura Nader. New York. Routledge.

Law, John, and John Hassard, eds. 1999. *Actor Network Theory and After*. Oxford: Blackwell Publishers/The Sociological Review.

Lee, Richard. 1992. "Readings in the 'New Science'. A Selective Annotated Bibliography." *Review* 15 (1):113–71.

Lévy-Leblond, Jean-Marc. 1996. *Aux Contraires. Exercice de la Pensée et Pratique de la Science*. Paris: Gallimard.
Lewin, Roger. 1992. *Complexity: Life at the Edge of Chaos*. New York: Macmillan.
Louçã, Francisco. 1997. *Turbulência na Economia: Uma Abordagem Evolucionista dos Ciclos e da Complexidade em Processos Históricos*. Porto, Portugal: Afrontamento.
Michael, Mike. 2000. *Reconnecting Culture, Technology and Nature: From Society to Heterogeneity*. London: Routledge.
———. 2002. "Comprehension, Apprehension, Prehension: Heterogeneity and the Public Understanding of Science." *Science, Technology and Human Values* 27 (3):357–78.
Nowotny, Helga, Peter Scott, and Michael Gibbons. 2001. *Re-Thinking Science: Knowledge and the Public in an Age of Uncertainty*. Cambridge: Polity Press.
Nunes, João Arriscado. 1998/99. "Para Além das 'Duas Culturas': Tecnociências, Tecnoculturas e Teoria Crítica." *Revista Crítica de Ciências Sociais* 52/53:15–60.
———. 2001. "Teoria Crítica, Cultura e Ciência. O(s) Espaço(s) e o(s) Conhecimento(s) da Globalização." In *Globalização: Fatalidade ou Utopia?* ed. Boaventura de Sousa Santos. Porto, Portugal: Afrontamento.
Nunes, João Arriscado, and Maria Eduarda Gonçalves, eds. 2001. *Enteados de Galileu? A Semiperiferia no Sistema Mundial da Ciência*. Porto: Afrontamento.
Oyama, Susan, Paul E. Griffiths, and Russell D. Gray, eds. 2001. *Cycles of Contingency: Developmental Systems and Evolution*. Cambridge, MA: MIT Press.
Ross, Andrew, ed. 1996. *Science Wars*. Durham, NC: Duke University Press.
Santos, Boaventura de Sousa. 1987. *Um Discurso Sobre as Ciências*. Porto, Portugal: Afrontamento.
1989. *Introdução a uma Ciência Pós-Moderna*. Porto: Afrontamento.
———. 1991. "Ciência." In *Dicionário do Pensamento Contemporâneo*, ed. Manuel Maria Carrilho. Lisboa: Dom Quixote.
———. 1992. "A Discourse on the Sciences " *Review* 15 (1): 9–47.
———. 1995. *Toward a New Common Sense: Law, Science and Politics in the Paradigmatic Transition*. New York: Routledge.
———. 1998a. "Oppositional Postmodernisms and Globalizations." *Law and Social Inquiry* 23 (1):121–39.
———. 1998b. "The Fall of the *Angelus Novus*: Beyond the Modern Game of Roots and Options." *Current Sociology* 46 (2):81–118.
———. 2000. *Crítica da Razão Indolente: Contra o Desperdício da Experiência*. Porto, Portugal: Afrontamento.
———. 2001. "Os Processos da Globalização." In Globalização: Fatalidade ou Utopia? ed. Boaventura de Sousa Santos. Porto, Portugal: Afrontamento.
———. 2002a. "Introdução Geral à Coleção." In *Reinventar A Emancipação Social. Para Novos Manifestos, vol. 1: Democratizar a Democracia. Os Caminhos da Democracia Participativa*, ed. Boaventura de Sousa Santos. Rio de Janeiro: Civilização Brasileira.
———. 2002b. *Toward a New Legal Common Sense*. London: Butterworth.
———, ed. 2004. *Semear Outras Soluções: Os Caminhos da Biodiversidade e dos Conhecimentos Rivais*. Porto, Portugal: Afrontamento.
Santos, Boaventura de Sousa, Maria Paula G. Meneses, and João Arriscado Nunes. 2004. "Introdução: Para Ampliar o Cânone da Ciência. A Diversidade Epistemológica

do Mundo." In *Semear Outras soluções: Os Caminhos da Biodiversidade e dos Conhecimentos Rivais*, ed. Boaventura de Sousa Santos. Porto, Portugal: Afrontamento.

Schiebinger, Londa. 1999. *Has Feminism Changed Science?* Cambridge, MA: Harvard University Press.

Scott, James C. 1998. *Seeing Like a State: How Certain Schemes to Improve the Human Condition Have Failed*. New Haven, CT: Yale University Press.

Singh, Rama S., Costas B. Krimbas, Diane B. Paul, and John Beatty, eds. 2001. *Thinking about Evolution: Historical, Philosophical, and Political Perspectives*. Cambridge: Cambridge University Press.

Sokal, Alan, and Jean Bricmont. 1997. *Impostures Intellectuelles*. Paris: Odile Jacob.

Stengers, Isabelle. 1992. "Les 'Nouvelles Sciences', Modèle ou Défi?" *Review* 15 (1):91–112.

——. 1996/97. *Cosmopolitiques*. 7 vols. Paris: La Découverte/Les Empêcheurs de Penser en Rond.

Strum, Shirley C., and Linda M. Fedigan, eds. 2000. *Primate Encounters: Models of Science, Gender, and Society*. Chicago: University of Chicago Press.

Taylor, Mark C. 2001. *The Moment of Complexity*. Chicago: University of Chicago Press.

Taylor, Peter J. 1995. "Building on Construction: An Exploration of Heterogeneous Constructionism, Using an Analogy from Psychology and a Sketch from Socioeconomic Modelling." *Perspectives on Science* 3 (1):66–98.

Theocharis, T., and M. Psimopoulos. 1987. "Where Science Has Gone Wrong." *Nature* 329:595–98.

Thrift, Nigel. 1999. "The Place of Complexity." *Theory, Culture and Society* 16 (3):31–69.

Toulmin, Stephen. 2001. *Return to Reason*. Cambridge, MA: Harvard University Press.

Wallerstein, Immanuel. 1999. *The End of the World as We Know It: Social Science for the 21st Century*. Minneapolis: University of Minnesota Press.

3

Science Wars

Whither the Two Cultures

Richard Lee

The ardent discussions and acrimonious debates of the 1990s that came to be known as the "science wars" have, to a great extent, calmed. Those defenders of "science" that were so outspoken a decade ago seem to have returned not only to what they do best, but what would seem to be the most efficient use of time and energy as well. And, on the other side, those who study the scientific enterprise itself, although continuing to develop their long-term intellectual project, seem too to have become less overtly polemical. Nonetheless, those fiery disputes of yesterday are at best banked embers today. The truth of the matter is that the issues were not resolved and the future of what was at stake—the yawning chasm between the arena of "truth," "objectivity," "reason," and "Western Civilization" and that of "values," "relativism," and "multiculturalism"—is still open. This is the very definition of the "two cultures," which has been a defining characteristic of the modern world over the past five centuries (see Lee and Wallerstein 2004). Indeed, the science wars are intimately associated with a larger series of concerns, the "culture wars."

The culture wars that have been much in the news over the past thirty years represent much more than a simple struggle between the old and the new, between the modern and the postmodern. They belong to a long history of debate and outright conflict in the modern world over how valid knowledge may be produced, what are the grounds and the domain of its authority, who may speak in its voice, and thus what courses of social action may be considered legitimate.

At the end of the seventeenth century, just as the questions of universalism and particularism, civilization and culture, were being argued in the "Battles of the Books" in what we would now consider the humanities in

France and England, Isaac Newton was working out a rationalist model, combining mathematical representation and empirical observation, for the analysis of physical reality based on universal natural laws applicable in the disciplines that would come to be known as the sciences. It linked the particular facts that were the product of observation through and with the universal facts of time-reversible mathematical equations. The solution allowed for prediction, and would eventually succeed thereby in jettisoning value-laden, thus relativist, explanatory modes. By the turn of the nineteenth century, Laplace could state of God that he did not need that hypothesis. Nonetheless, the adequacy of science *tout court*—its pretensions of truth and objectivity and their association with progress—remained an object of controversy. In the English-speaking world, one would recall the Tory Radical opposition to industrial society and the "dismal science" and the exchange between the liberal-humanist critic Matthew Arnold and the apologist for the physical sciences T. H. Huxley during the nineteenth century and a mid-twentieth-century dispute similar to this latter between the literary critic H. R. Leavis and the physicist C. P. Snow, who coined the term the "two cultures" (see Snow 1965) to describe the seemingly unbridgeable gulf between the sciences and the humanities.

Esteem for science, and its adoptive sibling, technology, reached its apogee after 1945, but began to falter by the 1960s. In 1962 Thomas Kuhn published his *The Structure of Scientific Revolutions* that, in spite of his later misgivings, by "suggesting the fallacy of viewing scientific knowledge as absolute, objective, and universal . . . opened the door to a new epistemological understanding of science that had the potential to undermine its privileged status." Up until the beginning of the 1960s, the norms of universalism, communism, disinterestedness and organized skepticism that Robert Merton had identified as hallmarks of science and scientific practice were accepted as an accurate description of the way science actually operated, moreover, operated most effectively and efficiently. Critics from a wide spectrum of disciplines and social locations now began to suggest that Merton's norms, if they had ever characterized scientific practice, no longer did, or should, for many scientists. Specifically, areas singled out included the "sacrifice of humaneness" (including indictments of the military-industrial complex and the antinuclear and ban-the-bomb movements), "environmental insults" (Carson's *Silent Spring* appeared in 1962), "male" science, political cooptation (a critique emanating from both the right and the left that cited science as not objective enough), peer review (its shortcomings) and evolution (by the scientific creationists) (Trachtman and Perrucci 2000, 8–9, 14–23).

In this atmosphere, it is well to remember that it was trained scientists like Kuhn, Paul Feyerabend and Stephen Toulmin who carried out much of the original work that led to the establishment of the new field in the history and sociology of science that provoked the ire of the defenders of science in the 1990s.[1] The resulting move to apply the scientific method to the scientific enterprise itself solidified in the form of the "strong program" in the sociology of scientific knowledge defined by David Bloor in 1976.[2] It would be causal ("concerned with the conditions which bring about belief or states of knowledge"), impartial ("with respect to truth and falsity, rationality or irrationality, success or failure"), symmetrical (with the "same types of cause explain[ing], say, true and false beliefs"), and reflexive ("its patterns of explanation would have to be applicable to sociology itself") (Bloor 1991[1976], 7). The SSK project "set out to construct an 'anti-epistemology,' to break down the legitimacy of the distinction between 'contexts of discovery and justification,' and to develop an anti-individualistic and anti-empiricist framework for the sociology of knowledge in which 'social factors' counted not as contaminants but as constitutive of the very idea of scientific knowledge . . . in opposition to philosophical rationalism, foundationalism, essentialism, and, to a lesser extent, realism" (Shapin 1995, 297). The dilemma of and challenge to dualism and the insistence on both the collective nature and local contingency of knowledge formation and reproduction figured prominently in science studies from the beginning.[3] It was highlighted in work that characteristically crossed multiple disciplinary boundaries to investigate at the most basic level how scientific knowledge is made (detailed ethnographic studies: e.g., Latour and Woolgar 1986[1979]; Knorr Cetina 1981; Lynch 1985) and how its claims are defended and institutionalized (including the structure and historical construction of the rhetoric and politics and power relations legitimating and sustaining public knowledge: e.g., Shapin and Schaeffer 1985; Shapin 1994). The field has now developed to include analyses that offer alternative visions of what "science" should or could be and how it might relate to the larger world of social relations.

By the early 1990s, with sciences studies well established, it had become "apparent that some scientists felt sufficiently threatened that they were impelled to go public with a defense of the rationality and the benevolence of science and an attack on what they viewed as uninformed, biased, and unwarranted criticism" (Trachtman and Perrucci 2000, 24). This "going public," and the chronology of the science wars proper, may well be taken to begin in 1992 with attacks directed primarily at the strong social constructivist and relativist bent of SSK. Two books were prominent in bringing to public attention the work of the group of historians, philosophers and sociologists who had been involved in reconceptualizing many common views about the nature of science: the physicist Steven Weinberg's 1992 *Dreams of a Final Theory: The Search for the Fundamental Laws of Nature*, and biologist Lewis Wolpert's 1993 *The Unnatural Nature of Science: Why Science Does Not Make (Common) Sense*. For Wolpert, the bone of contention is relativism, to which he replies that "[s]cientists can be very proud to be naïve realists" (1993, 117). This is also Weinberg's position. It should be noted, however, that the status of social constructivism was problematic even among some science studies scholars, a fact that is still often overlooked.

The issues came to a head with the "defense" mounted on a wide front in 1994 by the publication of *Higher Superstition: The Academic Left and Its Quarrels with Science*, by Paul Gross, a biologist at the University of Virginia, and Norman Levitt, a mathematician at Rutgers University. Their scattershot attack was directed at a broad "academic left"—of feminist theory, postmodern philosophy, deconstruction, deep ecology—and the way it "dislikes science." These dislikes were, it was argued, not just of the uses to which science is put, but "to the social structures through which science is institutionalized, to the system of education by which professional scientists are produced, and to a mentality that is taken, rightly or wrongly, as characteristic of scientists. Most surprisingly, there is open hostility toward the *actual content* of scientific knowledge and toward the assumption, which one might have supposed universal among educated people, that scientific knowledge is reasonably reliable and rests on a sound methodology" (Gross and Levitt 1994, 6, 2). The authors touched a chord in the scientific community despite the fact that the book itself stands as a monument to the very type of intellectual dilettantism it attacked. Nonetheless, many still make common cause with this line of reasoning (e.g., Baptista 2002). In an analysis of the "reading" given the work of Jacques Derrida by Gross and Levitt, Arkady Plotnitsky (who has both scientific and literary credentials) wrote that

> [i]n general, scholarly problems of monumental proportions are, to use the language of topology, found in the immediate vicinity of just about every point of *Higher Superstition*. It is not so much embarrassing errors even as egregious as that of the misreading of "*topique differantielle*" as differential topology, that are most crucial (we all make mistakes, sometimes absurd mistakes), but the intellectually and scholarly inadmissible practices and attitudes that pervade—and **define**—this sadly irresponsible book. Gross and Levitt's warning concerning "threats to the essential grace and comity of scholarship and the academic life" (ix) becomes, in one of many bizarre ironies of the book, its self-description. . . . The tragedy is that so many scientists, including some among the best scientists, have taken it seriously and accepted its arguments, and even adopted its unacceptable attitudes. (1997, par. 9, emphasis in original)

The exchanges that ensued made plain just how fundamental the issues were, and not simply that C. P. Snow's two cultures of scientists and nonscientists still could not communicate with one another.

The National Association of Scholars joined in the responses to the supposed "antiscience" threat (see Holton 1993 on the "dangers" of irrationalism) with two conferences arranged in 1994. The "Science in American Life" exhibit arranged by the American Chemical Society, which incorporated criticism of science, prompted an exchange between Tom Gieryn (1996) and Paul Gross (1996). Harry Collins and Lewis Wolpert confronted one another directly at the British Association for the Advancement of Science in 1994 and Steve Fuller organized a follow-up to this conference that sought some mutual understanding but ended with disappointing results (see Fuller 2000). Gross and Levitt themselves proved unforthcoming during the debate on the content of their book at the Society for Social Studies of Science Annual meeting in 1995. The meeting was punctuated, however, with an unseemly exchange over Donna Haraway's, and other feminist critics', scientific credentials. In 1995, the New York Academy of Sciences sponsored a conference entitled "The Flight from Science and Reason" that engaged the questions raised in *Higher Superstition* by a long list of defenders of science against such "science bashers" as constructivists, deconstructionists, creationists, feminists, Afro-centrists, and radical environmentalists ("ecosentimentalists") (Gross, Levitt and Lewis 1996).

What so far was missing was a concerted response to Gross and Levitt, and their "proscience" supporters. The editors of *Social Text* took on this project in 1996 with a special issue dedicated to the science wars. What the editors of the journal did not know was that the physicist, Alan Sokal, inspired by his reading of *Higher Superstition*, was involved in an "actively sustained" conspiracy to deceive the journal into publishing his "'Transgressing the Boundaries: Toward a Transformative Hermeneutics of Quantum Gravity" in which he "parodied postmodern stylistic conventions and derived politically correct conclusions from an esoteric subfield of science" (Segerstråle 2000b, n.p.).[4] Sokal exposed the hoax in his "A Physicist Experiments with Cultural Studies," which appeared almost concurrently in *Lingua Franca*, characterizing the article as combining "nonsense" and "silliness." For those who were taken in by Sokal and those they represented it was an extraordinary breech of intellectual ethics and scholarly integrity; for those making common cause with Sokal (and Gross and Levitt et al.), it forcefully demonstrated his point of declining "standards of rigor in the academic community" (Weinberg 1996, 11) and specifically the intellectual laxness of those it was designed to attack. This group proved notoriously difficult to define, however, given the diversity of the cultural studies community and the conflation of cultural studies with science studies as the debates played out.

In particular, the proscience fighters were concerned to protect some "true" idea of science against what they considered "false" representations. For Ullica Segerstråle, they seemed not "to recognize the right of other academics to do their own interpretations of science within the particular frameworks of their own disciplines"; the problem was, naturally, that this work never stayed within the confines of specialist disciplines but became a part of undergraduate curricula and the larger public understanding of science (2000a: 21). In another venue, Segerstråle relates this to an internal struggle within the political left (recalling the culture wars, as we shall see): the older left of Gross, Levitt and Sokal "equates science with reliable knowledge, a tool in the struggle for social justice. . . . In contrast, the cultural Left equates science with power that can be used for social oppression of minorities. For them, therefore, science criticism is a way to liberation" and it comes down to, as Bruce Robbins and Andrew Ross have asked, whether nonscientists would "have some say in the decision-making process of the professional scientific community" (2000b, n.p.).

Unfortunately, the fallout over the Sokal affair obscured Ross's own take on the science wars: in his "Introduction" to the *Social Text* collection, he treats the science wars as a "second front" in the "holy Culture Wars" (1996a: 6), and suggests a conjunctural explanation for the controversy, especially near-term decline in governmental support, including funding cuts, for big science (cancellation of the Superconducting Supercollider and closing of the Congressional Office of Technology Assessment) that others have noted. However, this is not just a turf-battle over the allocation of scarce research dollars and Ross's analysis also presents a longer-term interpretation of the science wars that situate the only real resolution in a shift in the process privileging formal rationality to one favoring substantive concerns:

> The rise in technoskepticism, then, parallels a crisis in industrialization which is often mistaken for a crisis of the environment. . . . [T]he remoteness of scientific knowledge from the social and physical environments in which it will come to be measured and utilized is as irrational as anything we might imagine, and downright hazardous when it involves materials that can only be properly tested in the open environment. . . . [D]emonstrating the socially constructed nature of the scientist's knowledge . . . may help to demystify, but it must be joined by insistence on methodological reform—to involve the local experience of users in the research process from the outset and to ensure that the process is shaped less by a manufacturer's interests than by the needs of communities affected by the product. This is the way that leads from cultural relativism to social rationality. (1996a, 2, 3–4)

Ross also sets the "relativism" argument in a material, spatial, context and links it positively with "diversity": "Once it is acknowledge that the West does not have a monopoly on all the good scientific ideas in the world, or that reason, divorced from value, is not everywhere and always a productive human principle, then we should expect to see some self-modification of the universalist claims maintained on behalf of empirical rationality. . . .

This is the way that leads from relativism to diversity" (Ross 1996a, 4). As Sandra Harding had earlier insisted: "There is plenty of science still to be done once physics is invited and permitted to step down and take its place as one human social activity among many others. What kinds of knowledge about the empirical world do we need in order to live at all, and to live more reasonably with each other on this planet from this moment on? Who should make up the 'we' who answers this question" (Harding 1992, 20)?

In their large-scale survey, Leon Trachtman and Robert Perrucci found that rather than any "General Cynicism" about science, work in science studies has not strained the estimation of the social worth of science to the general public: "despite the concerns of science's spokes people, the battle over social constructivism is a minor in-house academic quarrel with essentially no public following and is likely to have little impact on public understanding, public appreciation or public funding of science" (2000, 166, 167). However, this study does show that knowledge is socially situated and that "[w]hat may be emerging in this contentious debate about how knowledge is produced and used is a recognition that truth will have to be negotiated by an expanded pool of consequential actors with the right to speak" (2000, 168).

In this sense, the "science wars" may seem to be nothing more than a specifically focused instance of the more generalized "culture wars" and the two do share several characteristics. First, the themes of the debates constituting the science wars, like those of the culture wars, have a long history and resonate across national and linguistic borders. Second, as the institutions of knowledge formation opened up to previously excluded groups as a result of the upheavals of the 1960s, entrenched theoretical and methodological perspectives found it more and more difficult to dismiss the critiques and their overtly political agendas that were now mounted from within the institutions themselves. It became clear to many involved that any review of the premises of knowledge formation amounted to an evaluation of the structure of social relations as well. This dimension of the culture wars was replicated in the science wars. Finally, the disputes in both arenas, if indeed there are two, were polarized and the tone exacerbated by a series of publications defending what had been, up to the 1960s, the dominant positions.

Instances of debates over history and historical amnesia, language, education and the "canon," including their gendered and racial/ethnic foundations and the power relations ensuring their reproduction, may be observed throughout the world. These "cultural" mechanisms have become centers of controversy due to their common functions, especially as techniques through which "nation" and "nationalism" take form as consequences of the construction of "otherness" and its attendant struggles positioning groups on the hierarchy of the world division of labor. As Gregory Jay notes, "many nations are trying to find a way to balance the claims of individuality, ethnic or racial solidarity, democracy, economic development, women's liberation, and nationalism" (Jay 1997, 60, 62). Nevertheless, although the major themes of the contemporary "culture wars" are not limited to the United States, it has been here where these controversies have been grouped under a single rubric. Heated discussions associated with nationalism, patriotism, the role of religion and race, and ethnic and gender relations have developed around the same dilemmas over who-gets-what in terms of how-we-see-the-world as are evidenced abroad. In the United States, however, there has been a clear articulation of the mechanisms relating putatively universal ideals and the particularist context in which they function. Furthermore, the range of conflicts associated with the culture wars here, including the family, art, religion, education, law and politics, is such that according to John Davison Hunter, the culture wars touch "virtually all Americans" (1991, xi). Hunter's 1991 book, entitled simply *Culture Wars*, did much to popularize the term and crystallize the issues. And all of the parties to such debates as those focused on homosexuality and the family, abortion and health care (or the content of schooling in science), seem to have identified the issues as having a long-term component that speaks to the reproduction of social norms through education. What should be included in textbooks, especially how are specific narratives of the past associated with a particular vision of public culture legitimating specific hierarchies in the present and therefore what might be possible to imagine for the future? Should private schools be publicly funded and, in particular, what are the implications of imposing collective support for sectarian agendas?

All have understood that differing positions imply value-laden choices, thus are profoundly political in nature and speak to clear but contested power relations in a struggle for control, even though they may take the form of arguments over truth and relativism, tolerance and prejudice. As the agendas directing values, morality and acceptable conduct that are set at the top eventually redirect programs in the elementary and secondary schools, it should be of no surprise that debates have tended to implicate the content of higher education and the organizing and legitimating role of the university: as symbolic system (the "canon") and as material practice (access or gate-keeping, disciplinary/departmental structures, and pedagogical and research organization).

University reform, not the least of which involved the curriculum, stands as an enduring legacy of the 1960s. The culture wars have been bound up with the struggle over these reforms and especially the teaching of American literature, that is, the traditional canon reopened to new and, to some, often disconcerting interpretations, as well as the canon itself destabilized and redefined by a series of new inclusions and exclusions, when it has not been abolished altogether. Whether linked to the canon, classroom experimentation, or more generally to race and gender studies, the reforms have certainly come under attack in a whole series of doom-and-gloom writings. Two works stand out, however: Allan Bloom's 1987 *Closing of the American Mind: How Higher Education Has Failed Democracy and Impoverished the Souls of Today's Students* and E. D. Hirsch's 1988 *Cultural Literacy: What Every American Needs to Know*. These and their like presented an agenda for and outlined the purpose of shaping public schooling and higher education in ways that "abstracted equity from excellence and cultural criticism from the discourse of social responsibility" (Giroux 1992, 123). As many critics immediately realized, Hirsch's "culture" was unproblematic because he had removed it from the dynamics of its construction in struggle and power. Barbara Herrnstein Smith stated that from a position of high institutional authority,

> *Cultural Literacy* (book, list, term, and concept) does a very good job of obscuring the nation's very real educational problems and assuring the American public, many of whom are naturally happy to hear it, that those problems (along with other social and economic ills, both real and imaginary, from poverty and unemployment to "cultural fragmentation" and the "competitive edge" of the Japanese) are caused primarily by befuddled education professors and school administrators following what scientists have shown to be the incorrect principles of progressive education and, consequently, can be solved by school reforms that require no funding, entail no social or political changes, create no uncomfortable feelings for anyone except teachers and school administrators, and do not touch the structures of a single American institution, including its school system. (1992, 88–9)

It was in this context of the crusade to create "a narrowly specific cultural capital [as] the normative *referent* for everyone, but [to remain] the *property* of a small and powerful caste that is linguistically and ethnically unified" that the Stanford curriculum debate (and many such in universities all over the United States) was played out during the mid-1980s. At Stanford a limited reform was passed in 1988 but the vision that critics of reform presented proved to be well-entrenched: on the one hand, one of the decline of civilization and the loss of the shared values on which the unity of community, all diversities erased, rested, and on the other hand, the politicization of the "ivory tower," the erstwhile value-neutral ideal of the university. But any sense of contemporary cultural fragmentation can only be supposed on the presumption of a prior unity. What Hirsch, "in an illegitimate analogy between language and culture . . . refers to as 'the national culture' . . . is nothing but a *particular* (egregiously classbound and otherwise parochial) set of items of 'knowledge' that Hirsch himself privileges and that he *wants* the state educational system to make 'standard'" (Smith 1992, 77, 78). Remark, for instance, that who controls that specific representational form we call "history" and to what end has been a matter of serious controversy over the past decade, although one must acknowledge again that this is not new, and the university is again at the center of struggles over what, or whose, history is recognized as authoritative and propagated as grounds for legitimate social action.

The curriculum, in fact, is a historically specific narrative and pedagogy a particular form of cultural politics. However, when this historical dimension is taken into consideration, the contemporary culture wars appear as a post-1968 conjunctural moment (see Lee 1996) in a long-term trend (see Dejean 1997). Neither the intellectual arrangements nor the institutional organization of the structures of knowledge were able to successfully negotiate the outcomes of the challenges of the 1960s. And now, once students graduate, it is apparent that the promise of a better life for all who go through the system cannot be kept in a world ultimately ruled by the law of value. Thus, the university, in this case especially education in the humanities, has always been a political instrument.

The arguments over the politically suspect politicization of the university that pitted those that viewed the realm of knowledge and its institutions as loci of struggle against those that upheld "truth" achieved through value-neutrality as the ideal of scholarship seemed paradoxical in the light of actual practice. As Gerald Graff contends: "It is the academic left that has arguably made the major contributions to scholarship in literary studies over the past three decades, profoundly affecting fields as remote from literature as law, architecture, anthropology and the social sciences. Yet it is the right that defends the virtues of disinterested scholarship, while the left debunks disinterestedness in the name of politics and power. The producers of the best objective scholarship defend partisanship, while the defenders of objectivity produce mostly partisan political polemics" (1995, 308). The paradox is further compounded when the trajectories of "theory" and "identity politics" are considered.

Theory, whether loosely characterized as poststructuralism or deconstruction, unmasked the power relations inherent in the Arnoldian high-culture aesthetic, anathema on the right, but also tended to result in formal analyses of "texts," of whatever type, disengaged from concrete political agendas, roundly criticized on the left. According to Joan Scott, "[o]n the right, there are denunciations of the nihilism of theory, which, it is said, will leave us orphans without cultural patrimony. On the left, theory is indicted for its impracticality: it does not connect to 'real life' or 'lived experience' and so cannot lead directly to politics, to revolution, or at least to social reform." Scott goes on to point out that "[i]n the attack on 'theory,' right and left clear the field of all possible critiques of their foundational premises; with those intact, they can fight safely and familiarly among themselves" (1995, 301).

Unlike the antiessentialism of poststructuralism, identity politics presupposes foundations making the point that without presuppositions there can be no action on the part of the marginalized. According to Stuart Hall, the poor fit between theory and practice opens up an opportunity for rethinking the terms of the debate. For identity politics, even in the form of "strategic essentialism,"

> sees difference as "their traditions versus ours," not in a positional way, but in a mutually exclusive, autonomous and self-sufficient one. And it is therefore unable to grasp the dialogic strategies and hybrid forms essential to the diaspora aesthetic. A movement beyond this essentialism is not an aesthetic or critical strategy without a cultural politics, without a marking of difference. It is not simply rearticulation and reappropriation for the sake of it. What it evades is the essentializing of difference into two mutually opposed either/ors . . . replac[ing] the "or" with the potentiality or the possibility of an "and." That is the logic of coupling rather than the logic of a binary opposition. . . . The essentializing moment is weak because it naturalizes and dehistoricizes difference, mistaking what is historical and cultural for what is natural, biological, and genetic. (Hall 1996, 472)

At issue today is the epistemological basis of the representational apparatus that has characterized our understanding of the social world over the long term. As Stuart Hall's analysis suggests, the outcome of the contemporary controversies may have profound consequences for the dichotomous thinking, and particularly the divorce of facts from values, that has formed the basis of the epistemology, the "geo-culture" of the modern world-system as it has developed over the past five centuries (see Lee 2000 and Lee and Wallerstein 2001, 2004). In this case it is "civilization"—"culture"—the first universal, implicit in the term as it was deployed by Lynne Cheney (1988), and the second, particularist—that is in question. "What the political correctness debate and related phenomena display, symptomatically, is precisely the connection between representation in the field of knowledge and representation in the fields of society and politics." Far from the "orthodox" versus "progressive" elements that Hunter postulated, the protagonists of the culture wars simply cannot be shoehorned into "either/or" groups, of whatever stripe. Moreover, it is the impression that "academics are producing *a body of different truths* that threaten certain traditional value systems and institutions" (Jay 1997, 31) that has engendered the backlash of the culture wars. In this sense, Hall's analysis suggests that the conjunctural moment of the culture wars is also a point of structural crisis leading to a restructuring in the field of knowledge, that is, the de-legitimation of the principle of the excluded middle and the implicit call for a relational conception of human reality. But the epistemological status of dichotomous thinking is constitutive of and constituted by the role of the natural sciences in the modern world and their location at the privileged pole of the structures of knowledge that has become the object of contention in the science wars.

Thus, the science wars are more than just a special case of the culture wars and a plea for more democracy and value-consideration in decision-making in the context of the innate inequalities that are part of the all-encompassing corporate climate. They rather indicate the importance of the questions focusing on the epistemological status of the sciences that have come into play. This only becomes clear, however, when we realize that what the science warriors seem not to have adequately considered, and what surveys like those of Trachtman and Perucci (2000) fail to capture, is that "epistemological radicalism" now extends to the sciences themselves, especially to those involved in what has come to be known as "complexity studies" (see Lee 1992).[5] The combination of the conviction that there is a "real" world and that the future is "determined" by the past, but that that future is nonetheless unpredictable (time itself is thus "creative"), and the parallel assaults on dualism challenge the epistemological status of the sciences as discoverers, guardians and purveyors of authoritative knowledge, truth, by redefining what it means to describe the evolution of natural systems. Already in 1986 Sir James Lighthill, in his capacity as president of the International Union of Theoretical and Applied Mechanics, felt compelled to apologize on behalf of "practitioners of mechanics . . . for having misled the general educated public by spreading ideas about the determinism of systems satisfying Newton's laws of motion [implying complete predictability] that, after 1960, were to be proved incorrect" (1986, 38). Indeed, as Boaventura de Sousa Santos writes, "[i]n place of eternity, we now have history; in place of determinism, unpredictability; in place of mechanicism, interpenetration, spontaneity, irreversibility, and evolution; in place of order, disorder; in place of necessity, creativity and contingency" (1992, 25). This amounts to overturning the dominant model shaping our understanding of the natural world and the human world as well, renders moot the "contradiction" between determinism and free will, impinges directly on the manner in which scholars make claims for the legitimacy of their interpretations of social reality, and underscores the covert, long-term structural, nature of the debates, that have come to be known as the "Science Wars" and the "Culture Wars." Though "total," what Santos calls "post-modern knowledge . . . is knowledge about the conditions of possibility of human action projected into the world from local time-spaces" (1992, 38) to which the arena of "[p]lay, theater, text, or biography" (1992, 36) no longer seems so foreign, and suggests how António Baptista's (one of the latest in the line of those defending the traditional idea of science) exclusion of Weber, Pareto, Marx and Durkheim from a single pantheon that includes Smith, Ricardo, Lavoisier, Darwin, Humboldt, Planck and Poincaré (Baptista 2002, 53) has begun to look increasingly artificial and counterproductive.

In sum, on the one hand, the collapse of the two cultures is a reality that no rearguard action can check. On the other hand, this connotes an open future. This open future and the reintegration of facts and values in the construction of authoritative knowledge, a consequence of "complexity studies," entails the ethical imperative of scholars to imagine and work to bring into being, contingent on what Santos has called "conditions of possibility," a reorganization of the structures of knowledge that will lead to a more substantively rational world.

NOTES

1. Although not perfectly synonymous: SSK, "sociology of scientific knowledge"; SST, "science and technology studies"; "social studies of science"; or simply "science studies." The new field met with extraordinary success and rapidly developed a significant, and influential if controversial, literature (see Shapin 1995).

2. In company with the Empirical Program of Relativism (Harry Collins) and Actor-Network Theory (Bruno Latour).

3. Underlining the problematic role of the "social"—not to be conceived as "a 'dimension,' an 'influence,' or a 'factor' to be juxtaposed with the 'factors' of evidence and rationality," Latour and Woolgar removed the term from the subtitle of the second, 1986, edition of their 1979 *Laboratory Life: The Social Construction of Scientific Facts*. As Shapin further observes, SSK's "claim was that 'the social dimension' of knowledge needed to be attended to in order to understand what counts as a fact of discovery, what inferences are made from facts, what is regarded as rational or proper conduct, how objectivity is recognized, and how the credibility of claims is assessed. The target here was not at all the legitimacy of scientific knowledge but the legitimacy of individualist frameworks for interpreting scientific knowledge" (1995, 300).

4. The original article, the *Lingua Franca* revelation, responses, reactions in the press (domestic and foreign), and longer essays and colloquies on the "affair" are conveniently collected in Editors of *Lingua Franca* (2000); Ross (1996b) is an expanded version of the special issue of *Social Text* without the Sokal piece.

5. As Ilya Prigogine has noted, the "sciences are not the reflection of a static rationality to be resisted or submitted to; they are furthering understanding in the same way as are human activities taken as a whole" (Prigogine 1988, 3) and goes so far as to state that "I believe that what we do today depends on our image of the future, rather than the future depending on what we do today" (Prigogine in Snell and Yevtushenko 1992, 28).

REFERENCES

Baptista, António Manuel. 2002. *O Discurso Pós-Moderno Contra a Ciência: Obscurantismo e Irresponsabilidade*. Lisboa: Gradiva.

Bloom, Allan. 1987. *The Closing of the American Mind: How Higher Education Has Failed Democracy and Impoverished the Souls of Today's Students*. New York: Simon and Schuster.

Bloor, David. 1991. *Knowledge and Social Imagery*. 2nd ed. Chicago: University of Chicago Press. (Orig. pub. 1976.)

Cheney, Lynne. 1988. *Humanities in America: A Report to the President, the Congress, and the American People*. Washington, DC: National Endowment for the Humanities.

Dejean, Joan. 1997. *Ancients against Moderns: Culture Wars and the Making of a Fin de Siècle*. Chicago: University of Chicago Press.

Editors of *Lingua Franca*, eds. 2000. *The Sokal Hoax: The Sham that Shook the Academy*. Lincoln: University of Nebraska Press.

Fuller, Steve. 2000. "Science Studies through the Looking Glass: An Intellectual Itinerary." In *Beyond the Science Wars: The Missing Discourse about Science and Society*, ed. Ullica Segerstråle. Albany: SUNY Press, 185–217.

Gieryn, Tom. 1996. "Policing STS: A Boundary-Work Souvenir from the Smithsonian Exhibition on 'Science in American Life.'" *Science, Technology and Human Values*, 21, no. 1, Winter:100–15.

Giroux, Henry A. 1992. "Liberal Arts Education and the Struggle for Public Life: Dreaming about Democracy." In *The Politics of Liberal Education*, ed. Darryl J. Gless and Barbara Herrnstein Smith. Durham, NC: Duke University Press, 119–44.

Graff, Gerald. 1995. "A Paradox of the Culture War." In *PC Wars: Politics and Theory in the Academy*, ed. Jeffrey Williams. New York: Routledge, 308–12.

Gross, Paul R. 1996. "Reply to Tom Gieryn." *Science, Technology and Human Values*, 21, no. 1, Winter, 116–20.

Gross, Paul R., and Norman Levitt. 1994. *Higher Superstition: The Academic Left and Its Quarrels with Science*. Baltimore: Johns Hopkins University Press.

Gross, Paul R., Norman Levitt, and Martin W. Lewis, eds. 1996. *The Flight from Science and Reason*. New York: New York Academy of Sciences.

Hall, Stuart. 1996. "What Is This 'Black' in Black Popular Culture?" In *Stuart Hall: Critical Dialogues in Cultural Studies*, ed. David Morley and Kuan-Hsing Chen. New York: Routledge, 465–75.

Harding, Sandra. 1992. "Why Physics Is a Bad Model for Physics." In *The End of Science: Attack and Defense*, ed. Richard Q. Elvee. Lanham, MD: University Press of America, 1–21.

Hirsch, E. D. 1988. *Cultural Literacy: What Every American Needs to Know*. New York: Vintage Books.

Holton, Gerald. 1993. *Science and Anti-Science*. Cambridge, MA: Harvard University Press.

Hunter, James Davison. 1991. *Culture Wars: The Struggle to Define America*. New York: Basic Books.

Jay, Gregory S. 1997. *American Literature and the Culture Wars*. Ithaca, NY: Cornell University Press.

Knorr Cetina, Karin. 1981. *The Manufacture of Knowledge*. Oxford, UK: Pergamon.

Kuhn, Thomas. 1962. *The Structure of Scientific Revolutions*. Chicago: University of Chicago Press.

Latour, Bruno, and Steve Woolgar. 1986. *Laboratory Life: The Construction of Scientific Facts*. 2nd ed. Princeton, NJ: Princeton University Press. (Orig. pub. 1979.)

Lee, Richard E. 1992. "Readings in the 'New Science': A Selective, Annotated Bibliography." *Review* 15, no. 1, Winter:113–71.

———. 1996. "Structures of Knowledge." In *The Age of Transition: Trajectory of the World-System, 1945–2025*, by Terence K. Hopkins, Immanuel Wallerstein, et al. London: Zed, 179–206.

———. 2000. "The Structures of Knowledge and the Future of the Social Sciences: Two Postulates, Two Propositions and a Closing Remark." In "Festschrift for Immanuel Wallerstein—Part II," ed. Giovanni Arrighi and Walter Goldfrank. *Journal of World-Systems Research* 6, no. 3, Fall/Winter:786–96.

Lee, Richard E., and Immanuel Wallerstein. 2001. "Structures of Knowledge." In *The Blackwell Companion to Sociology*, ed. Judith R. Blau. London: Blackwell, 227–35.

Lee, Richard E., and Immanuel Wallerstein, coords. 2004. *Overcoming the Two Cultures: Science versus the Humanities in the Modern World-System*. Boulder, CO: Paradigm.

Lighthill, Sir James. 1986. "The Recently Recognized Failure of Predictability in Newtonian Dynamics." *Proceedings of the Royal Society of London*, 407, 1832, 8 September, 35–48.

Lynch, Michael. 1985. *Art and Artifact in Laboratory Science*. London: Routledge.

Plotnitsky, Arkady. 1997. "'But It Is Above All Not True': Derrida, Relativity, and the 'Science Wars.'" *Postmodern Culture* 7 (2):n.p.

Prigogine, Ilya. 1988. "The New Convergence of Science and Culture." *UNESCO Courier* May:9–13.

Ross, Andrew. 1996a. "Introduction." *Social Text*, 46/47, Vol. 14, 1 and 2, Spring/Summer:1–13.

———, ed. 1996b. *Science Wars*. Durham, NC: Duke University Press.

Santos, Boaventura de Sousa. 1992. "A Discourse on the Sciences." *Review*, 15, no. 1, Winter:9–47.

Scott, Joan W. 1995. "The Rhetoric of Crisis in Higher Education." In *Higher Education under Fire: Politics, Economics, and the Crisis of the Humanities*, ed. Michael Bérubé and Cary Nelson. New York: Routledge, 293–304.

Segerstråle, Ullica. 2000a. "Science and Science Studies: Enemies or Allies?" In *Beyond the Science Wars: The Missing Discourse about Science and Society*, ed. Ullica Segerstråle. Albany: SUNY Press.

———. 2000b. "History of Science: Stirred, Not Shaken." *Science* 290 (5497):n.p. at http://sciencemag.org.

Shapin, Steven. 1994. *A Social History of Truth: Civility and Science in Seventeenth-Century England*. Chicago: University of Chicago Press.

———. 1995. "Here and Everywhere: Sociology of Scientific Knowledge." *Annual Review of Sociology* 21:289–321.

Shapin, Steven, and Simon Schaffer. 1985. *Leviathan and the Air Pump: Hobbes Boyle and the Experimental Life*. Princeton, NJ: Princeton University Press.

Smith, Barbara Herrnstein. 1992. "Cult-Lit: Hirsch, Literacy, and the 'National Culture.'" In *The Politics of Liberal Education*, ed. Darryl J. Gless and Barbara Herrnstein Smith. Durham, NC: Duke University Press, 75–94.

Snell, Marilyn Berlin, and Yevgeny Yevtushenko. 1992. "Beyond Being and Becoming." *New Perspectives Quarterly* 9, no. 2, Spring:22–28.

Snow, C. P. 1965. *The Two Cultures and a Second Look*. New York: Cambridge University Press.

Trachtman, Leon E., and Robert Perrucci. 2000. *Science under Siege? Interest Groups and the Science Wars*. Lanham, MD: Rowman and Littlefield.

Weinberg, Steven. 1992. *Dreams of a Final Theory: The Search for the Fundamental Laws of Nature*. New York: Pantheon.

———. 1996. "Sokal's Hoax." *New York Review of Books*, August 8.

Wolpert, Lewis. 1993. *The Unnatural Nature of Science: Why Science Does Not Make (Common) Sense*. Cambridge, MA: Harvard University Press.

4

On Wars and Revolutions

Peter Wagner

There are periods in human history during which the perennial discussion about knowledge resorts to politico-military metaphors. Thus, from the middle of the twentieth century onwards, the term "scientific revolution" was increasingly used to refer to those changes in the human view of the world that emerged between the late sixteenth and the eighteenth centuries and brought about what we now tend to call "modern science."[1] More recently, we hear of "science wars," and although the major battleground seems to be in North America, some hostilities have also been observed in other parts of the world.

Between revolutions and wars, there is a deep connection. It is a feature of revolutions in the modern sense of the term to aim at spreading their impact beyond their place of origin, and they often do so by military means. Just as often, though, and only partly in response to revolutionary wars, the resistance to revolution also resorts to violent means. Such wars are then wars against revolutions, counterrevolutionary wars. Furthermore, it is also true that revolutions tend to occur at the end of wars, when there often is a desire to build some new order. But we will come back to this question only at the end of our reflections.

Given that both these terms, "scientific revolution" and "science wars," are creatures of the twentieth century and have somewhat uncertain temporal connotations, it is far from self-evident whether the wars in question are revolutionary wars, aiming at continuing the conquest of the world by scientific knowledge, or counterrevolutionary wars, bringing finally to a halt the devastation of the world under the reign of scientific knowledge. As often happens with wars, even the outbreak of hostilities is subject to a variety of interpretations. If I see it correctly, those who started to talk about a war, and thus may be seen as those who declared war, were the warriors of science aiming to continue on (what they see as) their revolutionary path. They tend to claim, though, that they were subject to an aggression, an undeclared war, so to say, and had not only to reciprocate in kind, but also to lay open to the public what indeed was going on: war on science.

Both the strong language and the explanation offered will take the more distanced onlooker by surprise. It is difficult to observe any significant hostilities with the naked eye; and scientific activities seem to pursue their ordinary path rather unobstructed. It is true, there are casualties: reputation damaged, promotion denied. But they are not only, after all, of a too minor and overall quite ordinary nature to justify this language, they also occur almost exclusively in the opponents' camp, not on the side of the science warriors.

Historians of wars and revolutions can often tell us about asymmetries in the sociopolitical constellations that were in the background of those events and go quite far in helping us understand why they occurred. Mostly, however, they are quite at a loss in explaining when they occurred, why hostilities started at a specific moment and not sooner or later. If, on the battleground of the recent so-called science wars, there was a provocation that required retaliation, namely combined effects of the radically empirical and "symmetric" turn in the social studies of science and the "linguistic turn" in the human sciences,[2] why is it that it took numerous years for the counterattack to happen? Only a close look is required to answer this question. There was no declaration of war at all; all that was going on was the breaking up of unfounded and self-serving assumptions about what science is by looking at the actual activities of scientists and at the concepts they use to describe them.[3] One contributor to this debate, the editor of this volume, Boaventura de Sousa Santos, was indeed arguing for nothing but "common sense" (Santos 1992 [1987]). And we will see in a moment that he even argued quite commonsensically for common sense. It needed quite some rhetorical armory to reinterpret the close analysis of scientific practices as a declaration of war on the sciences. One is tempted to say that a little bit more of common sense could have prevented the outbreak of hostilities. But it is one of the problems with wars that they are difficult to stop and that they tend to spread, involving ever more combatants as well as innocent civilians.

The warriors for the cause of science may object that, while they were not declaring war, their critics certainly were declaring revolution, for instance, as Santos did, by diagnosing an "emergent paradigm" that breaks with common scientific practices and aims at opening up broader perspectives. This claim certainly was provocative. Rather than revolutionary, established scientists may even have seen it as counterrevolutionary, as rolling back, namely, the accomplishments of the scientific revolution. Ironically, those accomplishments, as scientists tend to see them, can be described in terms that are structurally similar to the critical diagnosis.

Scientists claim to break with "traditional" knowledge practices and, thus, achieve a superior understanding of the world. An important element of their self-view is the systematic questioning of the world by means of empirical investigations and the subjection of their own findings to criticism by other scientists on grounds of results from further investigation. They get angry when they think somebody argues that their investigations should be halted and we should again accept the world as it appears to us, "traditionally." They are "reminded"—in the sense of a long cultural memory—of religious indictments against interrogating worldly phenomena. That is why, in search for more contemporary examples, Stalinism, Nazism and most recently Islamism come to their mind—all regimes, or aspiring regimes, by the way, that hardly ever left an opportunity of applying scientific insights for their purposes unused.

What the warriors on behalf of science persistently fail to understand is that nothing could be further from the contemporary agenda in the discussion about knowledge. If Santos was arguing for a revolution, in terms of an "emergent paradigm," it was not a counterrevolution against science, but a revolution to add on to the scientific revolution, to carry critical investigation further, and not to halt it. Enlightened scientists should join into this call, on the basis of their own self-understanding, and not fight against it. To understand this call better, at least as I see it, we need three brief forays into adjacent territories, into philosophy, into history and into politics.

SCIENCE AND THE QUESTION OF REPRESENTATION

Philosophers and scientists have always been concerned about the relation of their knowledge about the world to (the rest of) this world. This concern has sometimes been phrased as the question of the representation of the social world in and through the writing of scholars, or as the *adaequatio intellectus et rei*, the reaching of such adequacy being the objective of scientific inquiry. From Plato and Aristotle, and then Thomas of Aquinas, this *problématique* stretches into the positivism of the sciences of the nineteenth century, including the social sciences, as well as its critical alternative, scientific Marxism. With important qualifications, it was revived in the neo-positivism of the twentieth century and the unified science program that tried to (re)integrate the social sciences with the natural sciences.

Despite this continuity of concern, the very concept of representation of the world has also been heavily criticized. At several points in intellectual history it appeared as if the whole idea needed to be abandoned, because there just is no way to verify the relation between the world and thoughts about the world other than again by the use of thought (even though not exclusively by thought). Thus, certain knowledge of *adaequatio* is in principle unattainable. In most recent debate, during the closing decades of the twentieth century, the accumulation of such doubt has often been labeled a "crisis of representation," implicitly and misleadingly pretending novelty of these doubts. This crisis is in many respects the background to the "science wars." Through all of earlier such crises, however, the *problématique* keeps reviving, often with new qualifications, and those who revive it often argue that it cannot be entirely abandoned either, since otherwise the status of statements about the world would become entirely unclear. A war over this issue, thus, would be a very peculiar war: It cannot be won, not even by the attempt at annihilation of one of the belligerent parties, since such annihilation is impossible.

Even though this debate about representation has been going on virtually through all of the history of philosophy, the break that came with the radical skepticism characteristic of what can be called the philosophy of modernity is often seen as shifting the terrain. Between the middle of the seventeenth century, Descartes's time of writing, and the end of the eighteenth century, Kant's time, the ground on which representation could be possible was systematically explored. This period, of course, is precisely also the period of the so-called scientific revolution. Furthermore, it is also the period of the immediate aftermath of the religious wars in Europe—an issue to which we will return. Descartes's *cogito ergo sum* radically subjectivized knowledge claims. Nevertheless he was able to develop on that basis a rationalism according to which mathematical representation was adequate to the world. Kant's distinction between *noumena* and phenomena has sometimes been interpreted in a similar way, that is, by considering *noumena*, although they cannot be experienced, as theoretical entities the relations between which are governed by laws. A science developing such laws would then be truthfully representing the world. Arguably, however, the more significant impulse from these one and a half centuries of debate was an aggravation of the general problem of representation, at least within the philosophical mode of reasoning. This crisis of representation stemmed from the increasing gap between the perceiving and knowing subject and the world.

The combined outcome of the Cartesian and the Kantian moves, namely, was that, on the one hand, philosophical reasoning appeared to have exhausted its possibilities, at least for the time being. At the end-point of this reasoning, on the other hand, a space had been opened for a different form of inquiry. The gap between the subject and the world, or between the *noumena* and the phenomena, permitted, maybe even invited, attempts to fill it by positive, empirical knowledge, elaborated in systematic, conceptually guided ways. A terminological shift accompanied these debates. Natural philosophy was increasingly complemented by—some may want to say: gave way to—"empirical philosophy." And the latter inaugurates precisely that mode of investigation that we now know under the name of science. From the end of this period onwards, that is throughout the nineteenth century and into the twentieth century, these new sciences developed enormously. They achieved what we still consider major breakthroughs in knowledge, and they specialized and grew into the disciplines of the natural sciences as well as, more slowly and more contestedly, the social sciences. During this period of rapid development it was largely and easily forgotten that the turn toward empirical investigation had by far not answered the question of representation satisfactorily or, in other words, that the relation between philosophy and the sciences remained an important one, despite all claims on the part of the sciences to have emancipated from philosophy.[4] The explicit reopening of that question should only be a matter of time.

SCIENCE AND ITS CRITICS: A RECURRING DEBATE

"Man who lives in a world of hazards is compelled to seek for security." Thus starts a book that was published in the United States of America on 11 October 1929. Only a few days later, Wall Street crashed, and security was in short supply. The book was John Dewey's *The Quest for Certainty*, and the precise timing of its publication was only coincidentally related to the Wall Street crash. However, there is a broader context of crisis to which it indeed responds. For decades, concern had again been growing about the relation between the truth claims of philosophers, on the one hand, and, on the other, the kind of knowledge the sciences had been producing with increasing success throughout the nineteenth century. Those sciences, including the social sciences, were seen as overspecialized and fragmented while at the same time highly successful in generating new insights and in transforming the world through science-based technology. Having initially developed under the shelter of philosophy, they seemed to make philosophy—just like religion—both superfluous and unsustainable, since no—ontological or metaphysical—claims about any continuity of the order of things nor any ground for that order could be upheld in the face of rapidly changing scientific concepts and empirical insights.

Through the early decades of the twentieth century, however, novel conceptions of the relation between philosophy and the sciences were proposed that were based not least on a critique of what John Dewey (1984 [1929], 195) called the "spectator theory of knowing." With this term, he summarized all views, for him characteristic of most philosophy and science, that saw the world as exposed to the human gaze and assumed that it is exactly this distance between the knowing subject and the object to be known that allows for certain knowledge. The spectator theory of knowing thus starts out from a radical skepticism, from the attitude that nothing can be taken for granted. And it is in response to the doubts about the world around us as we experience it that the quest for certainty needs to proceed through a distancing from the world to gain an "outside" look at it. The distancing from the sensations of the world had been considered a precondition for the identification of those assumptions on which claims for valid knowledge could be erected. By the same move, the question of epistemology itself had emerged as the problem of the relation between the "reality" of the world and the "representations" that human beings provide of the world in their philosophy and science. The preference in the philosophy of science was for a long time, as Dewey points out, for an epistemological formalism that presupposes this distance between the knower and that which is to be known. Thus emerges what I would call the modernist conception of knowledge and of the sciences (Wagner 2001, ch. 1). It went along with the—quasi-sociological—idea of separating out a social realm in which the pursuit of knowledge was possible in independence from other concerns of social life. But, of course, as Dewey insists, human beings cannot step out of the world they inhabit and of which they are a part.

Similar turns were made in European philosophy during the same period. Ludwig Wittgenstein had hoped to clarify the problem of representation in the *Tractatus logicus-philosophicus*, in which he employs a picture theory of knowing, whereas in the *Philosophical investigations* he sees our languages as being part of our forms of life, and thus not looking at life from a distance. In *Holzwege*, Martin Heidegger analyzed the modernist theory of knowledge as representation comprehensively as the attempt "to provide oneself a picture" of the world by making a world that is placed in front of oneself. Rather than making "a picture of the world," however, this move "grasps the world as picture," relates to the world as if it were an object in front of us. In two intertwined processes, the world becomes a picture and the human being a subject. Richard Rorty tried to summarize this reinterpretation of the history of philosophy and the sciences provided by Dewey, Heidegger and Wittgenstein as the abdication of the attempt of seeing knowledge as "the mirror of nature" (Rorty 1980; Friese and Wagner 1997, 208–11).

Rorty's synthesis roughly coincided with Santos' *Discourse*. The 1980s became a new period of reflection about the sciences and the kind of knowledge they were producing. Dewey had already spoken of an emergent paradigm—to use Santos's term, even though Dewey conceptualized the shift somewhat differently. In the concluding lecture of *The Quest for Certainty*, he likened the development of the sciences of his time to a "Copernican revolution," borrowing the term from Kant, the implications of which for philosophy he aimed to have spelt out. "The old centre was mind knowing by means of an equipment of powers complete within itself, and merely exercised upon an antecedent external material equally complete in itself. The new centre is indefinite interactions taking place within a course of nature which is not fixed and complete, but which is capable of direction to new and different results through the mediation of intentional operations" (231–32). One of the key events in Dewey's history of the sciences is Heisenberg's work on the "uncertainty principle," just as it is for Santos. How similar is the earlier debate to the current one?

Most contributors to the debate in the early twentieth century worked with a strong distinction between philosophy and the sciences. In many of the European contributions, the loss of the hegemony of philosophy over the sciences was central to the problem. The initial observation of the fragmentation and compartmentalization of the sciences as a consequence of the loss of the philosophical umbrella led later to diagnoses such as the hegemony of instrumental reason—and thus the "eclipse" of reason in a more comprehensive sense—in Horkheimer, and the emergence of modern technology as a new metaphysics in Heidegger. The alternatives these authors offered in their early writings pointed into different directions for a rethinking, but in both cases they entailed a return to philosophy, even though in highly different guises.

The alternative approach was to reject any return to philosophy, however conceptualized, and to embrace the new sciences. In different, even directly opposed ways, this was what the (neo)positivism of the Vienna School and American pragmatism proposed. In Dewey's terms, positivism would transfer the—misguided, in his view—quest for certainty from philosophy to the sciences. Positivism basically works with a model of a mind-independent reality and draws a clear boundary between knowing and acting. As a philosophy of science, positivism tries to establish the claim for the certainty of superior scientific knowledge in a new form, based not least on logic and mathematics, which for Dewey are merely one pole of possible knowledge. Dewey himself, in contrast, also embraces the sciences, but only because he sees them undergoing a decisive change in orientation, his "Copernican revolution," as a consequence of Heisenberg's discovery of the principle of indeterminacy. Science will no longer want to discover laws of motion of antecedent objects, to which human beings have to conform, but will move to "constructing enjoyable objects directed by knowledge of consequences" (Dewey 1984, 217). Was it possibly John Dewey, whom nobody suspected, who declared war on the scientific establishment?

But a declaration of war, or of revolution, that passes unnoticed is no declaration of war at all. Five decades later, Richard Rorty (1980) should claim that Dewey, Heidegger and Wittgenstein were the most important thinkers of the twentieth century, and this not least because of their rethinking of knowledge, science and philosophy, all with a view to overcoming metaphysics and epistemology. But had these thinkers really been influential in this way at their time, then the question of epistemic certainty should long have been put to rest. However this was far from being the case. Half a century after *The Quest for Certainty*, another "report on knowledge" underlined the fact that the sciences of the twentieth century had continued to make unwarranted claims to superior knowledge. The author, Jean-François Lyotard (1979, 25–27), resorted to Wittgenstein's notions of language games and their irreducible plurality to explain why such a claim could not be upheld. He thus became one of those who triggered off a new debate on epistemology, often now referred to in the context of discussions about postmodernity. Similarly, Santos's observations form part of a renewed reflection on the time-honored debate about the ways in which the sciences relate to the world that they investigate.

The outraged reactions to the so-called postmodernist challenge to epistemology and to the sciences, though, operate as if no such debate had ever happened. Why is it then, we may ask, that such debates recur instead of having the "revolutionary" impact on our understanding of knowledge that both Dewey and Santos expected and hoped for. To give some answer to this question, we need to briefly review the issue of the relation between knowledge and the world, with a view to both its conceptualization and its actual practice in the history of modernity.

SCIENCE AND ITS APPLICATIONS: SCIENCE-INSPIRED TRANSFORMATIONS OF THE WORLD

Both the authors of the interwar period and those of the 1970s and 1980s look back at an immediate past during which the large-scale application of science-based technology considerably transformed the world they were experiencing. Dewey, Heidegger and Wittgenstein were writing after what was later called the second industrial revolution, with the development of electrical and chemical industry as well as a revolution in transportation, and the application of all of this during the First World War, of crucial importance at least for the Europeans. Santos, Lyotard and those historians and sociologists of science who have started to seriously study empirically the activities of scientists from the late 1970s onwards, in turn, write after the "*trente glorieuses*" (Jean Fourastié), those thirty glorious years of economic growth after the Second World War that transformed the face of the earth in an unprecedented way.

Contextually sensitive historians of ideas will agree that such observation helps understand the recurrence of topics. Unlike Dewey thought, no revolution in philosophy and science was taking place during the interwar years, so that a similar problematic phenomenon could reoccur. Nor was, we may add today, a new paradigm emerging during the 1980s, unlike Santos thought. Science-inspired transformations of the world have remained a topic of some urgency. However, scientists may rather easily reject any responsibility for those developments philosophers and historians of science were observing. Ironically, or maybe: significantly, what allows them to deny any such responsibility is precisely the self-understanding of science in relation to the world that is diagnosed by their critics. They will claim that their activity proceeds according to its own rules, at a certain distance from worldly affairs. These rules, as scientists tend to argue, both support their claim to epistemological superiority and the absence of any responsibility for "application," since this is what happens after they have done their work and is beyond their control.

The claim to superior knowledge, though, can be used to add a further twist to the reasoning: Despite denying responsibility for hazards resulting from application because their approach is neutral, objective, methodical and self-driven, scientists are often ready to accept praise for their useful knowledge (and funding to produce more it), which is useful precisely because it is superior to other knowledge. There is a saying to describe this attitude: to have one's cake and eat it. The point here is not the denial of the usefulness of some technical innovations that follow upon scientific discovery. The point is to understand what the relation of distancing and intervening is in the scientific-technological enterprise. Bruno Latour (1993)—one of the most favorite enemies of the warriors on behalf of science—has aptly described the apparent paradox as part of what he calls the modern constitution: Because the scientific attitude claims to take—objectifying—distance from the world, it is not only able to look at the world, but also to act upon it, apparently from the outside.

Some observers, especially during the 1930s, have moved from similar insights to the diagnosis of the reign of instrumental reason or of unfettered technology. In Horkheimer and Heidegger, for instance, such view was the direct outcome of their philosophy of knowledge and technology. From Dewey's standpoint such outcome is a mere historical-empirical possibility, and he was even optimistic that the "Copernican revolution" underway would lead to a new conception of knowledge. In the more recent debate, often known under the label of postmodernism, a strong view on historical direction is avoided (with Lyotard coming closest to having one, but remaining peculiarly undecided). In contrast, this debate has been the most explicit expression of generalized epistemological uncertainty.

This range of quite diverse interpretations suggests above all the need for a more sustained comparative-historical analysis of the sciences. Such analysis should not start out from a denunciation of the scientists' claim, but should see their practices as the search for alternative ways to certainty.

Exactly because—against the scientists' own claims—they entailed action, such as experimentation with nature, not just observation, the sciences could *create certainty* rather than merely *postulating* it. Even though no such historical analysis can be offered here in detail, some observations may indicate the direction in which it would have to go.

Descartes has long been accused of promoting the unachievable and ultimately damaging project of grounding certainty beyond the specificity of experience. More recently, Descartes's discourse on method has been read in context, that is as addressing the question of security in human social life at a time when the consequences of the reformation and the religious wars signaled one major step in the destruction of the foundations of certainty (Toulmin 1990). Similarly, the long neglected controversy between Hobbes and Boyle about the grounding of knowledge in experience—and, thus, scientific experiment—marks another case in which the fervor of the attempts to reestablish certainties can hardly be explained otherwise than through the impact of violent religious strife on the consciousness of the authors (Shapin and Schaffer 1985; Latour 1993; Stengers 1993). The scientific revolution was a revolution that followed a war; and it aimed at providing foundations for a new order. This order was, in the view of the revolutionaries, to be built on a belief that was of a different kind than any religious belief, on versions of which earlier orders had been grounded. What its protagonists failed to recognize was that any belief in a higher order is not only ultimately erected on uncertain grounds, it is also bound to create resistance. The scientific revolution may have aimed to end a war to make the world safe for rationality. What it indeed accomplished was to shift the terrain on which wars were led.

In such a contextual perspective, namely, the specificity of Descartes's, Boyle's and Hobbes' attempts is in the first instance a high awareness of an inescapable dilemma. On the one hand, there is a deep intellectual consciousness of contingency, of the lack of stability and certainty in the world. Radical doubt is indeed the starting point of the reasoning. In this sense, these episodes are rightly regarded as high points in the history of modernist thought, indeed inaugurating modernist discourse in philosophy and in political theory. On the other hand, Boyle, Hobbes and Descartes were also driven by the conviction that humankind could not live well without some categories of social and natural life, which impose themselves on everybody, and such categories would only impose themselves if and because they were undeniably valid. The uncertainty that was experientially self-evident in a context of devastating religious and political strife had to be overcome by appeal to an instance that in their view could only be outside such experience. The coexistence of these two, apparently incompatible convictions gives their work a character which one might label dogmatically modernist: radical in their rejection of unfounded assumptions, but inflexible in their insistence on some definable minimum conditions of cognitive and political order that could, and would have to, be universally established. Scientists up to this day have inherited much of this attitude.

Rather than interpreting the writings of Descartes, Boyle and Hobbes as distanced philosophical-scientific truths, we could however also regard them as activities that perform an operation on the world with a view to giving that world a new direction. This is especially clear with experiments on nature, as argued for by Boyle, which of course do not just observe the truth of nature but act upon nature. From such an angle, the authors would be seen as identifying a *problématique* in a given situation, would aim at problem-solving and would be driven by a situation-specific sense of urgency—rather than aiming to provide a timeless truth of philosophy or science (which, of course, was how they themselves described their activities). If history of philosophy and the sciences is read from such an angle, we become able to address issues that cannot be addressed by either principled defenders of modernist science or similarly principled critics of it.

As Boaventura de Sousa Santos points out in more recent writings (Santos 2001), the insistence on knowing as intelligent problem-solving in a given situation has to allow for different kinds of situations—not least: situations with greater or lesser degrees of (experienced) security—and for different proposals to solve the problems at hand—not least: solutions that work by generalization and solutions that work by specification, or as he calls it: small-scale, coarse-grain resolution versus large-scale fine-grain resolution knowledge. Modernism would then no longer fall into a separate category of knowledge-seeking, but would merely be that approach that insists on the need for generalization in—and because of—situations characterized by high insecurity.

VARIETIES OF EPISTEMIC CERTAINTY

As a consequence of such differentiation, this approach would allow for varieties in the search for certainty. More specifically, it would invite to reread the debate about "the crisis of the European sciences" (Husserl 1970 [1935]) as a dispute about various ways of engaging the world. The dispute was certainly marked by the diagnosis that an instrumental concept of knowledge lending itself to the transformation of the world by means of technology was about to become predominant. In the first instance, however, it relied on the possibility of making distinctions between various conceptions of knowledge.

Dewey (1984, 195) reserved a strong verdict for a philosophy that sticks to its quest for certainty: "Any philosophy that in its quest for certainty ignores the reality of the uncertain in the ongoing processes of nature denies the conditions out of which it arises. The attempt to include all that is doubtful within the fixed grasp of that which is theoretically certain is committed to insincerity and evasion, and in consequence will have the stigmata of internal contradiction." In the light of our earlier considerations this statement may be read as an indictment of the modernist sciences.[5] They rested on assumptions about the certainty of the world, and that is indeed where they failed—proved both contradictory and evasive, sometimes also insincere—when nature proved to be more uncertain than expected. Nevertheless the reference to "the reality of the uncertain" contains a significant openness. It invites the question about the degree of certainty that is attainable in different states of the real. Rather than attacking critics, scientists may better aim at understanding historically variable degrees of attainable certainty of their knowledge—or what Santos terms "epistemological confidence" (see also Wagner 1995).

Our questioning, though, cannot stop here. Epistemology is not overcome by merely, even if rightly, claiming that epistemic claims vary and may have variable validity. Certainly, once the assumption of a clear-cut separation of a mind-independent reality from the human knowledge about such reality is rejected, the traditional epistemological question is undermined. The fact remains, however, that human beings in the social world persistently make arguments for forms of knowledge that are held to be superior to others. The persuasive force of those arguments may have an impact on whether knowledge will be widely accepted and whether people will act on the basis of such knowledge and transform the social world. Even if epistemology in a traditional sense is overcome, actual claims for knowledge and their impact will have to keep being investigated. One cannot rest content with a general refutation of epistemically formulated validity claims—as, one needs to say, much of the history and sociology of the sciences during the 1970s and 1980s has done. Such claims should be read less as statements about the world, statements that could indeed be denounced by a critique of epistemology, than rather as ways of questioning the world, whose "truth" resides always in their nature as action on problems. Thus our question may now be reread as to why specific ways of questioning the world appeared particularly plausible at certain times. In this light, we can return to the history of the sciences and their epistemic claims, in particular during the twentieth century.

From our brief analysis two conclusions can now be drawn. First, this critique of science—both in the 1930s and in the 1970s and 1980s—had particular historical contexts of emergence. In the first case, it followed upon the rapid development of the scientific disciplines during the nineteenth century and the increasing application of their findings in what is often called the second industrial revolution at the end of that century and then, in the second case, in the even much wore widespread diffusion of these applications. Its problématique is circumscribed by the exigency to develop a critical perspective on the impact of modernism on the one hand, but the lack of plausibility of any recourse to "tradition" or "embeddedness" of social life, which would appear as an impossible return in time, on the other (on the rupture with tradition, see Wagner 2001, ch. 4).

Second, though, the strong critique of science as the one-dimensional imposition of instrumental reason is ultimately not entirely persuasive. It relies on strong presuppositions about both the nature of modernist science and its all-pervasiveness, on the one hand, and the society on which such conception can so apparently easily be imposed, on the other. The emphasis on varieties of ways of world-making, paraphrasing Nelson Goodman, and consequently variable degrees of certainty leads toward a more fruitful perspective. In turn, however, one should not forget the history of imposition of modernist science that started the critique of science in the first place. It is at this point, that Boaventura de Sousa Santos, if I see it correctly, aims to carry the debate further, in his *Discourse on the Sciences* and in his more recent writings (Santos 2001), on which my concluding discussion will concentrate.

EPISTEMOLOGY AND CRITIQUE: KNOWLEDGE AS REGULATION AND KNOWLEDGE AS EMANCIPATION

At a closer look, there are two views of knowledge in Boaventura de Sousa Santos's program; I will call them the strong view and the weak view. His conceptualization proceeds through the diagnosis of two epistemological breaks: The first break leads from common sense to science as we know it, a modernist science that develops knowledge-as-regulation and that supports colonization and order. The second break, yet to be accomplished, will then lead out of mainstream science to the new common sense, and that is to knowledge-as-emancipation, a knowledge that enhances solidarity. This is the strong view. There is, however, also a softer, weaker version of the program. In this version, the second break leads here to a variety of perspectives, to a "plurality of knowledges," to a new situation thus, in which knowledge appears in a multiplicity of forms.

The strong view draws a sharp line between regulation and emancipation and their respective knowledge forms. Thus, it follows many critical philosophies of social science, from Karl Marx to the early Jürgen Habermas. Those critical philosophies insist rightly—against objectivism and positivism, including the claims of what Santos calls "mainstream science"—on the indissoluble connection between knowledge, on the one hand, and the world that it is meant to be about, on the other. However, many, if not most, of these philosophies abandon their own questioning immediately after having developed it, namely by postulating to have already found the answer to it. Instead of one form of valid knowledge, there are then two: one that dominates because it supports domination, and another one that struggles to come into being because it undermines domination. This simple solution fundamentally neglects—or even denies—the ambivalence of modernity. If modernity, whether it actually exists or is an emancipatory and always unfinished project, is about freedom and autonomy, that is, self-determination, then it is also about self-regulation. There is no true knowledge of emancipation that would not in some sense also have to be a knowledge of regulation, of self-regulation, namely. The history of modernity certainly was a history of barbarisms, and science-based instrumental knowledge was enlisted in support of those. But if there is any way to read it fruitfully for the present, it needs to be read as a history of civilization failing. Otherwise, we would never see, let alone understand, the attempts made. There certainly has been what Santos calls a historical process of the "cannibalization of social emancipation." But we would need to know much more—in the form of fine-grain resolution, large-scale analyses, in Santos" terms—about how precisely this came about. An interpretation of the "first epistemological break" goes wrong if it totally denies the ambition of emancipation even in modernism and its science in the face of an otherwise appropriate critique of the rising asymmetry between "regulation" and "emancipation."

A certain lack of ambivalence with regard to the first break is repeated in the discussion of the second epistemological break. In the strong reading of his proposal, Santos appears to succumb to the time-honored inclination of rather markedly outlining a not-yet-existing alternative that solves all problems, an alternative in which knowledge and solidarity become one. But a quick historical observation suggests that he may be running into a trap that he himself posed. Is such a conception not precisely reviving the Enlightenment dream of a self-regulating society, a dream about which one can justifiably say that at least in some of its versions it has been converted from being emancipatory to being regulatory, and now in the sense of regulation by some imposed on others, and not as self-regulation? Why should we assume that something similar could not again be the case if we follow that same route again? In Santos's own, fruitful terms, the problem here is that one should not move from the absolute reign of experience over expectation, or from total indifference both to experience and expectation, toward a similarly unconstrained reign of expectation over experience. That is a recipe for disillusionment, at best, and was historically a recipe for disaster. The challenge of knowledge rather is to find the situation-wise appropriate relation of expectation to experience.[6]

Santos rightly underlines that there is conflict and struggle in the social world. However, even if we could assume that friend and foe could be clearly identified, there is struggle not merely between regulators and emancipators. If this were the case, then the struggle would be over once that fight is won. Under conditions of modernity, there will always be contestation, always struggle between different perspectives. And it is unlikely that there will be a knowledge that unequivocally is associated with emancipation and solidarity. Rather, even among those who support those goals, there will be variety of perspectives and, thus, a "constellation of knowledges" will emerge that does not settle a dispute, but provides it with resources for reasoning. Against indifference, one then would not just posit solidarity with friends. Rather, concern for others would show precisely in both solidarity and dispute, and both at the same time. This is what I see in the second, the weak program in Boaventura de Sousa Santos's work, and it is the one I fully agree with.

In his strong program, he continues the tradition of critique pointing to something that is not, but could be, and elevates this to a higher position, seen as immediately reachable, if not even reached, as soon as the obstacles are removed. Something of such a critical perspective will be needed as long as there is a strong asymmetry in the acceptance of knowledge claims in the world, as long as, in other words, modernist science can easily align itself with the powers-that-be on the terms of instrumental reason. But such critique should not overdraw its positive claims. Santos's weak program instead suggests a different understanding of critique, pointing also to something that is not, but that is always struggling to come into existence, thus is always present but in the form of failing. An epistemology of seeing, as he calls it, would not merely show the way to knowledge, it would also need to make visible the obstacles on that way. Since many of those obstacles will not disappear under any circumstances, the search for knowledge means the search for the different ways to overcome them.

In this light, we will have to conclude that there will be no easily established new order after the revolution. There is a need for a revolution, but maybe not in the form of a new paradigm that governs all our knowledge-seeking practices, rather in the form of putting the practices of the sciences into a place in this world, relativizing their claims to epistemic certainty by making them communicate with other ways of engaging the world. Because we are far away from such a situation, because of the strong asymmetries in the world of knowledge, there is a need for revolution. But this revolution does not necessitate a war, just a considerable dose of common sense, or maybe: new common sense.

NOTES

1. The term "scientific revolution," it appears, was made prominent by Alexandre Koyré in the 1930s and was widely used after the Second World War, but not much before (Shapin 1996, 2)

2. For the former the Edinburgh strong program in the sociology of scientific knowledge is the key source, though inspired by Thomas Kuhn's analysis of "scientific revolutions" (Kuhn 1962); for the latter, Richard Rorty's review of twentieth-century philosophy seems to have provided the term (Rorty 1967), though not the research program. For recent assessments, see Wittrock and Wagner 1987 for the former, and Wagner 2003 for the latter.

3. If there is one key aversion scientists seriously need to overcome, it is their refusal to see science as a human activity, with its specificity, but alongside others. Since "science" is what they are doing daily, this move should not be too difficult. It is their common refusal to make it, though, that turns them into dogmatics who claim to be right by virtue of their faith. And at that moment, war becomes a very likely occurrence.

4. Scientists today often may make fun of the alleged irrelevance of philosophy of science for their activity. But, to return the ball into their camp, the fact that the light goes on in my room when I turn the switch without me knowing what happens does not mean that physics is not important for understanding electricity (as Friedrich Dürrenmatt already pointed out).

5. This was not how Dewey saw it, who remained very positive about the sciences, but only because he saw them as undergoing the "Copernican revolution" toward the understanding of knowledge he thought to be adequate. Sixty years later, Santos, while also being—overly?—optimistic, cannot reason any longer in the same terms. His "emergent paradigm" takes distance not from philosophy, but from earlier—and still dominant—forms of scientific practice. It reflects the experience of another half century of ambivalent "progress" in the use of instrumental reason.

6. The concept of a relation between expectation and experience, borrowed from Reinhart Koselleck (1985 and elsewhere) can be translated into the metaphor of variable "degrees of distancing" that may be required/appropriate for different kinds of questions that may be asked. My thanks to Olav Korsnaes for making me think more about the metaphor of distance with regard to knowledge-seeking practices (which is also in the center of Carlo Ginzburg's recent collection of essays [Ginzburg 2002]).

REFERENCES

Dewey, John. 1984. *The Quest for Certainty.* In *The Later Works 1925–1953*, vol. 4. Carbondale: Southern Illinois University Press. (Orig. pub. 1929.)

Friese, Heidrun, and Wagner, Peter. 1997. "Bild und Wissen. Reflexionen zu Erkenntnis und Darstellung." In den Sozialwissenschaften, ed. Ute Hoffmann, Bernward Joerges and Ingrid Severin, *LogIcons. Bilder zwischen Theorie und Anschauung.* Berlin: Edition Sigma, pp. 207–22.

Ginzburg, Carlo. 2002. *Wooden Eyes. Nine Reflections on Distance.* London: Verso. (Orig. pub. 1998 in Italian.)

Heidegger, Martin. 1983. *Die Selbstbehauptung der deutschen Universität.* Frankfurt/M: Klostermann. (Orig. pub. 1933.)

Horkheimer, Max. 1931. "Die gegenwärtige Lage der Sozialphilosophie und die Aufgaben eines Instituts für Sozialforschung." Vol. 37 of *Frankfurter Universitätsreden.* Frankfurt/M: Englert und Schlosser.

Husserl, Edmund. 1970. *The Crisis of European Sciences and Transcendental Phenomenology*. Evanston, IL: Northwestern University Press. (Orig. pub. 1935.)

Koselleck, Reinhart. 1985. *Futures Past. On the Semantics of Historical Time*. Cambridge, MA: MIT Press. (Orig. pub. 1979 in German.)

Kuhn, Thomas S. 1962. *The Structure of Scientific Revolutions*. Chicago: The University of Chicago Press.

Latour, Bruno. 1993. *We Have Never Been Modern*. Hemel Hempstead: Harvester Wheatsheaf.

Lyotard, Jean-François. 1979. *La Condition Postmoderne*. Paris: Minuit.

Rorty, Richard, ed. 1967. *The Linguistic Turn. Recent Essays in Philosophical Method*. Chicago: The University of Chicago Press.

———. 1980. *Philosophy and the Mirror of Nature*. Princeton, NJ: Princeton University Press.

Santos, Boaventura de Sousa. 1992. "A Discourse on the Sciences." *Review*, 15 (1):9–47. (Orig. pub. 1987 in Portuguese.)

———. 2001. "Towards an Epistemology of Blindness. Why the New Forms of 'Ceremonial Adequacy' neither Regulate nor Emancipate." *European Journal of Social Theory* 4 (3):251–79.

Shapin, Steven. 1996. *The Scientific Revolution*. Chicago: University of Chicago Press.

Shapin, Steven, and Simon Schaffer. 1985. *Leviathan and the Air Pump*. Princeton, NJ: Princeton University Press.

Stengers, Isabelle. 1993. *L'invention des sciences modernes*. Paris: La découverte.

Toulmin, Stephen. 1990. *Cosmopolis. The Hidden Agenda of Modernity*. Chicago: The University of Chicago Press.

Wagner, Peter. 1995. "Sociology and Contingency. Historicizing Epistemology." *Social Science Information/Information sur les sciences sociales* 34 (2):179–204.

———. 2001. *Theorizing Modernity*. London: Sage.

———. 2003. "As Intellectual History Meets Historical Sociology. Historical Sociology after the Linguistic Turn." In *Handbook of Historical Sociology*, ed. Gerard Delanty and Engin Isin. London: Sage, 2003, 168–79.

Wittgenstein, Ludwig. 1984. *Tractatus logicus-philosophicus*. Frankfurt/Main: Suhrkamp (Werkausgabe, vol. 1).

5

Authorizing Knowledge in Science and Anthropology

Joan Fujimura

This chapter examines an incident in the history of geometry in an effort to articulate what is at stake in the science and culture wars that are currently being waged in academia. The incident is a mid-nineteenth-century "science war" that erupted around a challenge to Euclid's fifth postulate, a challenge that has since transformed our understanding of the circumference of a unit circle (or π) and our concepts of time and space. The historical outcome of this challenge is the now-accepted non-Euclidean geometry, and the events surrounding its genesis raise issues of central concern to us at the millennium.

The issue I address by discussing this early science war in the context of current debates is that of scientific authority. I argue that the science wars then and now are not about science versus antiscience, not about objectivity versus subjectivity, but about authority in science: What kind of science should be practiced, and who gets to define it? This war for authority has raged and currently rages between disciplines but also within disciplines and fields. There are similarities and differences between the warring camps that defy the often used dichotomous categories of "science" versus "humanism" into which many different positions and arguments between positions have been lumped. For example, within anthropology the debates for authority to define the "correct" approach for studying human communities take place within sociocultural anthropology, as well as between sociocultural and biological anthropology.[1] The language of "canons," "universals," and "generalized explanations" is often ascribed to the natural sciences, and the language of "variation," "specificities," and "contingencies" is used to describe humanistic studies.[2] Some anthropologists are currently authorizing themselves as scientists who speak for science and against other kinds of scholarly work that they denigrate as antiscience, art, moralizing, or politics.[3] For another example, some physicists and mathematicians are claiming the authority to denounce work in the humanities, as well as in the social and cultural studies of the sciences, mathematics, technologies, and medicines specifically (often abbreviated to "science studies"). Other physicists and mathematicians want no part of this offensive.

My effort here is to expand the terms of the discussion by asking several questions: What are the boundaries of science?[4] Who qualifies as a scientist, and who does not? What is at stake in the wars? I address these questions through an exploration of the complexities of science and mathematics that is intended to move the discussion away from simplistic categories and dichotomies. I hope that my central example from the history of geometry will add to the understanding of what science is and can be, in order to further discussion across camps. My argument is that science is more complex, diverse, and multiple than the understanding of science that is held as an ideal type model (or rhetorically used) by some anthropologists and scientists involved in the debates. This latter model of science has been found lacking by science studies writers. Rather than join either side of this debate constructed around polar extremes that represent nothing that any of us actually do, I want to discuss what science studies is attempting to accomplish through its analyses and critiques.

THE "SCIENCE WARS"

The movement often called the "science wars" has taken an active role in criticizing science studies, using textual, conference, corridor, and virtual (World Wide Web and e-mail) sites to advance its offensive. An early conspicuous salvo in the North American science wars was the book Higher Superstition, by Paul Gross and Norman Levitt (1994), which explicitly attacked social and cultural studies of science and technology as heretical.[5] A round of discussions followed, including the conference "The Flight from Science and Reason," organized in part by biochemist Gross and mathematician Levitt and hosted by the New York Academy of Science (May 31–June 2, 1995).[6] In 1996 a few literary theorists, sociologists, anthropologists, historians, philosophers, biologists, and population geneticists who are engaged in science studies published a set of responses to Gross and Levitt's work in a theme issue of the collectively run, non-refereed journal Social Text. [7] The issue was subsequently republished by Duke University Press in the volume Science Wars (Ross 1996).

Included in the Social Text issue was an article by Alan Sokal, a physicist who teaches at New York University, entitled "Transgressing the Boundaries:

Toward a Transformative Hermeneutics of Quantum Gravity" (1996b). Sokal's essay presented itself as a sincere, albeit naive, attempt at deconstructing physics using the conceptually transfiguring work of physicists Heisenberg and Bohr, "revisionist studies in the history and philosophy of science," "feminist and poststructuralist critiques," and cultural studies to try to develop a "future postmodern and liberatory science" (Sokal 1996b, 217–18). But Sokal (1996a) later revealed in a Lingua Franca article that his essay was a hoax, an attempt to parody what he considered to be incorrect constructivist and postmodernist arguments about science based on his interpretation of authors such as Jacques Derrida.[8] Sokal's article engendered more discussion and debate, including workshops around the country.

In this article I consider a passage about pi (π) in Sokal's hoax article that actually makes considerable mathematical sense. I use the history behind this passage to draw an analogy between today's "defenders" of "science" and nineteenth-century "defenders" of premodern geometry and number theory who reacted to new developments in mathematics with indignant polemics and exclusionary tactics.

The texts written by Gross, Levitt, and Sokal primarily aim their criticisms at "social and cultural studies of science" and "postmodernism," with "the academic Left" as an additional target for Gross and Levitt. (I place quotes around these terms because Gross and Levitt are imprecise in their use of them. For example, in each category they include diverse writings in the humanities and social sciences that would be parsed into many different categories by the authors of the writings and their other readers.) The major arguments made by Gross, Levitt, and Sokal are, in the barest terms, as follows:

1. Authors of major texts in the science studies are antiscientific, anti-objectivist, and antirealist.
2. These authors are nonscientists who have pretensions to know and criticize science when they do not understand science, except perhaps at the level of popularizations.
3. The argument by science studies writers that scientific knowledge is socially constructed is absurd and speaks of science as if it produces false claims.
4. Science is good, science is neutral, and science has the best available methods (for example, the testing of hypotheses) for producing credible claims.
5. While social science and feminist epistemologist writings may be wrong about science, postmodernist writings are worse. They are empty of any real substantive content and are merely decorative incomprehensible language masquerading as knowledge. (This last claim belongs more to Sokal than to Gross and Levitt.)

Some of these points have already been refuted. For example, Roger Hart (1996) has focused on Gross and Levitt's misrepresentations of the science studies literature through misquotes and quoting out of context. I want to discuss other issues raised in the work of Gross, Levitt, and Sokal. First is their misunderstanding of the notion of construction.

CONSTRUCTIVISM IN SCIENCE STUDIES

The term "construction," or "social construction," has been a point of contention in science wars rhetorics. Constructivism is generally employed in science studies to argue that scientific representations and objects are constructed by scientists engaged in particular kinds of scientific work.[9] These representations can take the form of theories, diagrams, drawings, photographs, autoradiographs, and plant and animal models. Scientific representations are created, mediated, and judged by sets of rules and procedures, often called protocols. Each community of scientists claims the license and mandate to construct and enforce such rules and procedures. Thus scientific work includes the methodological development of rules by which representations are constructed and confirmed, as well as the production of representations. Constructivist studies have provided an extensive body of historical and contemporary examples of how locally practiced sciences produce knowledge, how standardized practices and tools are constructed in particular scientific fields, how standards of judgment and verification are established, how specific controversies are resolved, and how social and cultural locations circumscribe scientific problem selections and constructions.[10]

Biologists, computer scientists, theoretical statisticians, mathematicians, physicists, and engineers are, in a different sense, constructivists. With particular desires, dreams, cultural resources, and technical competencies in their toolkits, they construct new realities, as well as new representations (e.g., Fujimura 1996; Haraway 1997). It is their job. I refer not only to their construction of theoretical statements about nature. Scientists also construct and create new entities. They create novel creatures, new life, new ways for atoms to exist and move, and new problems that create new solutions that are new entities. These, in turn, change our societies and cultural practices. Scientific constructions defy the human-nature binary, the social-natural binary, in their production and their effects.

In some cases scientists create new objects and new organisms in order to study a particular problem. These objects and organisms are meant to represent particular phenomena. For example, the "OncoMouse" is a living and breathing mouse genetically engineered by DuPont to produce a particular type of tumor consisting of cells containing *ras* oncogenes.[11] This mouse is an experimental model of cancer. Before the OncoMouse, biology and genetics had a long history of constructing experimental animal strains and cell lines according to specifications designed by scientists interested in particular causal models of cancer. For instance, they crossbred and re-crossbred mice to create strains that would reliably produce particular tumors for study by geneticists and biologists. Theories of cancer were based in part on the experiments conducted on these models, which in turn were designed with particular causal theories in mind. Biologists have been producing research tools of these types since at least the beginning of this century (e.g., Clarke and Fujimura 1992; Fujimura 1996).

Other examples of objects produced through science include the new creatures created by agricultural biologists and breeders for commercial purposes. Dolly, the cloned sheep, is only the most recent of these efforts. Dolly was the result of painstaking efforts, 434 attempts. A nonbiological example that comes to mind is the splitting of atoms in a linear accelerator. The amount of money, time, materials, and work (building and managing) that goes into splitting atoms by humans is phenomenal.

Science studies have taken as one of our tasks the analysis of how these productions and others (for example, theories of cancer causation, theories of organismal growth and development, and theories of time and space) are generated through the practices, instruments, and materials of scientific work. Alan Sokal has expressed a concern that constructivist studies of science imply a lack of faith in objective realities.

> What concerns me is the proliferation . . . of a particular kind of nonsense and sloppy thinking: one that denies the existence of objective realities, or (when challenged) admits their existence but downplays their practical relevance. . . . My concern about the spread of subjectivist thinking is intellectual and political. Intellectually, the problem with such doctrines is that they are false (when not simply meaningless). There is a real world; its properties are not merely social constructions: facts and evidence do matter. What sane person would contend otherwise? And yet, much contemporary academic theorizing consists precisely of attempts to blur these obvious truths. (Sokal 1996a, 63)

Whether or not the authors Sokal criticizes deny the existence of objective realities is a question for empirical investigation, that is, at least a careful reading of the literature. If Sokal refers to the reality of the objects or properties represented through scientific work, the science studies I have read are quite varied in their positions on this issue. But such analyses do not assume that theories and other abstract representations are mere ephemera, as Sokal implies in his crudest representation of science studies. "Anyone who believes that the laws of physics are mere social conventions is invited to try transgressing those conventions from the windows of my apartment. (I live on the twenty-first floor)" (Sokal 1996a, 62).[12]

Some science studies authors have argued that theories are real in their consequences.[13] They are interested in how people use abstract representations to organize their actions in particular ways. For example, people allow their bodies to be irradiated or injected with chemicals because they and their doctors subscribe to particular theories of cancer. Similarly, scientific theories of racial differences had real consequences in the history of the United States as justifications for particular forms of social organization, with slavery as just one example. For a less venal example, Einstein argued that space and time are not simply mediated and represented by rulers and clocks but, instead, are generated by these instruments. Indeed, notions of space and time represent "the humble and hidden practice of superimposing notches, hands and coordinates" (Mermin 1997b, 13). Those who take Einstein's view still can study how these notions generated from the work of rulers, clocks, and their makers have been used to organize human lives in many cultures.

Sokal states that when science studies does admit to the existence of objective realities, it "downplays their practical relevance" (1996a, 63). He perhaps intends to say here that social and cultural studies of science deny that the reality of natural phenomena have anything to do with how scientific controversies are resolved. Whether representations reign because they best represent the properties of the real world is something scientists and philosophers argue about constantly. In philosophy of science, this discussion would be parsed into, at the very least, discussions of realism and positivism, which are not the same.[14] In science studies there is no consensus on issues of realism, positivism, or relativism. I would argue that some are realists, some are positivists, some are both, some are neither, and others take no positions on these two issues; and the term relativism houses many different views.[15] For example, in contrast to Sokal's assumption, physicist Karen Barad (1996) sees no contradiction between realism and social constructivism, as the subtitle of her article attests. Using Niels Bohr's explication of Heisenberg's uncertainty principle, she discusses the production of the natural object through the "intra-action" of the instrument and the object it measures. For her, the natural object is both real and constructed.

Some science studies take relativist positions on the role of evidence in ending controversies and instead focus on examining the ways through which reliability and stability are produced in scientific and engineering work. There has been much interest in how evidence is constructed, argued for, ignored, marshaled, and so forth. It is also possible for those in science studies to examine the ways in which scientists argue for one position or another without taking positions ourselves. There are often good reasons for not taking a position. For example, given that mathematicians and physicists are engaged in endless and fascinating discussions about the "nature" of space, it would be presumptuous of science studies researchers to assume that we understand the reality of space enough to adjudicate among competing representations of space and to then use that "reality" to explain the state of geometry and physics. As physicist N. David Mermin reminds us,

> It was indeed a convention among scientists, buried so deep in their culture to be unrecognizable as such, that space and time were real objective entities measured by clocks and meter sticks. Einstein's profound insight was that, on the contrary, space and time are abstractions, serving to coordinate the results of such measurements. This is what sociologists of scientific knowledge have been saying for over two decades about all kinds of entities that scientists view as objectively real. (1997b, 13)

Instead of appealing to "realities," some science studies researchers observe the practices of scientists and the minutiae of their debates and controversies. They try to understand how scientists come to their representations of the world. And they want to understand how this work takes place within the world, within society, within history. Scientific work is not disassociated and detached from the world. Understanding it requires that we examine it in context. Some science studies writers argue that it is not productive to assume that we understand what nature is and then use that nature to explain science or society. There are also debates about whether we understand society and can use those understandings to explain science. Sokal's implication that the protocols in science studies imply a lack of faith in objective realities and his lumping of the diversity of protocols are simplistic and nonproductive.

SCIENTIFIC AUTHORITY AND THE POST-EUCLIDEAN REVOLUTION IN MATHEMATICS

A second issue raised by the work of Gross, Levitt, and Sokal is that of authority. In their criticisms of social and cultural studies, Gross, Levitt, and Sokal assume their right to determine *who gets to speak for and about science and whose work gets to be considered scientific and therefore credible.*[16] They claim to be able to determine what is true and false in the statements made not only in the sciences but also in social and cultural studies of science, and even more generally in the social science and humanities literatures. Either they do not recognize that "humanistic studies" have their own standards for determining the quality of scholarship, or they disapprove of those standards. Another topic eschewed in the Gross, Levitt, and Sokal critiques is the question of what scientific standards are. When one studies the actual practices of science, one sees many different standards practiced at different times for different stages of a research project as well as in different kinds of scientific research. Gross, Levitt, and Sokal assume that we all know what scientific standards are and that they apply equally in all sciences, social sciences, and humanities.

In this section I want to compare Sokal's rhetoric with that of historical antecedents who made similar authoritative statements about how science should be done. Sokal's hoax article referred me to the history I discuss here.

HISTORICIZING PI

I begin with the story of π (pi).[17] Pi is the circumference of a circle with unit diameter, that is, a circle with diameter of length one (of whichever unit of measurement one wants to use). I will refer to such a circle as a *unit circle*. What is a circle? It is a collection of points equidistant from a given point, such as an origin point. In Euclidean geometry, π is a constant, a transcendental number having a value (to five decimal places) of 3.14159.

In *Social Text* Sokal concocts what he thought would pass as a postmodernist (but nonsensical) emulation of a passage on mathematics by French philosopher Jacques Derrida. Derrida's statement, as translated into English and quoted by Sokal, follows:

> The Einsteinian constant is not a constant, is not a center. It is the very concept of variability—it is, finally, the concept of the game (jeu). In other words, it is not the concept of something—of a center starting from which an observer could master the field—but the very concept of the game which, after all, I was trying to elaborate. (Derrida 1970, 267 in Sokal 1996b, 221)

In an abstruse interpretation of this already abstruse passage, Sokal states that

> The π of Euclid and the G of Newton, formerly thought to be constant and universal, are now perceived in their ineluctable historicity; the putative observer becomes fatally decentered, disconnected from any epistemic link to a space-time point that can no longer be defined by geometry alone. (1996b, 222)

In addition to his play on Derrida's passage, Sokal added a statement that one can construct multiple π's, all of which could be valid in particular situations.

In subsequent discussions of his hoax, Sokal stated that "throughout the article, I employ scientific and mathematical concepts in ways that few scientists or mathematicians could possibly take seriously. . . . I intentionally wrote the article so that any competent physicist or mathematician (or undergraduate physics or math major) would realize that it is a spoof" (Sokal 1996a, 63). In a subsequent essay Sokal credits these "meaningless or absurd statements" to Derrida and other "postmodern Masters."

> The most hilarious parts of my article were not written by me. Rather, they're direct quotes from the postmodern Masters, whom I shower with mock praise. In fact, the article is structured around the silliest quotations I could find about mathematics and physics (and the philosophy of mathematics and physics) from some of the most prominent French and American intellectuals; my only contribution was to invent a nonsensical argument linking these quotations together and praising them . . .
>
> Now, what precisely do I mean by "silliness"? Here's a very rough categorization: First of all, one has meaningless or absurd statements, name dropping, and the display of false erudition. Secondly, one has sloppy thinking and poor philosophy, which come together notably (though not always) in the form of glib relativism. (Sokal in press, emphasis in original)

Since Sokal's article is in hoax mode and his revelation of the hoax in Lingua Franca was not specific about π, we cannot know whether he intended readers to believe that π is "constant and universal." Therefore, it is important to point out that "the π of Euclid, . . . formerly thought to be constant and universal," is "now perceived in [its] ineluctable historicity." The view of a "constant and universal" unit circle circumference (π) was refuted 170 years ago by non-Euclidean geometry, as is well documented in the history of mathematics. In twentieth-century mathematics and physics, the value of the non-Euclidean π now depends on motion, space-time, and gravity (Feynman 1997; Wolfe 1945).[18] Thus this mathematical statement that Sokal attributes to the "postmodern Masters" is well accepted by mathematicians, physicists, and engineers.

The reconstruction of geometry was part of major changes in mathematics and science that began in the 1820s.[19] As part of the revolution in geometry, the value of the circumference of a unit circle (or π) changed. Until that time, Euclidean geometry had reigned over the world of mathematics for 2,000 years. Then, in the 1820s, three mathematicians—Carl Friedrich Gauss in Germany, János Bolyai in Hungary, and Nikolai Ivanovich Lobachewsky in Russia—independently created new ways of doing mathematics by challenging Euclid's fifth postulate, on which Euclidean plane geometry is based.

I want to demonstrate the reconstruction of π in two ways: first, via a discussion of non-Euclidean geometry and, second, through a discussion of non-Euclidean distances.

NON-EUCLIDEAN GEOMETRY

> Out of nothing, I have created a strange new universe.
>
> —János Bolyai, letter to his father, November 1823.[20]

How could it be that the circumference of a unit circle is not 3.14159. . . ? According to high-school geometry, it seems obvious that this is the only possible value for this circumference. To answer this question, we must consider the alternative geometries invented by mathematicians when they questioned Euclid's fifth postulate.

Euclid's fifth postulate for geometry in the plane states that "if a straight line falling on two straight lines makes the interior angles on the same side less than two right angles, the two straight lines, if produced indefinitely, meet on that side on which are the angles less than the two right angles" (see figure 5.1). The lines, according to Euclidean geometry, must intersect. Gauss, Bolyai, and Lobachewsky challenged this postulate by inventing alternative plane geometries where the lines do not necessarily meet (see figure 5.2). Others considered the possibility that all lines intersect.

This challenge to Euclid's fifth postulate has had dramatic consequences for geometry. For instance, in Euclidean plane geometry, there is exactly one line through a point P parallel to a line L. This is the familiar world of Euclidean plane geometry. But in one example of non-Euclidean plane geometry, there can be infinitely many lines through a point P parallel to a line L (see figure 5.2). This geometry is called hyperbolic plane geometry.

At first glance (as well as second, third, fourth glance) figure 5.2 seems nonsensical, and for 2,000 years mathematicians and logicians attempted to prove this. Gauss, Bolyai, and Lobachewsky proved that figure 5.2 did make sense, by imagining and inventing alternative plane geometries, thus challenging Euclid's fifth postulate. Mathematicians gradually came to accept these new geometries in part by invoking mathematical logic to show that they were just as logically consistent as Euclidean geometry was within its system of postulates. In these new geometries, the circumference of a unit circle (that is, π) is no longer constant and universal. It can be a value different from 3.14159. . . and is one of the interesting topics of study in post-Euclidean geometry.

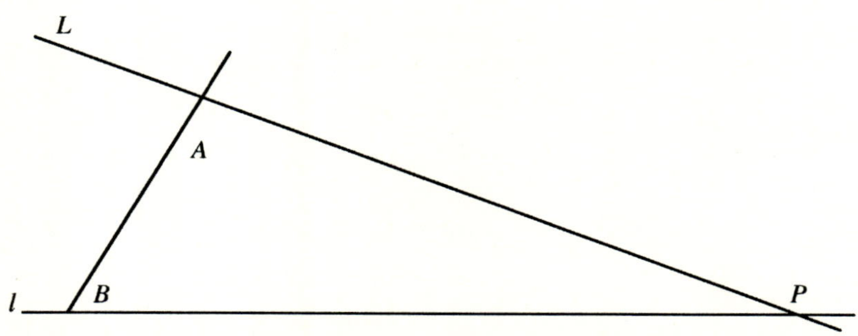

Figure 5.1. Euclid's fifth postulate. If angles A and B add to less than 180 degrees, then the lines L and l when produced indefinitely meet at some point P.

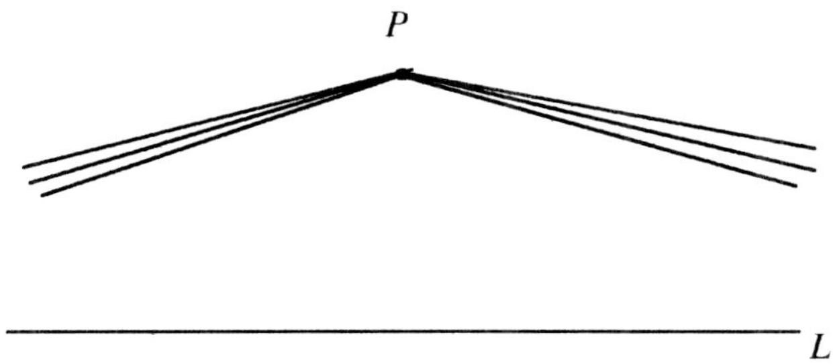

Figure 5.2. Hyperbolic geometry. Infinitely many lines parallel to line L through point P.

Those of us who are used to discussing only Euclidean geometry intuitively assume that there can be only one line drawn through point P that lies parallel to line L. But this intuition does not apply, for example, when we consider distant galaxies and the curvature of space.[21] Resemblances between non-Euclidean geometry and geometry on spheres and other curved surfaces are useful in understanding non-Euclidean geometries. For instance, imagine the lines through P in figure 5.2 as small circles on a ping-pong ball. Each "line" will meet up with itself before it intersects with the line L. Thus there are infinitely many "parallels" through the point P to the line L. This highly simplified example is useful for imagining infinitely many parallels to a line, but it is not equivalent to the more subtle non-Euclidean geometry in a plane, for which I do not have adequate space here to explain fully (Wolfe 1945,155, 178).

Mathematician Marvin Jay Greenberg explains that Einstein had to use non-Euclidean geometry to develop his ideas about distances, the curvature of space-time, and mass, all of which were crucial to his work on the theory of relativity.[22]

> We must question the nature of our instruments—aren't they designed on the basis of Euclidean assumptions? We must question our interpretation of "lines"—couldn't light rays travel on curved paths? We must question whether space, especially space of cosmic dimensions, cannot be described by geometries other than these two (Euclidean and hyperbolic).
>
> The latter is in fact our present scientific attitude. According to Einstein, space and time are inseparable and the geometry of space-time is affected by matter, so that light rays are indeed curved by the gravitational attraction of masses. Einstein said, "To this (non-Euclidean) interpretation of geometry I attach great importance, for should I not have been acquainted with it, I never would have been able to develop the theory of relativity." (Greenberg 1974, 249)

Euclidean geometry, then, was not sufficient to allow the development of Einstein's theory of relativity.[23] Sokal's postmodern master is on the mark when he states "mathematically, Einstein breaks with the tradition dating back to Euclid,. . .and employs instead the non-Euclidean geometry developed by Riemann" (1996a, 221).[24]

NON-EUCLIDEAN DISTANCES

For non-Euclidean geometry, the circumference of a unit circle can have many values, but it is too complicated to demonstrate here how one computes these values. Instead, I provide another example, that of non-Euclidean distances, in which computing alternative values is easier to explain.

A circle is a collection of points equidistant from a given point, such as an origin point. What is distance? For a long time, distance meant Euclidean distance. Since the post-Euclidean revolution, however, mathematicians have constructed infinitely many distances, many of which are used daily in scientific research and applications. For each one of these distances, there will be a different π.

For example, figure 5.3 presents an example of a non-Euclidean distance called the maximum distance d_m, or $d_m = \max\{ |x|, |y| \}$, where maximum stands for the larger of the two numbers in the parenthesis. For instance, the maximum distance from the origin to (.5,1) is the larger of the two numbers .5 and 1, that is, 1. For this distance, the value for the circumference of a unit circle is 4, rather than the more familiar 3.14159.

But what about real problems in the world? Surely Euclidean distance must be used. Theory is fine in research, but in the practical world there can be only one metric. Not so. For an example of the maximum metric, consider the problem of the reliability of the materials that go into building bridges. These materials are tested for qualities such as the maximum stress

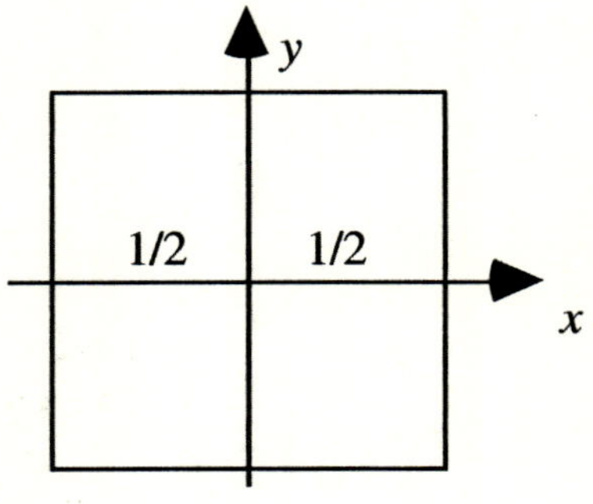

Figure 5.3. The maximum metric circle with unit diameter and radius 1/2 is the set of points (x, y) with dm = 1/2. The circumference of this circle is 4.

that they can withstand without breaking. That is, reliability engineers who test the materials that go into building bridges use the maximum metric, since it is the maximum stress to which a bridge cable will be subjected that matters.

CANONS AND "PURITY" CONTROL IN NINETEENTH-CENTURY GEOMETRY

Sokal's "postmodern Masters" claim that π is no longer constant and universal, then, is now accepted. This so-called postmodern claim was first made in the early 1800s and presented what was then considered to be a heretical challenge to Euclid's firmly established fifth postulate. Indeed, in 1829 Gauss predicted resistance to his non-Euclidean geometry in the form of the "howl from the Boeotians" and would not allow his findings to be published until after his death.[25]

> It is amazing that despite his great reputation, Gauss was actually afraid to make public his discoveries in non-Euclidean geometry. He wrote to F. W. Bessel in 1829 that he feared "the howl from the Boeotians" if he were to publish his revolutionary discoveries. He told H. C. Schumacher that he had "a great antipathy against being drawn into any sort of polemic." (Greenberg 1974, 146)

In the 1820s Gauss was probably correct in his evaluation of the potential reactions to his ideas, should he have published them. In a footnote to Gauss's reference to the Boeotians, Greenberg explains that the critique that Gauss expected actually came in response to Riemann, Helmholz, and Hilbert's work on non-Euclidean geometry published in the 1860s and, a century later again, in response to Einstein's ideas published in the 1920s.

> [The term Boeotians was] an allusion to dull, obtuse individuals. Actually, the "Boeotian" critics of non-Euclidean geometry—conceited people who claimed to have proved that Gauss, Riemann, and Helmholz were blockheads—did not show up before the middle of the 1870s. "If you witnessed the struggle against Einstein in the Twenties, you may have some idea of (the) amusing kind of literature (produced by these critics). . . . Frege, rebuking Hilbert like a schoolboy, also joined the Boeotians. . . . Your system of axioms, he said to Hilbert, is like a system of equations you cannot solve (Freudenthal, 1962)." (Greenberg 1974, 146)

Referring to a potential source of resistance in a letter to F. A. Taurinus, who had also investigated Euclid's fifth postulate, Gauss wrote, "But it seems to me that we know, despite the say-nothing word-wisdom of the metaphysicians, too little, or too nearly nothing at all, about the true nature of space, to consider as absolutely impossible that which appears to us unnatural" (Wolfe 1945, 47). As Greenberg explains,

> The "metaphysicians" referred to by Gauss in his letter to Taurinus were followers of Immanuel Kant, the supreme European philosopher of the late eighteenth century and much of the nineteenth century. Gauss' discovery of non-Euclidean geometry refuted Kant's position that Euclidean space is *inherent in the structure of our mind. In his Critique of Pure Reason* (1781) Kant declared that "the concept of (Euclidean) space is by no means of empirical origin, but is an inevitable necessity of thought." (1974,146)

According to mathematician Harold E. Wolfe, the transformation in mathematical thinking that allowed Gauss to determine that Euclid's fifth postulate could not be proven was a transformation of mathematics from a metaphysical study to an experimental science.

> In that day, it required not only perspicacity, but courage, to recognize that geometry becomes an experimental science, once it is applied to physical space, and that its postulates and their consequences need only be accepted if convenient and if they agree reasonably well with experimental data. . . . The discovery of Non-Euclidean Geometry led eventually to the complete destruction of the Kantian space conception and at least revealed not only the true distinction between *concept and experience* but, what is even more important, their *interrelation*. (Wolfe 1945, 44; emphasis added)

So great was Gauss's fear of his critics that, upon receiving correspondence from Wolfgang Bolyai, János's father, urging him to support the junior Bolyai's work on non-Euclidean geometry, Gauss responded that his intention was to withhold publication of his work until after his death.

> My intention was, with regard to my own work, of which very little up to the present has been published, not to allow it to become known during my lifetime. Most people have not the insight to understand our conclusions and I have encountered only a few who received with any particular interest what I communicated to them. In order to understand these things, one must first have a keen perception of what is needed, and upon this point the majority are quite confused. On the other hand, it was my plan to put all down on paper eventually, so that at least it would not finally perish with me. (Gauss's letter to Wolfgang Bolyai, quoted in Greenberg 1974, 144)[26]

Gauss so disappointed János Bolyai that, after his initial publication of a twenty-six-page appendix to a book published by his father, Bolyai never again published any more of his work on non-Euclidean geometry.

Lobachewsky was much bolder than either Bolyai or Gauss in his attempts to promote non-Euclidean geometry. He published the first account of this new work in Russian in 1829 and another account in German in

1840. But his attempts to have this paper reviewed met with the ridicule and rejection that Gauss had feared. In 1832 the paper "On the Principles of Geometry," sent to the Petersburg Academy by the university council of Kazan University, where Lobachewsky was then serving as the rector, met with the following negative review by Mihail Vasil'evic Ostrogradskii.

> Having pointed out that of the two definite integrals Mr. Lobachewsky claims to have computed by means of his new method one is already known and the other is false, Mr. Ostrogradskii notes that, in addition, the work has been carried out with so little care that most of it is incomprehensible. He therefore is of the opinion that the paper of Mr. Lobachewsky does not merit the attention of the Academy. (Kagan 1905–1907, 247, quoted in Rosenfeld 1988, 209)

In 1834 another review of this work, "an insulting lampoon" entitled "On the Principles of Geometry, A Work of Mr. Lobachewsky," was published in the Petersburg literary journals *Syn otecestva* (Son of the fatherland) and *Severnyi arhiv* (Northern archive). The authors of the review were probably Ostrogradskii's students S. A. Buracek and S. I. Zelenyi.

> Glory to Mr. Lobachewsky who took upon himself the labor of revealing, on the one hand, the insolence and shamelessness of false new inventions, and on the other, the simpleminded ignorance of those who worship their new inventions.
>
> However, while I realize the full value of Mr. Lobachewsky's work, I cannot but hold it against him that, having failed to give his book an appropriate title, he forced us to think for a long time in vain. For instance, why not write, instead of *On the principles of geometry, A satire on geometry* or *A caricature of geometry* or a similar thing. (quoted in Rosenfeld 1988, 209)

Gauss's correspondence articulating his ideas was published after his death in 1855. It helped to persuade the mathematical world to take non-Euclidean ideas seriously.

> Some of the best mathematicians (Beltrami, Cayley, Klein, Poincaré, and Riemann) took up the subject, extending it, clarifying it, and applying it to other branches of mathematics, notably complex function theory. In 1868 the Italian mathematician Beltrami settled once and for all the question of a proof for the parallel postulate: he proved that no proof was possible! He did this by proving that non-Euclidean geometry is just as consistent as Euclidean geometry. (Greenberg 1974, 150)

Even after Beltrami had "settled once and for all" that non-Euclidean geometry is just as consistent as Euclidean geometry, non-Euclidean geometry still faced resistance. "As late as 1888, [mathematician] Lewis Carroll was [still] poking fun at non-Euclidean geometry" (Greenberg 1974, 150).

DISCIPLINE AND AUTHORITY IN SCIENCE

Mathematicians and scientists have changed the world by challenging canons, universals, and constants. The challenge to Euclid's fifth postulate presented by Friedrich Gauss, János Bolyai, and Nikolai Ivanovich Lobachewsky is just one example from the history of science. They and their supporters transformed mathematics and produced a new world where the circumference of a unit circle does not have to be 3.14159. . . . In his statement that "the π of Euclid, . . . formerly thought to be constant and universal, [is] now perceived in [its] ineluctable historicity," Sokal's parodical author refers (without citation) to the historical reconstruction of π that had occurred 170 years ago. To argue that π is 3.14159 . . . in Euclidean geometry is correct; to argue that 3.14159 . . . is the only value possible for π, as the circumference of a unit circle, is to lack imagination and to ignore radically interesting possibilities. The challenge to Euclidean mathematics that began in the 1820s changed the circle, mathematics, science, and the world in thoroughly radical and provocative ways.

PARODY AS SOCIAL CONTROL

Sokal specifically wrote his Social Text article as a hoax, an attempt to parody what he considered to be incorrect constructivist and postmodernist arguments about science by authors such as the philosopher Derrida. He also cited "postmodernist" and "cultural studies" writings as empty, inane, decorative, incomprehensible, nonsense, and "gibberish" strung together using politically correct keywords.

It is notable that Euclidean geometry "metaphysicians" also used satire to ridicule and mock work that they did not understand. In his 1832 review comments on the incomprehensibility of Lobachewksy's "On the Principles of Geometry," Mihail Vasil'evic Ostrogradskii of the Petersburg Academy assumed that his inability to understand the paper was due to Lobachewsky's writing with little care rather than to his own closed-mindedness. "Mr. Ostrogradskii notes that, in addition, the work has been carried out with so little care that most of it is incomprehensible" (quoted in Rosenfeld 1988, 209). In 1834 another review of Lobachewsky's work, an insulting lampoon probably written by one or two of Ostrogradskii's students was published in two Petersburg literary journals.

The similarity between the critical styles of nineteenth-century attacks on non-Euclidean geometry and contemporary attacks by Sokal, Gross, Levitt, and others is striking. Why do these critics lampoon, parody, and satirize the works of their adversaries? Assuming that they are sincere in their desire to challenge bad scholarship, one would expect serious attempts to obliterate or nullify their adversaries' points and not mockery and ridicule.

Ostrogradskii and his students most likely did not understand the revolutionary ideas expressed in Lobachewsky's paper and therefore could not argue the points presented in the paper. Similarly, Sokal may not understand the writings he criticizes in social and cultural studies of science and in cultural studies more generally. This misunderstanding may in part be due to his unfamiliarity with the language and context of discussion. Reading, especially across disciplines and across theoretical approaches (sometimes called "paradigms"), is not a transparently facile activity. Understanding the writing and reading conventions of another author, much less another discipline or approach, takes time, effort, and care. As in any field, not every writer in social and cultural studies of science is as careful as he or she should be, and some of the literature may be unclear or "rhetorically excessive." But the best science studies take the time, effort, and care to explore the sciences before writing about them. These studies involve patient labor, scholarship, rigor, and careful argumentation. Language is the major instrument for conceptualization and argumentation; thus the best work calibrates language much as scientists calibrate instruments for measuring the phenomena they attempt to inscribe. Criticism, then, should be based on a serious effort to engage with the material on its own terms rather than on the assumption that the reader and writer employ the same tools of analysis, description, and representation. This statement applies to both science studies and science studies critics.

Arkady Plotnitsky, who writes on literature and mathematics, argues that Sokal misunderstands Jacques Derrida's writings on Einstein's ideas because he does not understand the language of argumentation in philosophy. Derrida is concerned with the philosophical interpretations and implications of Einstein's theory of relativity. According to Plotnitsky, part of the reason for this misreading is Sokal's ignorance of the literature in philosophy on relativity and quantum physics and on the conventions of reading and writing in philosophy. Plotnitsky contends that should Sokal want to engage in philosophical debates or critiques, it is incumbent on him to read the literature within the conventions of practice in philosophy.

> Neither Derrida's more substantive discussions of mathematics and science . . . nor his caution in this respect, are considered by his recent critics in the scientific community. These critics instead appear to base their views of Derrida's ideas . . . on indiscriminately extracted, isolated references to science or on snippets of his texts, without placing such statements in the context of his work. . . . At stake here are . . . the most elementary and most traditional norms of reading. Such norms would be routinely applied by scientists in reading scientific texts but are massively disregarded by most scientists who commented on Derrida and other authors mentioned. (Plotnitsky 1997, #1)[27]

Rather than discuss or challenge his philosophical ideas, Sokal's hoax distorts and parodies Derrida's statements. But perhaps Sokal is applying the rules of physics to the study of Derrida. Here Plotnitsky employs a quote from Derrida to show that "competent" physicists would not argue in such a fashion, even in cross-disciplinary debates.

> The most competent scientists and those most committed to research, inventors and discoverers, are in general, on the contrary, very sensitive to history and to processes which modify the frontiers and established norms of their own discipline, in this way prompting them to ask other questions, other types of questions. I have never seen scientists reject in advance what seemed to come from other areas of research or inquiry, other disciplines, even if that encouraged them to modify their grounds and to question the fundamental axioms of their discipline. (Plotnitsky 1997, 27)

Science makes good use of skepticism, as Plotnitsky notes. Sokal is not acting as a scientist interested in arguing on the basis of evidence, informed by an awareness of potential errors in calibrating instrumentation, different interpretations of data, or, according to Plotnitsky, the fundamental principles of physics, biochemistry, or mathematics in any historical period.

What is Sokal doing then, if he is not arguing the merits of the science or of the philosophy? I suggest that Sokal uses ridicule and mockery for four interrelated reasons. First, he cannot engage with the specific arguments and empirical work of social and cultural studies of technoscience because he does not understand them. Second, again perhaps because he does not understand them, he constitutes cultural studies and social studies of science as the "Other," as different, wrong, and even harmful, much as early Euro-American missionaries and anthropologists viewed peoples of other cultures on their initial encounters. Third, in so judgmentally representing science studies as wrong, he constructs himself as the authority over all sciences, social sciences, and humanities. That is, he constructs his language, method, and set of understandings as the universally correct set of practices and procedures and then uses his epistemology to represent other ways of doing things as inferior and incorrect. In the pose of defender of universal truth, he becomes the arbiter of truth. Fourth, like early missionaries and anthropologists who brought news of the "Other" back to their home countries, Sokal passes his (mis)understandings and (mis)representations of these literatures on to those scientists who read only his expositions and incites their disapproval. In short, Sokal uses parody as a tool for social control, as a form of discipline.

CONCLUSION: CRAFTING KNOWLEDGE IN SCIENCE AND ANTHROPOLOGY

This chapter addresses issues of authority and control in knowledge production. I examine the science wars in nineteenth-century geometry to create a comparative case for the current wars being generated by critiques of contemporary science studies. My aim is to introduce a historical perspective into the methodological and epistemological debates that have recently rent the academy.

Writings in the history, sociology, anthropology, and literary studies of science document and analyze the constellation of instruments, practices, logics, cultural arrangements, metaphors, and other tools through which complex objects like pi, oncogenes, neutrinos, gravity waves, and airplanes emerge, stabilize, and then produce a wide variety of real effects in our technosociocultural world. This work produces complex accounts of the objects of technoscientific knowledge and of scientific and technological cultures and practices rather than clichés, labels, and stereotypes.[28]

Attacks on this work have their predecessors in the 1800s in "the Boeotians" who opposed the thinking that led to the novel non-Euclidean geometry. Despite these attacks, mathematicians began to question cannons and established practices and to reexamine and change the principles behind those canons and methods. We have expanded our conceptions from Euclidean mathematics to non-Euclidean mathematics; from Newtonian mechanics to quantum mechanics to Einstein's theory of relativity (all of which are still used); from dimorphic sexual reproduction to cloning. In comparison to all this, science studies, for all its spirited discussions and contentious debates, are relatively conventional.

The stakes in the authoritarian battles are high. They include institutional resources such as research funds, university faculty positions, promotions and tenure, and publication outlets. More than just personal gains for individual academics, these institutional resources are the means for reproducing different intellectual positions through the education of undergraduate and graduate students.

But the battle is not just intellectual or "academic." It also has consequences for people's lives, for knowledge making, and for political actions. The power that science, engineering, and medicine have over people's bodies and lives and over the environment certainly extends beyond the bodies and environments of scientists. For that reason alone, it seems most appropriate that the study of science, technology, and medicine be of concern to a broad spectrum of society.

ACKNOWLEDGMENTS

This chapter has benefited from comments by four referees and the journal editors. I am also grateful for the comments of Karen Barad, Howard S. Becker, Herbert Bernstein, Kathy Chetkovich, Kjell Doksum, Troy Duster, Paulla Ebron, Michael M. J. Fischer, Michael Fortun, Clifford Geertz,

Michael Goldhaber, Roger Hart, Bruno Latour, Michael Lynch, Donald MacKenzie, Arkady Plotnitsky, Roddey Reid, Sharon Traweek, Anna Tsing, Sylvia Yanagisako, and Mei Zhan. I also acknowledge the Henry R. Luce Foundation for their support of this work.

NOTES

1. As histories of anthropology demonstrate, debates about what anthropology should be have existed since its inception. The arguments, positions, and alliances have been complicated and have defied neat dichotomies throughout its history. And there has been no consensus about what is scientific anthropology and what is not.

2. This chapter was first presented at the 119th Annual Meeting of the American Ethnological Society, Seattle, WA, March 6–9, 1997, whose theme was "Anthropology and the Canon."

3. See, for instance, the group discussion "Science in Anthropology," which appeared in the Anthropology Newsletter from March 1996 through September 1996 (Skomal 1996a, 1996b, 1996c, 1996d).

4. See, e.g., Nader 1996 on the boundaries of science.

5. Science wars had erupted earlier in the United Kingdom with a critique of science studies by a science popularizer (see Lynch 1996).

6. The New York Academy of Sciences is not associated with the National Academy of Sciences. A volume of papers presented at the conferences was later published as The Flight from Reason (Gross et al. 1997). See science historian Paul Forman's (1997) review of this volume.

7. The Social Text editors are Stanley Aronowitz and Andrew Ross. See Robbins and Ross 1996 on why they published Sokal's article. They should not have published the manuscript, but mistakes are common in all human endeavors. That this mistake has become the standard bearer and rallying point of a strange alliance of political conservatives and supposedly progressive academics is fascinating.

8. Although I question Sokal's judgment of scholarship in what he calls postmodernism and science studies, I do not assume that all the writings that he lumps into these categories are examples of good scholarship. Neither do I assume that all writings in physics are examples of good scholarship. I do assume that each work should be evaluated on its merits and not on the category to which it is assigned.

9. There are many science studies writers who do not subscribe to constructivist ideas, but I focus on constructivism here because it is a major bone of contention for critics of science studies.

10. But it is important to realize that they represent in fact a diverse group of perspectives that share some commonalities and have many differences. These differences have been the source of creative debates in the field. See Fujimura 1996 and Pickering 1992 for more discussion.

11. "Ras" is a technical term that refers to a particular class of oncogenes, or cancer-causing genes.

12. It is distressing to see such misrepresentations applauded by other academics who usually demand careful investigation and argumentation.

13. But the relationship between a representation and its interpretation or use, and therefore its consequences, is not fixed. Representations change, depending on who is using and interpreting them, when, and for what purposes. Thus the study of representations should incorporate their diverse interpretations, renderings, and uses. This highlights the active role of audiences. The consequences of representations (that is, realities) are the products of cooperation, conflict, negotiation, and sometimes power struggles between audiences and producers, between readers and authors. To understand science as a way of producing realities, we also have to understand this interaction.

14. Philosophers of science take diverse positions on the issues of positivism, realism, and constructivism. For discussions from different perspectives, see Smith 1997 and articles in Boyd et al. 1991.

15. I do not have the space here to articulate the many positions and differences. See Smith 1997 for an excellent extended discussion of current controversies over truth, objectivity, realism, relativism, and constructivism in science studies.

16. Michael Lynch (1996) similarly points out that, for Gross and Levitt, the battle is actually about how "Science" (that mythical, nonexistent entity) should be portrayed, about who controls its portrayal, and not about what scientists or social scientists actually do. See also Lynch 1997 on Sokal.

17. The notation *pi* was introduced by William Jones in 1706 in his Synopsis Palmariorum Matheseos; or, a New Introduction to the Mathematics. According to Beckman, "Jones' introduction of the symbol in his book strongly suggests that he used the letter *pi* as an abbreviation for the English word periphery (of a circle with unit diameter)" (1970, 141). This usage of the symbol *pi* was later popularized by Euler in his Variae observationes circa series infinitas (1737).

18. According to Goldhaber,

> To Pythagoras and Euclid, pi was . . . the ratio of perfect circles' circumferences to their diameters. . . . Now in Einstein-Riemann geometries, as you will find spelled out in Landau and Lifshitz's classic, The Classical Theory of Fields, a spinning perfect circle turns out to have a circumference which is not 3.14159 . . . times the diameter; but rather the ratio varies according to the velocity of rotation. Using the original definition of pi, you find it is not a constant. (1997)

19. See especially chapter 3 of Wolfe 1945.

20. This statement was published in an appendix to Bolyai 1832 and quoted in Wolfe 1945, 51.

21. For an illuminating discussion of this point, see chapter 6 of Feynman 1997.

22. Greenberg refers the reader to George Gamow's article "The Evolutionary Universe" (1956), "which tells how Einstein developed a non-Euclidean geometry appropriate to general relativity from the ideas of Georg Friedrich Bernhard Riemann (1826–1866)" (Greenberg 1974, 249).

23. For a more detailed account, including a discussion of three-dimensional non-Euclidean geometry, see Feynman 1997, chapter 6.

24. In a later paper, Sokal (in press) also attempts to parody Bruno Latour's 1988 article on Einstein's theory of relativity. In contrast to Sokal's (1997:4-5) critique of Latour's article, physicist Mermin (1997a) gives Latour an A+ for his understanding of relativity and (1997b, 13) chides Sokal for not understanding the ideas of Einstein discussed by Latour.

25. Joan Richards (1979) has argued that resistance in nineteenth-century England to Reimann's and Helmholz's non-Euclidean geometry also came from outside mathematics due to the use of Euclidean geometry to shore up theological authority.

26. See also Rosenfeld 1988, 216, for a slightly different translation of Gauss's letter in German to Wolfgang Bolyai.

27. The numbers in the Plotnitsky quotation citations are the argument numbers in Plotnitsky's Internet journal article.

28. See Traweek 1996 for a discussion of complexity in science.

REFERENCES

Barad, Karen. 1996. "Meeting the Universe Halfway: Realism and Social Constructivism without Contradiction." In *Feminism, Science, and the Philosophy of Science: A Dialogue*, ed. Lynn Hankinson Nelson and Jack Nelson. Boston: Kluwer Academic Publishers, 161–94.

Beckmann, Petr. 1970. *A History of π (pi)*. Boulder, CO: Golem Press.

Bolyai, Wolfgang. 1832. *Tentamen juventutem studiosam in elementa matheseos purae, elementaris ac sublimioris, methodo intuitiva, evidentiaque huic propria, introducendi, cum appendice triplici*. n.p.: Maros-Vasarhelyini.

Boyd, Richard, Philip Gasper, and J. D. Trout, eds. 1991. *The Philosophy of Science*. Cambridge, MA: MIT Press.

Clarke, Adele, and Joan H. Fujimura, eds. 1992. *The Right Tools for the Job in Twentieth Century Life Sciences*. Princeton, NJ: Princeton University Press.

Derrida, Jacques. 1970. "Structure, Sign, and the Play in the Discourse of the Human Sciences." *In The Languages of Criticism and the Sciences of Man: The Structuralist Controversy*, ed. Richard Macksey and Eugenio Donato. Baltimore: Johns Hopkins University Press.

Feynman, Richard C. 1997. *Six Not-So-Easy Pieces: Einstein's Relativity, Symmetry*, and Space-Time. New York: Addison-Wesley.

Forman, Paul. 1997. "Assailing the Seasons. Review of The Flight from Reason." Science 276 (May 2):750-52.

Fujimura, Joan H. 1996. *Crafting Science: A Socio-History of the Quest for the Genetics of Cancer*. Cambridge, MA: Harvard University Press.

Gamow, George. 1956. "The Evolutionary Universe." *Scientific American* (September)136-54.

Goldhaber, Michael. 1997. "A Preface to a Contribution to the Critique of Sokal's Hoax." Paper presented at the meeting of the American Philosophical Association, March, Berkeley, CA.

Greenberg, Marvin Jay. 1974. *Euclidean and Non-Euclidean Geometries: Development and History*. San Francisco: W. H. Freeman and Co.

Gross, Paul R., and Norman Levitt. 1994. *Higher Superstition: The Academic Left and Its Quarrels with Science.* Baltimore: Johns Hopkins University Press.

Gross, Paul, Norman Levitt, and Martin W. Lewis. 1997. *The Flight from Reason.* Baltimore: Johns Hopkins University Press.

Haraway, Donna. 1997. *Modest_Witness@Second_Millennium: FemaleMan©_Meets_OncoMouse(TM): Feminism and Technoscience.* New York: Routledge.

Hart, Roger. 1996. "The Flight from Reason: Higher Superstition and the Refutation of Science Studies." In *Science Wars,* ed. Andrew Ross. Durham, NC: Duke University Press.

Kagan, V. F. 1905-1907. Foundations of Geometry [in Russian]. Vols. 1 and 2. Odessa, Russia: n.p.

Latour, Bruno. 1988. "A Relativistic Account of Einstein's Relativity." *Social Studies of Science* 18:3-44.

Lynch, Michael. 1996. "Detoxifying the 'Poison Pen Effect.'" In *Science Wars,* ed. Andrew Ross. Durham, NC: Duke University Press.

———. 1997. "A So-Called "Fraud": Moral Modulations in a Literary Scandal." *History of the Human Sciences* 10 (3):11-23.

Mermin, N. David. 1997a. "Reflections of an Earthling." Paper presented at the workshop Current Debates on Science, Science Studies, and Their Critics, May 10-11, University of California, Santa Cruz.

———. 1997b. "Reference Frame: What's Wrong with This Reading?" *Physics Today* 50 (October):11-13.

Nader, Laura, ed. 1996. Naked Science: *Anthropology Inquiry into Boundaries, Power, and Knowledge.* New York: Routledge Press.

Pickering, Andrew, ed. 1992. *Science as Practice and Culture.* Chicago: University of Chicago Press.

Plotnitsky, Arkady. 1997. "'But It Is above All Not True': Derrida, Relativity, and the 'Science Wars.'" *Postmodern Culture.* Available at http://muse.jhu.edu/journals/pmc/v007/7.2plotnitsky.html (accessed 29 Aug. 2004).

Richards, Joan L. 1979. "The Reception of a Mathematical Theory: Non-Euclidean Geometry in England, 1868-1883." *In Natural Order: Historical Studies of Scientific Culture,* ed. Barry Barnes and Steven Shapin. London: Sage Publications.

Robbins, Bruce, and Andrew Ross. 1996. "Mystery Science Theater." *Lingua Franca* 6 (5):54-57.

Rosenfeld, B. A. 1988. *A History of Non-Euclidean Geometry: Evolution of the Concept of a Geometric Space,* trans. Abe Shenitzer. New York: Springer-Verlag.

Ross, Andrew, ed. 1996. *Science Wars.* Durham, NC: Duke University Press.

Skomal, Susan N., ed. 1996a. "Conclusion." *Anthropology Newsletter* 37 (6):14-17.

———. 1996b. "Definitions." *Anthropology Newsletter* 37 (3):48-52.

———. 1996c. "Transformations in Science and Society." *Anthropology Newsletter* 37 (5):9-12.

———. 1996d. "What Is the Problem?" *Anthropology Newsletter* 37 (4):10-12.

Smith, Barbara Hernnstein. 1997. *Belief and Resistance: Dynamics of Contemporary Intellectual Controversy.* Cambridge, MA: Harvard University Press.

Sokal, Alan D. 1996a. "A Physicist Experiments with Cultural Studies." *Lingua Franca* 6 (4):62-64.

———. 1996b. "Transgressing the Boundaries: Toward a Transformative Hermeneutics of Quantum Gravity." *Social Text* 46-47, 14, 1-2 :217-52.

———. In press. "What the Social Text Affair Does and Does Not Prove." In *A House Built on Sand: Flaws in the Cultural Studies Account of Science*, ed. Noretta Koertge. Oxford University Press.

Traweek, Sharon. 1996. "Unity, Dyads, Triads, Quads, and Complexity: Cultural Choreographies of Science." *Social Text* 46-47, 14, 1-2 :129-39.

Wolfe, Harold E. 1945. *Introduction to Non-Euclidean Geometry*. New York: Henry Holt and Co.

6

The Structures of Knowledge or How Many Ways May We Know?

Immanuel Wallerstein

The Report of the Gulbenkian Commission bears the title, *Open the Social Sciences*. The title bears witness to the sense of the commission that the social sciences have become closed off, or have closed themselves off, from a full understanding of social reality, and that the methods which the social sciences had historically developed in order to pursue this understanding may themselves today be obstacles to this very understanding. Let me try to summarize what I think the report says about the past 200 years, and then turn to what this implies for what we should now do.

The commission saw the enterprise of the social sciences as an historical construction, institutionalized primarily in the period 1850–1945. We emphasized that this construction was therefore quite recent, and that the way in which social science was constructed was neither inevitable nor unchangeable. We tried to explain what elements in the nineteenth-century world led those who constructed this edifice to make the decisions that were made concerning the distinctions that were created between a named list of "disciplines." We sought to outline the underlying logic that accounted for why the multiple disciplines adopted various epistemologies and why each chose certain practical methodologies as their preferred ones. We also tried to explain why the post-1945 world found this logic constraining and set in motion a series of changes in the academy, which had the effect of undermining the distinctions among the disciplines.

The picture that we drew of the history of the social sciences was that of a U-shaped curve. Initially, from 1750 to 1850, the situation was very confused. There were many, many names being used as the appellations of proto-disciplines, and none or few seemed to command wide support. Then, in the period 1850–1945, this multiplicity of names was reduced to a small standard group that clearly distinguished the ones from the others. In our view, there were only six such names that were very widely accepted throughout the scholarly world. But then, in the period from 1945 on, the number of legitimate names of fields of study has been once again expanding and there is every sign that the number will continue to grow. Furthermore, whereas in 1945 there still seemed to be clear demarcations that separated one discipline from another, these distinctions have in the subsequent period been steadily eroded, so that today there is considerable de facto overlap and confusion. In short, we have in a sense returned to the situation of 1750–1850 of a large number of categories, which do not provide a useful taxonomy.

But this overlap and confusion is the least of our problems. This process of defining the categories of the social sciences has been occurring within the context of a much larger turmoil that goes beyond the social sciences and implicates the entire world of knowledge. We have been living for 200 years in a structure of the organization of knowledge in which "philosophy" and "science" have been considered distinctive, indeed virtually antagonistic, forms of knowledge. It is salutary to remember that this was not always so. This division between the so-called two cultures is also a rather recent social construction, only a bit older than that which divided up the social sciences into a specified list of disciplines. It was in fact virtually unknown anywhere in the world before the middle of the eighteenth century.

The secularization of society, which has been a continuing feature of the development of the modern world-system, expressed itself in the world of knowledge as a two-step process. The first step was the rejection of theology as the exclusive, or even the dominant, mode of knowing. Philosophy replaced theology; that is, humans replaced God as the source of knowledge. In practice, this meant a shift of locus of the authorities who could proclaim the validity of knowledge. In place of priests who had some special access to the word of God, we honored rational men who had some special insight into natural law, or natural laws. This shift was not enough for some persons, who argued that philosophy was merely a variant of theology: both proclaimed knowledge as being ordained by authority, in the one case of priests, in the other of philosophers. These critics insisted on the necessity of evidence drawn from the study of empirical reality. Such evidence, they said, was the basis of another form of knowledge they called "science." By the eighteenth century, these protagonists of "science" were openly rejecting "philosophy" as merely deductive speculation, and proclaiming that their form of knowledge was the only rational form.

On the one hand, this rejection of philosophy seemed to argue a rejection of authorities. It was in that sense "democratic." The scientists seemed to be saying that anyone could establish knowledge, provided he (or she) used the right "methods." And the validity of any knowledge that any scientists asserted could be tested by anyone else, simply by replicating the empirical observations and manipulation of data. Since this method of asserting knowledge seemed to be capable of generating practical inventions as well, it laid claim to being a particularly powerful mode of knowing. It was not long therefore before "science" achieved a dominant place in the hierarchy of knowledge production.

There was one major problem, however, in this "divorce" between philosophy and science. Theology and philosophy had both traditionally asserted that they could know *two* kinds of things: both what was true and what was good. Empirical science did not feel it had the tools to discern what was good; only what was true. The scientists handled this difficulty with some panache. They simply said they would try only to ascertain what was true and they would leave the search for the good in the hands of the philosophers (and the theologians). They did this knowingly and, to defend themselves, with some disdain. They asserted that it was more important to know what was true. Eventually some would even assert that it was impossible to know what was good, only what was true. This division between the true and the good constituted the underlying logic of the "two cultures." Philosophy (or more broadly, the humanities) was relegated to the search for the good (and the beautiful). Science insisted that it had the monopoly on the search for the true.

There was a second problem about this "divorce." The path of empirical science was in fact less "democratic" than it seemed to claim. There rapidly arose the question of who was entitled to adjudicate between competing scientific claims to truth. The answer that the scientists gave was that only the community of scientists could do this. But since scientific knowledge was inevitably and increasingly specialized, this meant that only subsets of scientists (those in each subspecialty) were deemed part of the group that had a claim to judge the validity of scientific truth. In point of fact, these groups were no larger than the group of philosophers who had previously claimed the ability to judge each other's insights into natural law or laws.

There was a third problem about this "divorce." Most persons were unwilling truly to separate the search for the true and the good. However hard scholars worked to establish a strict segregation of the two activities, it ran against the psychological grain, especially when the object of study was social reality. The desire to reunify the two searches returned clandestinely, in the work of both scientists and philosophers, even while they were busy denying its desirability, or even possibility. But because the reunification was clandestine, it impaired our collective ability to appraise it, to criticize it, and to improve it.

All three difficulties were kept in check for 200 years, but they have returned to haunt us in the last third of the twentieth century. The resolution of these difficulties constitutes today our central intellectual task.

There have been two major attacks on the trimodal division of knowledge into the natural sciences, the humanities, and the social sciences. And neither of these attacks has come from within the social sciences. These attacks have come to be called "complexity studies" (in the case of the natural sciences) and "cultural studies" (in the case of the humanities). In reality, starting from quite different standpoints, both of these movements have taken as their target of attack the same object, the dominant mode of natural science since the seventeenth century, that is, that form of science which is that based on Newtonian mechanics.

To be sure, in the early twentieth century Newtonian physics had been challenged by quantum physics. But quantum physics still shared the fundamental premise of Newtonian physics that physical reality was determined and had temporal symmetry, that therefore these processes were linear, and that fluctuations always returned to equilibria. In this view, nature was passive, and scientists could describe its functioning in terms of eternal laws, which could eventually be asserted in the form of simple equations. When we say that science as a mode of knowing became dominant in the nineteenth century, it is this set of premises of which we are speaking. That which could not be fit into this set of premises, for example, entropy (which is the description of necessary transformations in matter over time), was and is interpreted as an example of our scientific ignorance, which could and would eventually be overcome. Entropy was seen as a negative phenomenon, a sort of death of material phenomena.

Since the late nineteenth century, but especially in the last twenty years, a large group of natural scientists has been challenging these premises. They see the future as intrinsically indeterminate. They see equilibria as exceptional, and see material phenomena as moving constantly far from equilibria. They see entropy as leading to bifurcations, which bring new (albeit unpredictable) orders out of chaos, and therefore the process is not one of death but of creation. They see auto-organization as the fundamental process of all matter. And they resume this in two basic slogans: not temporal symmetry but the arrow of time; not simplicity as the ultimate product of science, but rather the explanation of complexity.

It is important to see what complexity studies is and what it is not. It is not a rejection of science as a mode of knowing. It is a rejection of a science based on a nature that is passive, in which all truth is already inscribed in the structures of the universe. What it is rather is the belief that "the possible is 'richer' than the real"(Prigogine 1996, 67). It is the assertion that all matter has a history, and it is its sinuous history, which presents material phenomena with the successive alternatives between which each "chooses" throughout its existence. It is not the belief that it is impossible to know, that is, to understand how the real world operates. It is the assertion that this process of understanding is far more complex that science traditionally asserted that it was.

Cultural studies attacked the same determinism and universalism under attack by the scientists of complexity. But for the most part those who put forward these views neglected to distinguish between Newtonian science and the science of complexity, or in many cases to be aware of the latter. Cultural studies attacked universalism primarily on the grounds that the assertions about social reality that were made in its name were not in fact universal. It represented an attack against the views of the dominant strata in the world-system, which generalized their realities into universal human realities, and thereby "forgot" whole segments of humanity, not only in the substantive statements but in the very epistemology of their research.

At the same time, cultural studies represented an attack on the traditional mode of humanistic scholarship, which had asserted universal values in the realm of the good and the beautiful (the so-called canons), and analyzed texts internally as incarnating these universal appreciations. Cultural studies insists that texts are social phenomena, created in a certain context, and read or appreciated in a certain context.

Classical physics had sought to eliminate certain "truths" on the grounds that these seeming anomalies merely reflected the fact that we were still ignorant of the underlying universal laws. Classical humanities had sought to eliminate certain appreciations of "the good and beautiful" on the grounds that these seeming divergences of appreciation merely reflected the fact that those who made them had not yet acquired good taste. In objecting to these traditional views in the natural sciences and the humanities, both movements C complexity studies and cultural studies C sought to "open" the field of knowledge to new possibilities that had been closed off by the nineteenth-century divorce between science and philosophy.

Where then does social science fit in this picture? In the nineteenth century, the social sciences, faced with the "two cultures," internalized their struggle as a *Methodenstreit*. There were those who leaned toward the humanities and utilized what was called an idiographic epistemology. They emphasized the particularity of all social phenomena, the limited utility of all generalizations, the need for empathetic understanding. And there were those who leaned toward the natural sciences and utilized what was called a nomothetic epistemology. They emphasized the logical parallel between human processes and all other material processes. They sought to join physics in the search for universal, simple laws that held across time and space. Social science was like someone tied to two horses galloping in opposite directions. Social science had no epistemological stance of its own and was torn apart by the struggle between the two colossi of the natural sciences and the humanities.

Today we find we are in a very different situation. On the one hand, complexity studies is emphasizing the arrow of time, a theme that has always been central to social science. It emphasizes complexity, and admits that human social systems are the most complex of all systems. And it emphasizes creativity in nature, thus extending to all nature what was previously thought to be a unique feature of homo sapiens.

Cultural studies is emphasizing the social context within which all texts, all communications, are made, and are received. It is thus utilizing a theme that has always been central to social science. It emphasizes the nonuniformity of social reality and the necessity of appreciating the rationality of the other.

These two movements offer social science an incredible opportunity to overcome its derivative and divided character, and to place the study of social reality within an integrated view of the study of all material reality. Far from being torn apart by horses galloping in opposite directions, I see both complexity studies and cultural studies as moving in the direction of social science. In a sense, what we are seeing is the "social scientization" of all knowledge.

Of course, like all opportunities, we shall only get *fortuna* if we seize it. What is now possible is a rational restructuring of the study of social reality. It can be one that understands that the arrow of time offers the possibility of creation. It can be one that understands that the multiplicity of human patterns of behavior is precisely the field of our research, and that we may approach an understanding of what is possible only when we shed our assumptions about what is universal.

Finally, we are all offered the possibility of reintegrating the knowledge of what is true and what is good. The probabilities of our futures are constructed by us within the framework of the structures that limit us. The good is the same as the true in the long run, for the true is the choice of the optimally rational, substantively rational, alternatives that present themselves to us. The idea that there are "two cultures," *a fortiori* that these two cultures are in contradiction to each other, is a gigantic mystification. The tripartite division of organized knowledge is an obstacle to our fuller understanding of the world. The task before us is to reconstruct our institutions in such a way that we maximize our chances of furthering collective knowledge. This is an enormous task, given the inherent conservatism of institutional authorities and the danger such a reconstruction poses to those who benefit from the inegalitarian distribution of resources and power in the world. But the fact that it is an enormous task does not mean that it is not doable. We have entered a bifurcation in the structures of knowledge, which appears in many ways to be chaotic. But of course we shall emerge from it with a new order. This order is not determined, but it is determinable. But we can only have *fortuna* if we seize it.

REFERENCES

Prigogine, Ilya. 1996. *La fin des certitudes*. Paris: Odile Jacob.

7

Beyond the Great Divide

Becoming Civilized?

Isabelle Stengers

INTRODUCTION

In *Um Discurso sobre as Ciências*,[1] Boaventura de Sousa Santos states that the reason we now give priority to a form of knowledge based on the prediction and control of phenomena has nothing to do with science. There are no scientific reasons for considering scientific explanations better than any other alternative explanations. This, then, is a value judgment.

We should add, obviously, that everything depends on the phenomenon in question. When it is an "experimental phenomenon," the procedure followed by the person who produced it in the laboratory has no rival. No other knowledge can, for instance, rival biology in what concerns the role of the DNA molecule in protein synthesis. None of these words—molecule, DNA, synthesis, protein—has significant meaning outside the laboratory. In contrast, from the moment we deal with a human being, with what s/he receives at birth, his/her "belonging," the DNA "genetic" explanation ceases to be paramount. The problem is, thus, that of the value judgments that enable scientific knowledges to "come out" of the sites in which they are produced and seek to answer questions that are important to all human beings, not just scientists, or rather, seek to replace questions that matter with questions that matter to scientists. This is a political, cultural and social problem, which does not mean that scientists are "innocent," devoid of dreams of power. But, by themselves, they have no power to make real their dreams of power. Other protagonists have an interest in such dreams, and this is a very old story indeed. In a faraway past, sovereigns called on the knowledge of priests and astrologists to reinforce their power. Legend has it that Hieron, the tyrant of Syracuse, received a new kind of offer. Archimedes, the engineer, told him of the interest that he, the sovereign, would have in helping the engineer make his dream come true. . . .

ARCHIMEDES'S COUP AND THE CHINESE SMILE

"We should never again let ourselves be taken in by Archimedes's coup," Bruno Latour wrote in *Nous n'avons jamais été modernes* (1991, 194). "Give me a fulcrum"—thus supposedly wrote Archimedes to King Hieron—"and I will lift the world." Archimedes's fulcrum is what is needed by a lever, which allows a weak man to lift a heavy weight. Tell me the weight you want to lift, the strength you can apply to it, and I will calculate the length of the lever and the exact place for the point of support. This is obviously a fiction; it assumes that the lever is perfectly rigid, and there are a few other "management" problems, but the idea is clear: Archimedes's science is a "knowledge that should matter" to the king, since it would allow him to invert the relations of force. Equip my laboratory, give me the researchers I need—innumerable scientists today ask their funders—and . . . I will invert the relations of force that make you feel so weak, so unarmed facing such enormous, multiple or conflicting problems. Take starvation in the world, for instance: I will produce a new type of being, a superb GMO . . . Give me a laboratory, and what will come out of it will allow you to prevent conflicts, price problems, distribution problems, problems of social and economic inequality.

This vignette points immediately to what Jacques Mirenowicz (2000) has so well described, the Vannevar Bush program, in which a disinterested science saw its recognition as a source of general and fundamentally apolitical progress for human beings. Industrial applications of scientific discoveries would bring an increase in the wealth produced, which would gradually benefit all. Who would continue to worry about an unequal distribution of parts if each one's part kept increasing? Growing abundance would make the antagonisms and conflicts arising from poverty meaningless. It is true that this vignette has lost credibility, but it continues to haunt our universities because it corresponds to what many scientists nostalgically call the lost Golden Age, while denouncing the loss of faith in science as the "growth of irrationality."

Second vignette. The story goes that the Chinese smiled at the naiveté of the Christian missionaries who wanted to interest them in the wonders of Galileo's and Newton's science, in the discovery of the "laws of nature." For them, the famous inclined plane—thanks to which Galileo, for the first time, turned a natural phenomenon into reliable testimony, into a fulcrum on which one could prove one's reason against general opinion—was child's play. No knowledge of this kind could, by definition, produce true understanding, since true understanding would have to harmoniously relate what happens in heaven, in the fields and in the government of the empire. A rooster with two heads or a new star in the sky, this indeed would be of consequence, a possible presage of dire political instability, of a revolt or revolution at court.

This vignette points directly to the remarkable stability of the Chinese Empire, with its palace revolutions and unchanging bureaucracy. For talented young people, it was out of the question to attempt Archimedes's coup, or to seek, like Galileo, to overthrow the authority of the church and of philosophical tradition with a "new science." Such young people would be chosen through a contest, hired and turned into knowledgeable officials placed throughout the immense empire, scrupulously keeping the harmonic order of things and people everywhere. They would learn to smile sweetly at the kind of naïve spirit of enterprise[2] that the missionaries who had arrived to convert them unwittingly evinced.

I need to resort to this second vignette in order to situate myself in relation to the question of "knowledges that count." Like the missionaries, I am the daughter of a history marked by a spirit of enterprise that gambles on the instability of any form of distribution between what matters and what doesn't: the entrepreneur is someone who sees distribution as a matter for speculation and risk; he is concerned with "making things count"—even if that means, as in the case of Archimedes's coup, directly addressing the sovereign's interests. But it may also mean, as in the case of social and political movements (earlier trade unionism, feminism, gay liberation, etc.), seeking to subvert heaven and earth, daily practices and timeless value judgments in order to change the register of what is taken into account, the evaluations of what counts. I am a daughter of this history. In order to avoid being struck again by Archimedes's coup, I will be concerned with the resources of this history, and not with some order in which each one would have to fill his proper place.

IN THE SAME BOAT

We know that philosophy does not mean wisdom. A "friend" of wisdom, maybe, but within an unstable urban world, where one always has to pose the question of how to take into account diverging interests. I insist on this in order to bring up the following: I have little nostalgia for the fact that the knowledges that began to matter from the emergence of the modern sciences displaced philosophy to the status of somewhat frivolous and arbitrary chatter. As it happens, we philosophers were played the trick that Plato had played on the Sophists: these, according to Plato's accusation, resorted to seduction, to the art of words and appearances, and were unable to base themselves on a truth that could make each and every one agree on what counts and how it should count. In his own way, Plato was already a small entrepreneur who made the advantages of Ideas count, proposing to extend to human matters the certainties of mathematical demonstration. Whereas Galileo opposed the world of facts, in whose the name he spoke, to the world of paper, in which conflicting certainties confront each other inconclusively, those who follow him can legitimately be seen as Plato's true heirs. Scientists are reserved the extensive domain of "how" things can be decided, and, if philosophers wish to survive, let them quarter themselves in "why." To the extent that, today, they realize, horrified, that this was a hoax, and that strange scientific-looking "hows" assume the right to make decisions about thought (human intelligence), morals (sociobiology), justice (economics), beauty (evolutionary psychology), human identity (cloning, etc.), and so on.

No nostalgia, therefore: as Whitehead would say in *Concept of Nature* (1964), all our resources are "in the same boat, to sink or swim together." The question "what is it that counts?" is our question, without transcendence, with no *a priori* argument, no reserved domain, either on the side of "values" or on the side of "facts," and thus of the knowledges that present themselves as "authorized by facts."

And it is here, of course, that we must play it safe, since there is a ready-made accusation: relativism. If nothing is entitled to count by itself, transcending our histories and permitting the judgment of our errors and our crimes, then what counts is simply relations of force. Plato's heirs, on the side of values, and Galileo's, on the side of facts, suddenly hold hands. It is true that they fight each other, but this fight about what counts must have the tragic grandeur of that which stages the Ends of Man. The contenders must be capable of mobilizing multitudes, appealing for their involvement, some in the name of reason, others in the name of salvation. The space that they polarize must be uninhabitable, a no man's land.

Those in the know have already understood, of course, that I am thinking here of Bruno Latour. Among today's thinkers (yesterday it was Foucault, but only on the side of the moralists, for he didn't much dwell on Galileo's heirs), he is the one who has most violently suffered this double rejection, in a constant manner, determined to hear nothing, discuss nothing. His crime was having attempted to inhabit the "middle kingdom" (ours, where hybrids proliferate, not millennial China), having made the two poles that want to disrupt it into products of that space where a multitude of small entrepreneurs busily operates.

If I mention Latour here, it is not only as a victim, but because his own trajectory, as I see it, today leads us to a proposition which goes far beyond the diagnosis of our disputes, toward a thought of the possible, which is, of course, easy to deride: his "political ecology" is a dream, a utopia. But dreams and utopias are important when they yield ideas, when they yield words to think in another way about what happens to us, to "make count in another manner" what other words would lead us to ignore or disqualify. And, in particular, those words, so numerous, so insistent, that repeat the Great Divide between "us," who can distinguish facts and values, the how and the why, the rational and the symbolic, etc., and the others, who mix everything. It was these words that, from the beginning, Latour sought to resist, but what I want to pursue here is the way in which an enterprise that could, in the beginning, be placed in the category of "critical demystification" turned into a thinking of creation and becoming. What is at stake, for me, is to "make count" the trajectory of this transformation, because it points to our own resources to "civilize" the amateur entrepreneurs of "Archimedes's coups" without appealing to a world order which we are supposed to have criminally transgressed.

3. IN FACT, WE ARE NOT THAT DIFFERENT FROM OTHERS

From the moment we accept the way in which the sciences are usually presented, with a hierarchy between a pure, disinterested science and its applications, which engender a progress with a universal vocation (benefiting the whole of humanity), the trick is played, and we are no longer like others. No doubt other civilizations aspired to purity, the ascesis that offers detachment from the world of interests and passions, but they were unacquainted with the idea of useful "relapses," which leads to the assertion that purity is also the condition for the technical appropriation of the world. However, it is not enough to denounce that presentation as connected to a political lie. Vannevar Bush did not invent this representation of the sciences when he turned it into the key of apolitical progress. We find it, for example, about a century before in the writings of the great chemist Liebig. Liebig attacked, in a ferocious manner, "Baconian" inductive science—"the result one reaches through his method should always be 'zero'" (Liebig 1894, 111)—and, in general, condemned every kind of science defined as working for the useful. He argued that it was necessary to let scientists deal freely only with those questions in which thought can precede experience—"an experience that is not beforehand connected with a theory, that is, with an idea, is as similar to a true investigation as the noise of a child's rattle is to music" (Liebig 1894, 114). He was one of the first who turned the question of the autonomy of science into a political question, heralding a long line of descent (it is unnecessary to recall Einstein's diatribe against those who only come to the Temple of Science for utilitarian interests, and whom the Angel of the Lord would expel if he had to draw the line between true builders and parasites). Indeed, once they stopped having to fear the authority of tradition, and began to confront the pressing interests of modern states and their industries, many scientists adopted the same *leitmotiv*: let us be the judges of what counts for us, feed our autonomous research, otherwise you will kill scientific progress. This means, of course, asking the public to make, not a disinterested effort, but an interested investment, which, in the fullness of time, will yield more than any demand for immediate returns could: let us create fulcrums, and you will then be able to put your levers in place.

What we need to do is change this conception of scientific knowledges without destroying it in the process. If the phrase "we are not that different from others" is to have an interesting meaning, it is not by diminishing everyone, us and them, relegating the difference to questions of belief: the Bolobolo (an imaginary people) believe that earthquakes are caused by the great rabbit Nabozozo, geologists believe they are an effect of tectonic plates, and all the differences that we could establish between these two beliefs would be the translation of our adherence to geologists' beliefs. There is something of a caricature in what I just said, but this is exactly what scientists perceive when they get angry about "social constructivism." They feel insulted, and not without reason, because the comparison is of a polemical nature. Indeed, the way in which I was able to evoke an imaginary people proves it: I could have just cited any example of a "belief." But nothing is just "anything" when we have a serious interest in the ways of doing and thinking of concrete peoples, and what then collapses is the possibility of placing on the same plane, for the sake of impartiality, knowledges that cannot be so placed because they count in a different manner and make what seemed to allow their comparison count in a different way, particularly the adjective "social."

We cannot expect changes if we offend, if we issue a polemical challenge in advance: none of the differences that interest you will matter to me, I make a point of relegating them to the undecidable and of explaining my decision on the basis of social determination. Bruno Latour has broken with the so-called strong program of the sociology of science because he has rejected every kind of asymmetry between a "socially constructed" nature and a society that answers to categories such as "social interest" or "relations of force," which explain differences between beliefs. The strong program is "ours" in the sense that its explanatory ambitions presuppose a separation between what is related to the social and what is related to the things that are the "object" of a knowledge, and also in the sense that the concern with this separation is itself our own. What counts cannot be explained in social terms, because the question of knowing whether we can explain an earthquake, an epidemic or rainfall, "in themselves," independently of their social consequences, this question is ours and has been one of the defining vectors of what we call society. "Let us define the fulcrum that constitutes an explanation 'independent from society,' and we will give you the means to change that society."

Actually, there is a double insult here. On the one hand, a knowledge that was not produced according to a method that includes resistance to tests, which, for us, verify that the definition of what it is concerned with is independent from its social consequences, will be destroyed by a confrontation that does not suit it at all. That is why the school is a great destroyer, proposing to the children of "nonmodern" peoples to consider that their ancestors are defined by their superstitions, that they do what they do because they don't know any better. On the other hand, a knowledge that was produced by actors united by their desire to valorize above all the eventual ability of what is studied to make a difference among the different possibilities of interpretation that can be produced about it, this knowledge will be insulted by the verdict that this is a vain effort, that nothing we deal with has the power to make such a difference.

This is why Bruno Latour abandoned the subject/object opposition for a human/nonhuman opposition. The subject/object opposition is ours; it points to questions of knowledge, questions that valorize in the first place the possibility of establishing a difference between what belongs to the subject that knows and to the object that is known. Humans and nonhumans, in turn, are not meant to be opposed. This opposition would be very strange indeed, with nonhumans making up the most peculiar group imaginable, including ancestors and deities, as well as rain, the pluviometer, automobiles, genetic resources, observation satellites and neutrons. What matters is the association, and this association places, first of all, the one who deals with it in a situation of deliberate agnosticism: we do not know what a human is independently of the whole of the nonhumans that matter to him, or on which he counts to act, think, predict, etc. Where is the subject, where the object? Could we conceive of an American without the right to bear arms, an author without copyright, a search without a warrant, an experimenter indifferent to the opposition between fact and artifact, a philosopher who only refers to texts published in the last five years, or an unemployed person who does not concentrate entirely on getting a new job? We perceive in these questions the whole fabric of a world that hesitates, that poses questions about the "habits" that could be changed without too many consequences, about what would perhaps be destroyed, what would not longer hold. . . .

It is from such questions that the problems raised by what counts and what doesn't, or counts less, or shouldn't count, or should count differently, are formulated, in a way that does not lend itself to the Great Divide. The possibility of changing the conception of scientific knowledges as autonomous in relation to social interests without doing away with the concerns of scientists, is part of such questions. This is what actor network theory, of which Bruno Latour was one of the inventors, has attempted to do.

4. THE FIVE HORIZONS OF RESEARCH

In "Joliot et la vascularisation des faits," Latour (2001a) describes the scientific work of Frédéric Joliot Curie in a way designed to oppose the distinction between "scientific work," on the one hand, and its social, economic and political context, on the other. It is not a question of stressing the suspicion that the statements about the role of neutrons in an eventual nuclear chain reaction would have been "influenced" by social, economic and political interests, which would amount to social constructivism. Rather, the point is to show how Joliot himself actively constructed that without which this role would have been reduced to a beautiful idea, to use Liebig's terms, an idea he would not have had the means to translate into experiment, and much less of making it count. Rather than summarize this description, I will briefly transpose it to the case of Justus Liebig, the defender of a science independent of utilitarian values.

From Latour's point of view, the position taken by Liebig belongs to one of the horizons that an innovative and ambitious researcher should construct in order for his work to "count": the horizon of "public representation." But, at the same time, Liebig also had to find "allies," in his case industrialists prepared to take risks with a new kind of industrial production that corresponded to Liebig's laboratorial practices, and to hire researchers trained in them. Liebig had to "build the autonomy" of his discipline, which meant having the new training he proposed for chemists recognized (a curriculum that dispensed with everything that was related to the traditional procedures of chemistry), publishing a journal which would be an authoritative reference in "organic chemistry," and so on. And he also had to "mobilize a world," that is, produce the protocols, instruments, and experimental mechanisms that allowed him to delocalize the results, guaranteeing that a chemist in London or Paris could reproduce what Liebig had obtained in his laboratory at Giessen.[3] When Liebig states that his only fruitful investigations are those guided by a theory, by an idea, his readers know that he is not a chemist of the Enlightenment, but rather one whose name becomes associated with the triumph of the German chemical industry, one who trains young chemists from all over the world, one whose fertilizers (but that's another story . . . [4]) promise to revolutionize the methods of agricultural production.

Public representation, alliances, building autonomy and mobilizing are not enough, though; what holds these together, but derives its own import from the fact of being able to hold them together, is what Latour calls "links" and "knots." This is what the history of science would retrospectively single out as objective knowledge proper, treating everything else as context. If Joliot or Liebig, in their respective laboratories, had not been able to establish, with the nonhuman beings with whom they dealt, solid, reliable links, able to withstand the tests and trials of demanding partners, everything else would have become undone, the results, unable to be reproduced, would have been ignored by colleagues who would not have read the published articles, the captains of industry would not have been there, the chemists trained in the laboratory at Giessen would not have found employment, the public would have read, with interest, texts explaining Liebig's failure and concluding, perhaps, that it is impossible to "professionalize" chemistry, to separate it from the heterogeneous mass of artisanal know-how in relation to which it becomes necessary to train a "true chemist," useful to industry, not a dreamer, a lover of "ideas."

This redescription is not an insult to Liebig. In fact, it allows us to show the extent to which the public position he assumed, with the aim of making others accept a definition of society as respectful of the autonomy of scientific research, was vital to him. The success of professional chemistry depends on not asking it to understand "everything," on granting it the freedom to ignore opaque artisanal procedures, which often depended on local raw material, on a set of non-standardizable operations. It was necessary that everything that entered the laboratory could be defined in the terms of the laboratory, in order to produce what Bruno Latour (2001b) has described, in connection with Pasteur, as an inversion of the relations of force. While the range of chemical procedures, like a spreading epidemic, exceeds, because of the multiple questions these procedures entail, the power of an "idea," the laboratory is the place where the adversary can become visible—if one is able to bring him there—, accessible to questions, susceptible to the confrontation of ideas.

In order not to insult scientists, it is necessary to insist on "if," that is, on the tests and controversies that make scientists gather around what matters to them: have we managed to bring the adversary to the laboratory, is that which we have interrogated a simple artifact, which will have no influence on what is produced outside the laboratory? And, in order not to reproduce the Great Divide, we need to insist on the fact that the information thus obtained will only be clearly pertinent to what is produced outside the laboratory if Pasteur, Liebig, or Joliot, like good entrepreneurs, are active in mobilizing a world, in getting the interest of those who oppose their ideas, in changing habits—in short, in constructing a society in which the knowledge that they acquired in the laboratory can count. In other words, the "coups of Archimedes" which Vannevar Bush made into the principle of apolitical human progress, entail, in fact, the successful creation of a "fulcrum," or, more precisely, a "knot" that joins scientists together around a being redefined in experimental terms, which does not fail to satisfy new tests and to articulate new "objective" properties. But, when all is said and done, links with nonhumans and knots between colleagues " do not explain the "coup" or the Great Divide itself, which designates a collective divergence between modern societies and "other" societies, since the question is that of the "lever," of that which permits the successful inversion of the relations of force inside the laboratory to change the "world," to "modernize" it. And here the perspective gets inverted: links and knots only count due to the successful performance of the operations included in the other four horizons of research.

Thus we return to Needham's thesis: it is not scientific rationality, neutrality, and objectivity that define the possibility of the Great Divide, but the freedom of "small entrepreneurs" who do not, Archimedes-like, "lift" the world, but who engage themselves in mobilizing it, in recruiting allies, in getting them interested in the screening that they propose to make, within a question of general interest, of what should be taken into account, what might or should be done as a consequence, what, acting as an obstacle, can be denounced as obscurantism, inertia, irrational resistance, etc. And it is this freedom of enterprise that the experts in "network actors" describe with appetite, following the course of an innovation, the way in which recruiting operations—which make distinct *a priori* interests converge—weave long and hybrid networks, in which an increasing number of humans become associated with an increasing number of nonhumans, in which the strategic question "what can we make him/it do?" is asked of both humans and nonhumans, without, of course, denying the difference, which becomes, every time, a risky and demanding gamble.

Those who refuse to abandon the accusation of constructivism always yell at this point, because for them humans are easy to recruit: this does not represent a success, and therefore it is an insult to place on the same footing a scientific breakthrough and an operation of transformation-seduction-alienation that is "merely social." The former implies knowledges that count, the latter involves only a bit of cunning, a bit of prestige, a bit of relations of domination. The concern of actor network theory lies precisely in breaking up this dissymmetry: it is not easier to recruit human allies, who know what matters to them,[5] or to discipline humans, making them acquire new habits, than it is to recruit nonhumans in an experimental demonstration, or to discipline them in such a way that they can cooperate in a standardized procedure, in the user-friendly manner demanded by the nonexperts who are going to manage it (for instance, a magnetic resonance imaging device or an ultrasonograph at a hospital, or a pregnancy test at home).

This is why Bruno Latour has never stopped calling attention to the fact that the question of the technoscientific networks provide a great opportunity for the social sciences, because the construction of networks puts actors on the stage who do not submit to a theory about what a society should be, because they themselves, be they innovating scientists, industrialists, jurists, technicians or others, never cease to speculate about what they can change in that society, about the possible, about the interesting, about what counts and what doesn't.

POLITICIZING NETWORKS?

If the task of the social sciences is to describe the construction of our sciences and of our world "as they are made," the actor network approach fits the bill. The fact that it has scandalized the partisans of the "great causalities" (theories of domination, critiques of technoscience) seems to me to have much to do with the fact that this approach does not provide a privileged vantage point from which one can diagnose the way in which humans are determined without being aware of it, as well as denounce what determines them. The sociologist "follows" those who are active, who speculate, who experience constraints to which they give "interesting" meanings; he does not deal with submissive beings, but with opportunistic entrepreneurs, with a multitude of small Archimedeses as well as small kings Hieron, some of whom might even be interested in the possibilities of the "great causalities." Indeed even a sheer theoretical vision (for instance, genetic determination) can become a "knowledge that counts" as long as it offers an authoritative equivalent of what a fulcrum provides, the means for a screening operation of what will be mobilized and what will be ignored.

The actor network approach can be said to be scientific because it is situated on the side of "how" and is perfectly indifferent to "why." What actually concerns it, the issue of associations, involves beings, both human and nonhuman, who "go about their business," sidestepping the great questions, both the existential ones, for instance the reasons why humans alienate their freedom, and the epistemological ones, for instance the reason why nonhumans are authorized to make a difference in human interpretations. What matters is the way in which tests are made and what, resisting the test, will be said to be real; how relations of obligation are established and dissolved (if you want . . . then you should), asymmetries (it is enough to . . .), commensurabilities (and therefore . . .), differentiations between what counts and what doesn't (measure and representation). And this approach is also scientific in the sense that it exposes the actor network researcher: it exposes him, not only to the laughter of those he interrogates, if his questions are naïve, but also to fascination, and, therefrom, to cynicism. In effect, in the world they can mobilize at their own risk, a world of active entrepreneurs, what prevails is *Vae victis*: woe to those who do not count enough to be considered, those whose demands can be bypassed or disqualified. It is true that the latter are not vanquished, either by Reason or by Method—there is no Great Divide—, but they were unable to make their position into a "compulsory passage point," through which others are forced to pass in order to carry out their own business. Correlatively, the great absence is, of course, politics, supposedly situated at the level of the "common good": networks are a way of doing politics differently, of fabricating a landscape marked by the differences between those who are active and those who suffer, groan and despair.

In 1991, *Nous n'avons jamais été modernes* posed the problem of politics (the Parliament of Things), although it didn't make very clear whether this was more than a "thought experiment" [6] making explicit the consequences of the fact that what counts is not the whole of "naked" citizens, united by their common humanity, but the set of associations between humans and nonhumans. However, in 1995, Latour published a text in which he deals with the question of ecology and, more precisely, poses the question of what would be the "greatness" appropriate for an eventual "seventh City," which would join the six cities defined by Boltanski and Thévenot (1991) (inspirational, domestic, of renown, civic, market, industrial).

> What would a man be without elephant, without plant, without lion, without cereal, without ocean, without ozone and without plankton, a lonely man, much lonelier than Robinson on his island? Less than a man. Certainly not a man. The city of ecology does not tell us that we need to pass from the human to nature [. . .]. The city of ecology simply says that we don't know what makes the common humanity of man, and that, without Amboseli's elephants, without the water that flows from the Drôme, without the bears of the Pyrenees, without the pigeons of the Lot, without the phreatic water of Beauce, maybe he would not be human. (Latour 1995, 19)

Here, there is an echo—suiting the "greatness" of the seventh city—of a "fear" that cannot be appeased by any general rule. The one who poses the problems of the City on the basis of his certitudes, on the basis of "nothing forbids it, so I can" that characterizes the entrepreneur, is "small," in the sense used by Boltanski and Thévenot. "Great" will certainly not be the one who always doubts, who is afraid of everything, and who turns the first disagreement that appears into a drama, but rather the one who knows that a question is only well constructed on the basis of uncertainty.

The question of the Great Divide has just changed direction. It is true that we are not different from others, but we are much more dangerous, because, as befits the "greatness" of entrepreneurs, we have thrown fear into the closet of obsolete accessories. But the new direction taken by the question does not constitute a regression, a call to submit to intangible values that should govern what counts to us. This time, the point is to start from the political question, in the sense that it becomes also an adventure. And an adventure that defines its first risk as the fact that elephants, bears, the freedom of research, climate variations, do not have representatives capable of making other representatives think. In *Politiques de la nature* (Latour 1999), the first question around which the Parliament of Things gathers is "how many are we?" How many should be gathered here in order that the problem that brings us together can be well formulated? It is a moment of perplexity, which opposes the fascination for Stalinist pragmatism: "How many armoured divisions does the Vatican have?"

To block the possibility of silencing a knowledge which, if it counted, would have forced others to organize themselves differently, to prevent the advantages derived from relegating a not fully articulated protest to the level of opinion, to oppose screenings that decide about what counts and what should be done thereafter, this is the first task that, from now on, Latour assigns to politics. Not only will we no longer be taken in by Archimedes's coup, in the sense that it will not be assumed that the world lets itself be legitimately lifted by the fulcrums provided by disinterested sciences, but we will actively undo the connection between mobility, the capacity of the world to "move," and "mobilization," based on relations of force, deciding what should be taken into account and what should "manage for itself." If that on which their humanity depends is destroyed, humans will no longer be only associated with nonhumans (associations are made and unmade, according to the tests they undergo), but will be tied to and might be destroyed by, reduced to, the inhumanity of hate, cynicism and despair (Latour 2001c). All the same, entrepreneurs are not called to respect the immobility of a world in which everything would be connected to everything else. Nevertheless, they are subjected to the collectively recognized obligation of "paying attention," which hampers their ideal of free enterprise: the world should not be thought only from the point of view of the opportunities for mobilization that it offers, since these opportunities ratify, first of all, the inequalities between those who have the power to make their objections prevail, imposing themselves as compulsory passage points for all innovation, and those who are defined as unimportant, unqualified to raise obstacles to the great march of progress. The latter were addressed in the following immortal words by Pascal Lamy, a European Commissioner who was our representative in the negotiations with the World Trade Organization: one cannot stop a clock.

DEMOCRATIZING DEMOCRACY?

I refer the reader to a recent book by Michel Callon, Pierre Lascoumes and Yannick Barthe, *Agir dans un monde incertain* (2001), for a eulogy of the "democratization of democracy" to which correspond "dialogic" practices that actively imply, at the earliest stage of a question of public interest, what the authors call "open-air research," in contrast with "confined research" (which includes fieldwork, for this produces data made to be translated into professional statements in a confined space). "Open-air" research involves all the groups interested in a question (including, obviously, the different groups of researchers) and all the knowledges that "might count"; in addition, the verdict about what should count, or may not count, has to follow procedures that redefine the obligations and demands associated with the notion of democracy. These procedures, in fact, bring to the fore the dynamics of emerging identities, in other words, those that characterize groups that become capable of making, in a determined and articulated way, what counts to them and of entering the composition of the collective, participating in the "dialogic" dynamic thanks to which the disparate ensemble (hybrid forum) of representatives of the question being debated will have to take one another into account and negotiate the relevance and significance of their respective positions.

But I would like to take up a "provoking" dimension of Bruno Latour's position, which was somewhat smoothed over by the "serious" version that the authors of *Agir dans un monde incertain* give of the ideas they share with Latour. The contrast is not unimportant, although it might be reduced to a question of "rhetoric." Since the question concerns what counts, rhetoric is important, because it designates what is presented as that which makes one think. In this case, whereas by using the adjective "dialogic" the authors of *Agir dans un monde incertain* intend to show a tranquilizing respect for humanistic values, Latour's exhortation, "let's seek to treat humans as well as we treat nonhumans," brings to the fore the radical character of the change proposed for our adventures both in knowledge and in politics.

We do not treat all nonhumans well: think of the animals in industrial farming which are dosed with antibiotics to prevent them from dying. We take advantage of the fact that cows, chickens and pigs make it possible to assimilate, in practice, the animal to butcher's meat with feet, to ignore the animal, to reduce it to our ideas of profitability of meat production (even though those in charge of that reduction, the animal breeders, "feel" that something is wrong—but this is a knowledge that does not count). In contradistinction, in order to make history in a pertinent manner with electrons or bacteria, we know it is better to abandon this kind of "idealism": it is this need to negotiate relevant links with them and not treat them only according to our interests that Latour calls "recalcitrance," and it is because they are "recalcitrant" that we have to treat them "well." This brings up a "great question": are we treating our children in schools more like cows (in an idealistic manner) or more like bacteria? And their teachers? And the unemployed? And the peoples we seek to modernize? What are the situations in which we take advantage of a nonrecalcitrance to a treatment which destroys "attachments" and produces "demoralized" beings (a little less than human since what that which was "their element" has been disqualified, relegated to the "irrational" or the "outdated")? What would it mean to extend the question of action in an uncertain world to the questioning of every situation in which our certainties are tied to the fact that certain protagonists let themselves be "badly" treated, enabling a distinction between the "normal" (for instance, the "good" student or the "motivated" teacher, according to pedagogues) and the pathological variations that demand police-like operations (to pacify, explain, treat, take care, recycle . . . exclude)? But if democratizing democracy entails making the production of recalcitrance into the precise condition for the adequate formulation of a problem, could we not then talk of a "cultural revolution"?

However, we cannot. The reference to a cultural revolution is too connected with a thinking of alienation, directed at the chains that need to be shaken and that it suffices to shake in order that humans find again that which was theirs by right. If the question of attachments—always related with the confusion and horror raised by the iconoclastic gesture that intends to liberate the other from the "beliefs" that alienate him—may be found in cultures which contrast in everything, from the caste system (Latour 2001c) to experimental practices, it is because what is at stake is not the struggle against something so general as alienation. Rather, what is actually at stake is the old question of civilization, and the following definition of it retroactively turns the Great Divide into a veritable declaration of war against all civilizations: a civilized group is a group which can present itself before other groups in such a way that is not insulting to them, which is able to affirm its divergent identity without pretending to base that identity on reasons which should compel general adhesion, in other words, without judging the others as lacking something "everybody" should possess.[7] The upshot of this definition is that no civilized human being can present himself as "free" or "rational," since this would imply that those to whom he presents himself might not be so. Everyone should be presented as "attached," compelled to think and feel by that which binds them, capable of becoming and learning through that which makes them human.

Hence, the entrepreneur who thinks himself free and without bonds, working for the "revolution," for a humanity free at last from superstitions and alienation, is not civilized. Not that he isn't, in fact, bound or doesn't have obligations, but rather because he appropriates the task that binds him as if it were at one and the same time his and naturally universalizable to all humans. In contrast, that which constitutes recalcitrance as a condition for a problem to be well formulated transforms the kind of freedom that entrepreneurs use to advantage (there isn't a natural order determining what counts) into radical obligation: we do not know what counts, we have to learn it, and in order to do that we need those who are insulted and endangered by what counts to us to be recalcitrant, to have the power to force us to "pay attention." We, children of entrepreneurs, have a vital need to be forced to think by and through others. Which is, after all, a very civilized way of presenting ourselves.

Thus, behind the "emerging identities" of Callon, Lascoumes and Barthe, stands a risky politics of belonging and the need to leave behind the current (in France at least) association of belonging and "isolationism," with the alternative that derives from this: either something unites, by right, all humans, guaranteeing the possibility of their convergence, or each group will state its own identitarian conditions (it's this or nothing), and then we will be in the realm of relations of force. If this alternative holds, farewell hybrid forums, the entrance of science into democracy, "dialogics." But of course this way of presenting belonging is an insult affiliated with the thesis of the Great Divide: those who belong are presumed to be identified with their "beliefs," whereas "we" are free, subject only to the universal constraints of rationality. And it is true that belonging means that those who belong are not free, as they cannot subject themselves to a cause that requires detachment, forgetting that which binds them, without becoming "less human" (desperate, "unmotivated," hateful, stupid, etc.). The question of what can be accepted, therefore, can never be posed in terms of the Right, with its "and therefore you should." But it can be posed in terms of becoming, of event, always local and circumstantial. Such becoming and event is possible because those who belong do not master that which binds them, do not have the power to utter "this or nothing." They are the human voice of a knowledge that counts, that binds them, but that, in binding them to nonhumans (DNA, or divinities, or mathematical abstractions), does not confer them the all too human power of determining by themselves how it should count, how the obligations derived therefrom should be formulated.

The problem of the dual theme of universality and progress, in whose name so many "Archimedes's coups" have been accomplished, so many operations of mobilization, screening and modernization have been justified, is that it leaves entrepreneurs free to become the sole privileged representatives of this theme ("the market," "the voters," "the future," "the facts" will decide, that is, will do the screening, among "us"). I believe this is the reason why Bruno Latour associates political ecology with an obligation that no one can claim for himself, excluding others, the obligation of having to work for the "progressive composition of a common world." Such an obligation makes immediately vibrate, as the risk par excellence, that which was the legitimate success of screening operations: if there are breeders who are desperate because they have to reduce their cows to the status of meat with feet, if a voice is reduced to silence, defeated by a majority vote, if a knowledge is disqualified as unworthy of being considered, the composition will be impoverished, a failure. Every knowledge that counts needs the others, without which the manner in which it counts cannot have full significance, destroying that with which it had to be articulated.

"Democratizing democracy," in other words, civilizing the entrepreneurs for whom it is primarily a means, or an obstacle, to reach their own ends, situates us, thus, entirely "beyond the Great Divide." Indeed the risk of failure, as I have defined it, unites the hypercontemporary—I'm thinking, namely, of "consensus" decision-making techniques, which activist groups today require to fabricate themselves, to engender the dynamics of empowerment without which activists could not avoid sacrificial militantism—together with the old—I'm thinking about the techniques of *palabre*,[8] which ask of the order of the world, about which it is accepted that each of the participants has some legitimate knowledge, but that none is its privileged spokesperson, that it "passes among them," that it gives its meaning to the case that divided or perplexed them. One might say that this is a utopia, but the one who protests in this way has already been transformed. He has to deny a possible there, where previously he thought he had no choice. Utopia can be a knowledge that counts.

NOTES

1. Also published in English as "A Discourse on the Sciences," *Review*, 15(1), 1992, 9–48.
2. It was Joseph Needham (see, namely, Needham 1973) who stressed, in contrast to China, the relationship between the development of the modern sciences and the spirit of enterprise in the European world.
3. See Morrel (1972).
4. See Rossiter (1975).
5. This is what Steven Epstein (2001) has superbly shown in relation to AIDS patients who refused to take part in clinical tests that had been conceived "in the name of science."
6. It is important to stress that this thought experiment, in itself, is not an anodyne operation. In the case mentioned, Latour connected it with the hope of a "slowing down" in the construction of the network by making its operations visible. But this hope could be read from a scientist's perspective: it would be enough to dispel the small error that makes us think of ourselves as modern in order to restore a politics worthy of the name.
7. I have used this definition in Stengers 1996.
8. The concept of "palabre" derives from Spanish, where it originally meant "a gift offered to a black chieftain for the purpose of falling into his good graces"; later, it came to mean the long conversation that accompanied this ritual. The first sense, archaic, may have been the origin of the word "palabrer," which was recorded in 1890. The modern meaning is that of "an idle, long and useless conversation." The author uses the word here in a clearly positive sense, pointing to one of the possible techniques that she calls elsewhere cosmopolitical relation (Stengers 1996). Thanks to João Maria André, Ana Cristina Macário Lopes and Joana Santos for helping to clarify this concept (Portuguese translator's note).

REFERENCES

Boltanski, L., and L. Thévenot.1991. *De la justification. Les économies de la grandeur*. Paris: Gallimard.

Callon, Michel, Pierre Lascoumes, and Yannick Barthe. 2001. *Agir dans un monde incertain*. Paris: Le Seuil.

Epistein, Steve. 2001. *La grande révolte des malades. Histoire du Sida 2*. Paris: Les Empêcheurs de penser en rond.

Latour, Bruno 1991. *Nous n'avons jamais été modernes*. Paris: La Découverte.

———. 1995. "Moderniser ou écologiser? A la recherche de la 'septième cité'." *Ecologie Politique* 13:5–27.

———. 1999. *Politiques de la nature*. Paris: La Découverte.

———. 2001a. "Joliot et la vascularisation des faits." *L'espoir de Pandore*. Paris: La Découverte.

———. 2001b. *Les microbes: guerres et paix, suivi de Irréductions*. Paris: La Découverte, 1984.

———. 2001c. "Faits, fétiches, faitiches, la divine surprise de l'action." *L'espoir de Pandore*. Paris: La Découverte.

Liebig, Justus. 1894. *Lord Bacon*. Paris: Librairie J-B Baillière et Fils.

Mirenowicz, Jacques. 2000. *Science et democratie: Le couple impossible?* Paris: Editions Charles Léopold Mayer.

Morell, J. B. 1972. "The Chemist Breeders: The Research Schools of Justus Liebig and Thomas Tmomson." *Ambix* 19:1–43.

Needham, Joseph. 1973. *La science chinoise et l'Occident*. Paris: Le Seuil.

Rossiter, M. W. 1975. *The Emergence of Agricultural Science. Justus Liebig and the Americans, 1840–1880*. New Haven, CT: Yale University Press.

Santos, Boaventura de Sousa. 1987. *Um discurso sobre as Ciências*. Porto, Portugal: Afrontamento.

Stengers, Isabelle. 1996. *Cosmopolitiques*. Paris: La Découverte.

Whitehead, Alfred North. 1964. *Concept of Nature*. Cambridge: Cambridge University Press, 1920.

8

The Self, Psychoanalysis, and Epistemology

A Dialogue

Frank Summers and Leonard Kaplan

It is the contention of the present chapter that neither side of the argument between objectivism and relativism recognizes the unique nature of psychoanalytic knowledge, and, therefore, neither position provides an epistemological basis for psychoanalysis. Both the objectivist and relativist viewpoints will be examined in detail to illustrate the limitations of each position. It will be argued that only a specifically human science standpoint avoids the irresolvable problems of the other two positions and provides a valid epistemological basis for psychoanalytic science.

Leonard Kaplan: *Before I ask a question or comment I think we should make clear that even in the face of a vast and a parasitic critical literature attacking psychoanalysis, it still remains the best map we have in the West to understand the intrapsychic aspect of the self, particularly in the way in which it understands the impact of forces outside of reflective consciousness on the aware mind. A general question: To what extent do any clinicians depend on epistemological decisions, that is, what is at stake for clinical practice as opposed to the appropriation of clinical findings for larger critical thought, for example, in the Frankfurt School or Norman O'Brown?*

Frank Summers: The relationship between clinical praxis and epistemology is a large and complex question to which I could not possibly do justice in this context. What I can say briefly is that epistemological assumptions do seem to have a great influence on the analyst's clinical strategy, although the type of impact it has differs widely among analysts and often in unpredictable ways so that there are no clear principles of therapeutic action that follow a given epistemological stance. For example, Hoffman (1998) contends that his relativism allows him greater freedom in suggesting ideas to patients because he does not hold the illusion that he will someday uncover the ultimate truth about the patient. However, Aron (1997), adopting the same epistemological position draws virtually opposite conclusion: He makes few interpretations because he does not believe he has the right to claim an understanding of the other. His view of analysis is a "meeting of minds," contact between two subjectivities, more akin to Bubers' I-Thou relationship than Freud's authoritarian claim to knowledge. Despite this difference, it is fair to say that relativists tend to be more tentative in their understanding than objectivists who follow Freud in believing they are "reading nature." The latter group feels a greater authority to insist on their understanding. The issue is most sharply drawn around clinical moments when the patient does not concur with the analyst. The objectivist is much more likely to regard the patient's opposition as resistance, to be interpreted forcefully to break down the patient's defenses. One does tend to hear less about resistance in the clinical discussions of relativists. However, it is also true that Gill (1981), one of the most widely respected analysts of his time and an avowed relativist who contended that reality must be negotiated between patient and analyst, used resistance as a cardinal principle of his clinical approach. For me, this example demonstrates that there is often a disconnection between epistemological position and therapeutic stance.

LK: *So does a good clinician even need to make an epistemological commitment? It seems perhaps not.*

FS: Let us start with objectivism. According to objectivism, psychoanalytic inquiry differs from that of the natural sciences only in the aspect of nature being investigated so that the methods and causal principles applicable to psychoanalytic data are no different from that of the natural sciences (e.g., Holt 1972; Wallerstein 1986, 1988). The objectivist argument tends to rely heavily on Popper's (1962) critique of positivism and his alternative view of scientific method as "conjecture and refutation" (Blight 1981, Holzman 1985, Wallerstein 1986, 1988). The positivist critique is that psychoanalysis has not proven its hypotheses because it has not compiled observations to demonstrate their truth (Grunbaum 1984). Psychoanalytic objectivists point out that Popper (1962) in his thorough critique of positivism showed that no amount of observations can ever conclusively prove a theory because future observations could theoretically disprove it and that there are no observations without theories. Popper concluded that science proceeds not by compiling random lists of observations, but by conjectures, leaps of imagination, that are then tested. Scientific theories are demarcated from nonscientific statements by virtue of their falsifiability.

Objectivist psychoanalysts argue that psychoanalysis is a series of propositions, conjectures, awaiting tests that could refute them, and, therefore, is no different from any other science. Some objectivists view the analyst as a natural scientist who uses the scientific method within the analytic setting (Bachrach 1989; Rubenstein 1976, 1980). According to other objectivists, the analyst understands the patient's motives at the experiential level, but invokes nonexperiential mechanisms to explain via causation, and this level of explanation is the natural scientific aspect of psychoanalysis (e.g., Blight 1981; Eagle 1980). For example, the patient's excessive sympathy may be understood clinically as a defense against underlying feelings of cruelty, but the scientific level explains this defense by the *mechanism* of reaction formation

Another group of objectivists, like Holzman (1985) and Wallerstein (1986, 1988), believe the psychoanalytic setting is too contaminated for objective investigation and, therefore, the ultimate validation of psychoanalytic hypotheses must come from controlled extraclinical experimental research. Despite these differences, the fundamental principle of objectivism in all its forms is that the subject matter of psychoanalysis is the natural world, and the task of the analyst is to explain its workings via the methods of the natural sciences.

LK: *Is the study of history scientific? I don't think so? Isn't the doing of history legitimate? So picking up from above, why do people like Morris Eagle who knows better think that the patient can be so reified that he becomes an object for making universal claims? Can even a sensitive analyst as Eagle ever offer a definitive claim that could be studied that sympathy in one was excessive and in another reaction formation? And even if this is correct or plausible in the instant case, how could one ever establish the parameters for a differential claim in another case? In other words, on its own terms, was the positive defense of psychoanalysis bad faith from the beginning of the project? In fact isn't this true for much of psychology as well? I might add that it would be impossible to construct a model of the person out of the disparate findings of research in academic psychology. Perception, cognition, learning theory, etc., are not constructed to integrate into an embodied understanding of an individual. So academic psychology is no more an objective science than is psychoanalysis.*

FS: In my understanding of objectivism, the view would be that history is a legitimate endeavor to the extent that it follows the scientific method. Historians who do not propose testable hypotheses nor provide evidence to prove them are not scientists, but those who do are practicing science as valid as that of any physicist. Their view is that the good historian has a hypothesis, which he then tests against the historical data for its accuracy, and they believe the analyst operates in the same fashion. Objectivists such as Eagle believe that criteria for reaction formation and differentiating it from sympathy are, in principle, possible. However, one can legitimately ask: If so, why do we not have such criteria? It seems to me Eagle is trying to walk a tightrope by claiming that he is not objectifying the person, but he can apply objective mechanisms to understand the person. I don't think the distinction works, but I think Eagle is trying to achieve scientific legitimacy for psychoanalysis, in the way he understands it, while maintaining a humane sensitivity to the patient. For Eagle, it is the use of mechanisms like reaction formation that can be applied to different patients in different circumstances, and criteria for judging behavior as "reaction formation" across patients and situations is, in principle, possible.

LK: *I have a different question then: Given the changes in scientific epistemology after Freud, people like Toulmin, Kuhn, Laktatos, Feyerabend, why aren't psychoanalytic thinkers like Eagle shifting to the newer and, in fact, now I think hegemonic models?*

FS: It is difficult for me to enter the minds of others, but my inference from reading people like Eagle and Holzman is that they fear that if they adopt an alternative model of science, psychoanalysis would be confined to a "soft science" without much justification despite the fact that the philosophers of science you mention apply their thinking to the natural sciences.

LK: *I do not think that this is true of Toulmin. We know that he considers psychoanalysis a significant knowledge base for understanding the self and world. In a longer elaboration we would have to ask if even the Aristotelian divide between physical science and politics is accurate. Perhaps each endeavor relies on the prudence that Aristotle specifies as the best we can do for politics. So the objectivist position may be naïve even where practitioners feel most secure.*

FS: There are several problems with the objectivist position. First, the objectivist contradicts himself in the very statement of his position because objectivism relies on a truth claim (that all scientific knowledge is testable) that is not founded on the testability it claims is the only basis for scientific knowledge.

Second, Popper's critique of positivism does not justify a "unitary" view of science. The positivist position is that there is only one method for science (Ayer, 1946), so Popper's attack on the positivist view of science is not relevant to the distinction between natural and human science. Second, the fact that Popper believed in one method for all science does not in any way imply that psychoanalysis fits that method; in fact, Popper concluded that psychoanalysis is not scientific because it is hopelessly unverifiable.

This point leads to the third, and perhaps most psychoanalytically significant, flaw in objectivism. Rather than fitting the methods of the science to the domain of inquiry, the objectivist predetermines his method and then attempts to fit the data to it. In this sense, the objectivist is in the position of the drunk in the old story who looks for his watch under the lamppost in the alley because the light is there although he lost it up the street. Absent any demonstration that psychological reality is material reality, or, at least, is best understood by the methods used to investigate the latter, there is no reason to believe the natural science model fits the analytic process, and the equation of psychoanalytic data with the material world is an unfounded theoretical prejudice.

If psychoanalysis claims to have objective knowledge, it must conform to the standards of the objectivist standards of falsifiability and the logic of explanation. Any less stringent criterion for objective truth does not fulfill the objectivist standard of verifiability, and, therefore leaves psychoanalysis open to scathing attacks from philosophers of science (e.g., Nagel 1959; Scriven 1959; Grunbaum 1984) who point out that the uncontrolled nature and lack of independent investigators in the analytic process reduce any truth claims generated in this setting to unfounded speculation. If psychoanalysis is regarded as an objectivist science, there is no answer to these charges because psychoanalytic investigation fails to meet objectivist criteria for knowledge.

LK: *Is the use of the term "science" at this point in history any better than talking in terms of critical analytical thought that is, isn't "science" itself now a rhetorical term? I take it for example astronomy fits what would be called natural science?*

FS: The claim that analysis is a natural science because the analyst uses "nonexperiential mechanisms" in his understanding is precisely the sort of "pseudoscience" that objectivist philosophers have so aptly criticized. Employing the concept, reaction formation, is not an explanation because it does not add to the clinical understanding that the patient uses over solicitousness to hide his cruelty. As long as psychoanalysis claims to be an explanatory science of observables, such attacks are warranted because any putative explanatory "mechanism" can be no more than a label for phenomena already understood.

LK: The critique of psychoanalysis by objectivist philosophers cannot be dismissed as the complaint of an outmoded positivist philosophy of science. Nonpositivist philosophers such as Popper (1962) and Scriven (1959) are no less relenting in their attacks on psychoanalysis than positivists such as Grunbaum (1984). If psychoanalysis must claim to be a natural science, these philosophers win the argument. It is at this point that we must turn to the viewpoint of the relativists who have provided the most commonly voiced epistemological alternative to this view of psychoanalysis.

Positivism itself is part of a broader historical ideology to break finally with any metaphysical assumptions still remaining at that point in history, isn't it? Doesn't this connect with Kant's attempts to anchor philosophy in Newtonian physics? Wouldn't a claim about universality of human psychology of any type undercut any claim about human choice, freedom, etc? You rightfully point out that such claims are philosophic and not open to a narrow meaning of science. Wasn't positivistic science itself muddled, pointing only to the Humean claim that the empirical is relevant if not independent?

FS: Comte, the founder of positivism, believed that efforts to understand the world have gone through three historical phases: the religious, the metaphysical, and the scientific. By "science" is meant the method of natural sciences. The social sciences must employ such methods or they are no more valid than the speculations of metaphysics, which we can now abandon due to the discovery of the scientific method. Such an application, endorsed by Freud and other psychoanalytic objectivists, would seem to preclude a role for human freedom, and Freud was not reluctant to draw exactly this conclusion. I believe any objectivist stance would have to view the human as a biological mechanism reducible to the same explanations as nature. Objectivism sees humankind as a machine with the illusion of freedom and difference from nature, but, in reality, a naked ape. One might well ask how such a stance could be applicable to the analytic setting. How can an analyst practice without the assumption of human freedom? If people are biologically determined machines, why should understanding be expected to have an impact?

It is not my understanding that Kant tried to anchor philosophy in Newtonian physics. I understand Kant to have grounded philosophy in the transcendental conditions for knowledge of any type. Kant is a prototypical antipositivist in the philosophical sense that he would have opposed the positivist attempt to ground knowledge in empirical observation. You would have to explain your comment on Kant for me to reply further.

Positivism, in my view, has been thoroughly debunked as a philosophy of science by Popper and others, and that is why it is dead in philosophical circles; however, objectivism is alive and well. Popper, for example, is an objectivist who has been at the forefront of the critique of positivism, as I say. So, while one may dismiss positivism as an outmoded and discredited philosophy of science, the same cannot be said of objectivism, which is why I use objectivism rather than positivism in this chapter.

LK: *My sense of Kant is that he did operate under the shadow of Newton and as such was caught in an antinomy between freedom, which was a central value for him as you state and a position that all nature was determined by deeper law. But in any case I am taken with the distinction between positivism and objectivism. I am not sure that the distinction is so clear to everyone and that the positivist wasn't trying for an objectivism. Any points of clarification here?*

FS: The distinction I make is as follows. Objectivism is the thesis that there is an objective world that includes all known phenomena, both physical and mental, that can be known by the method of the natural sciences. From this viewpoint, there is no reason to use any other method and to do so would be unscientific and no such results have validity. By positivism, I mean the work of A. J. Ayer, Rudolph Carnap, and the whole Vienna group, represented more currently by philosophers of science like Nagel and Grunbaum. The contention of these philosophers is that only the observable is real and science is a process of accumulation of observations. Anything not observed is "not sensed," therefore, literally "nonsense." To the extent that any sentence has meaning, it refers to observational events. If it does not, it is meaningless. This positivist view is not held by all objectivists; in fact it is specifically rejected by many of them. Popper is the philosopher most responsible for debunking positivism. Popper contended that science proceeds by conjecture and refutation so that hypothesis building, imagination, and creativity are crucial to the scientific endeavor. What makes a hypothesis scientific, for Popper, is its testability by the scientific method. So, he, like many, is an objectivist while rejecting positivism as a hopelessly inadequate scientific view. (Interestingly, Feyeraband, whom you mentioned, was one of Popper's prized students; he simply took Popper to another level.) I think the confusion between objectivism and positivism is that Comte, the founder of positivism, made as his major point the idea that the social world could be investigated by the same methods as the natural sciences, so positivism came to be mistakenly equated with the view that the social world is subject to the same methods as any other science. That is an objectivist view shared by positivism, but that does not mean anyone holding it is a positivist.

LK: *So Comtian positivism may have been a strategy to break the hold of metaphysical or teleological thinking?*

FS: Let us now move on to relativism. In reaction to the difficulties of the objectivist stance, the relativist regards the analyst's viewpoint as relative to her perspective. There are several variations of epistemological relativism. The most well known is Schafer's view (1983, 1992) that the essence of psychoanalysis consists of analyst and patient creating a narrative together. The patient's verbalizations are a narrative performance out of which the analyst selects some aspects to retell the story based on her "precritical assumptions," theories, and ways of understanding. Because other narratives are always possible, there are no objective psychoanalytic data that compel definite conclusions, and the chosen narrative is always subject to revision and new interpretation as analyst and patient "coauthor a script." The role of the patient becomes greater as she becomes a more reliable interpreter. There is nothing beyond the narration that can be used to adjudicate analytic truth. While narrations cannot be regarded as objective or definitive, they are useful.

Well aware that his position may be construed as solipsistic, Schafer (1983) attempts to answer such an objection with his view that psychoanalytic constructions are subject to the rules of verification, coherence, consistency, and completeness. The interpretation that fulfills these criteria and fits the events outside the analysis is preferred to alternatives that do not.

LK: *Isn't Schafer caught here with a modern notion of narrative, that is, beginning, middle, end, and coherence? Also how does narrative construction, according to Schafer or others who use narrative theory avoid problems of bad faith? What is the mechanism in the construction in the patient, analyst, or analytic field that points toward truth criteria?*

FS: Schafer is, indeed, caught in the modern view of narrative, but this does not seem to bother him; he believes coherence is a truth criterion. The deeper problem with his position lies in your second point: he attempts to avoid the problem of pure relativism or "bad faith" in your terms by applying the criteria of coherence, verification, consistency, and completeness. His unacknowledged problem is that to utilize such criteria is to undermine relativism. If one can appeal to verification, one goes beyond narrative to independent criteria for assessing the validity of the narrative. Schafer seems to be unaware of this blatant contradiction. I take this unresolved contradiction to mean that relativism is not a viable epistemological analytic stance. He appeals to verification precisely to respond to your question of how one can assess the truth of the narrative, and thereby undermines his relativist epistemology.

LK: *If analysts schooled in medicine and not in psychology and philosophy as you are catch up the postmodern literary thinking, won't they eschew Schafer's modernistic narrative thinking if only as a matter of fashion? To continue this point Schafer is a psychologist not medically trained if I remember correctly. Is medicine no longer capable of epistemological advance in psychology and psychoanalysis?*

FS: I do not know if medicine was ever capable of epistemological advance because medical training is an objectivist model. All the analysts I mentioned who propose alternatives to objectivism, including Schafer, Spence, Hoffman, and me are psychologists. While not all objectivists are physicians, I cannot think of a single physician-analyst who is not an objectivist. I think it is difficult to go through medical school and then throw away that education to be an analyst. Physicians fear a hermeneutic model because they sense it will dissuade physicians from pursuing psychoanalytic education. It may just well be that the nature of psychoanalytic knowledge is simply too different from medical training to make sense to those already trained and deeply steeped in the medical model. Most analytic candidates are from fields other than medicine.

The other major narrative theorist is Donald Spence (1982) who bases his hermeneutic position on a detailed criticism of analytic reasoning to show that the analyst has no claim to "historical truth," knowledge of events as they occurred. Spence argues that the patient's experience is not directly accessible to the analyst. Consequently, the analyst inevitably uses the "haze of his own experience," to "fill in" the missing pieces of the patient's communications, resulting in "unwitting interpretations" that become the nonverbalized basis for constructions that are inevitably distortions. Furthermore, analysts mistake created structural, linguistic similarities between current happenings and the past for an actual, causal relationship between the past and present. The "link" with the past is not a causal connection, but a linguistic creation of the analyst to develop an aesthetically pleasing story.

In his early work Spence believed such stories possessed "narrative truth," a consistent, coherent, comprehensive account, which has both aesthetic and practical components. Aesthetically, a good analytic interpretation finds a "home" for an "anomalous happening" by fitting the event into the patient's life story. Therapeutic effectiveness is a product of the linguistic and narrative closure effected by reducing life to organizing principles, rather than correspondence to actual events.

In his later work, Spence (1993, 1994), no longer finding rhetorical persuasion an acceptable alternative, attacks psychoanalysis for its reliance on rhetoric rather than evidence. He calls on psychoanalysis to overcome the relativity of the observer, separate out the influence of historical and cultural bias, and found psychoanalysis on evidence from the "bottom up," rather than from a "top down" theoretical bias in order to become a science.

LK: *Contemporary theory of memory in thinkers like Daniel Schacter make it clear to me that Wittgenstein's assertion that we can never prove what we remember actually happened, let alone in the way we recall, should be relevant here. Isn't the first move in the early work of Spence correctable by understanding what is going on in a good construction is, in fact, a reordering of life not only linguistically but affectively. So a successful outcome would be a new (re)solution of the earlier fragments, jump starts, avoidances, traumas, mistakes in the patient's life experience.*

FS: In my view, the best recent work on memory is Edelman's theory that memory is an active categorization process rather than the passive registration assumed by Freud and his contemporaries. Modell has used Edelman's theory to define *transference* as affective categorization. I do not believe "proof" applies to memory, but it does seem that narrative coherence does provide the criterion we use to assess what we remember. When we recall events, we see how the memory fits into the narrative we have of temporally contiguous happenings. If it does not fit, we begin to question the veridicality of the memory. To say that we cannot assess in any way the validity of a memory would be abolish all history as a fantasy. It is the fitting together of data that we use as the criterion to evaluate newly uncovered historical facts or memories. If we do not believe that memory can be trusted, would we not have to throw out of the courtroom all eyewitness testimony?

LK: *In fact, this might make for something more truthful. As an aside, years ago the one time I taught evidence in a law school, I asked the question: If Wittgenstein was right that we cannot generally prove that which we remember happened, how could we be certain about what constitutes evidence of past events? I think we all have had experiences of so-called screen memories where we have been certain of the occurrence of a past event but become aware that the recalled event never took place. The courtroom analogy is a good one and one where eyewitness testimony has been known to be defective. Pragmatism and not justice has ruled the day with the continued relevance of this testimony as truth.*

FS: I know I cannot come close to doing justice to the complex problem of the veracity of memory, but I can say this: if you accept that a screen memory never happened, how do you know it never happened? Any such evidence is based on memory, evidence of the past, so are you not contradicting your position? To state that an event happened or did not is to assume the veracity of knowledge of the past, precisely what you propose to question, so it seems you are privileging one view of the past over another, but why? On what basis is one retrospective preferred to another? In any case, you seem to be assuming you can have knowledge of the past by disputing a memory.

I believe you are right about Spence: a good construction is not simply linguistic, but affective, and affects provide the truth criterion analysts routinely use. If the patient cannot feel the interpretation, then its truth is questioned. It is one thing to say that Freud made invalid inferential leaps without sufficient evidence and quite another to say such evidence is in principle not possible. So, my response to your question is: What does "resolution" mean? Are there any criteria by which one can claim the superiority of any particular construction? Are all reorderings equal? The issue comes down to criteria: Is there any way to adjudicate between two different interpretations? All relativists have the problem of how to escape the conclusion that all interpretations are equal without contradicting their relativism.

LK: *Where then in this discussion would the tension between Kohut and Kernberg play out? What theory would either have to take to say that a borderline or narcissistic personality was either someone who had not developed psychological structures (Kohut) or was someone with repressed rage needing confrontation and interpretation (Kernberg)? If the two are not relevant here, what does that say about epistemology and theory in psychoanalysis generally where a decade (the 1970s) become irrelevant?*

FS: Kernberg is clearly an objectivist in that he sees the patient as having a diagnosable disorder that can be assessed just like any medical condition, and treatment prescribed on the basis of diagnosis. Kohut (1959) very early on defined psychoanalytic tools as empathy and introspection, thus defining psychoanalysis as a domain of interpersonal experience. Whether he was aware of it or not, by making empathy and introspection the tools of the trade, Kohut defined psychoanalysis as a hermeneutic science. I do believe that Kohut's view of the psychoanalytic method is inseparable from his theory, not so much of the structure of narcissism, but the emphasis in his clinical method on empathy and diminishing the role of interpretation. Kernberg, on the other hand, takes an interpretive stance toward narcissism, and this view cannot be dislodged from his medical model of psychoanalysis. Interpretation is the treatment for the pathology; the relationship has little clinical importance. Kohut (1984), in the end, has the opposite view: interpretation is effective because of the type of relationship it forms. He opposed what he called the "truth morality" of classical analysis in which the highest value is confrontation with the truth about oneself. In opposition, Kohut believed what matters is the development of the relationship because that is what helps the patient most. In general, objectivists tend to advocate interpretation because they believe that observation of reality must be given to the patient. Those with a more hermeneutic bent tend, like Kohut, to see the emotional resonance between patient and analyst as the key to therapeutic action. So, there is definitely a relationship between epistemology and clinical theory, but it is not about understanding of pathology as much as the clinical stance of what is expected to help the patient.

A recent version of the relativist position originates with the relational model of psychoanalysis, which is based on the principle that the psychoanalytic situation is a relational matrix mutually constructed by analyst and patient (e.g., Mitchell 1988). The analyst's embeddedness makes it impossible for her to perceive the relationship from "outside," so the analyst's understanding has no claim to special knowledge or objective truth (Mitchell 1993). Analysis is a meeting of two subjectivities, neither of which has any greater claim to the truth than the other (e.g., Aron 1997, Stolorow, Brand chaft, and Atwood 1987).

A version of the relational epistemological position is Hoffman's (1991, 1998) dialectical, constructivist paradigm, a model based on the principle that "the personal participation of the analyst in the process is considered to have a continuous effect on what he or she understands about himself or herself and about the patient in the interaction. The general assumption in this model is that the analyst is always a function of his or her perspective at the moment . . ." (Hoffman 1998, 136). He contrasts this paradigm to the view that therapists are "capable of standing outside the interaction with the patient, so that they can generate rather confident hypotheses and judgments. . . ." (Hoffman 1998, 165), a view he regards as positivist. Hoffman's model is based on the concept that interpersonal experience is distinguished from knowledge of the physical world by its inherent ambiguity so that the analysts understanding is necessarily "skewed in keeping with their personal participation in the process (1998, 138).

Hoffman opposes his epistemological position to unqualified relativism because "ambiguity and uncertainty do not connote the disappearance of an objective reality and the rule of unqualified relativism. On the contrary it is objectively the case that experience is intrinsically ambiguous" (1998, xxii). In opposition to "radical relativism," Hoffman believes in an "objective framework within which constructive activity takes place" consisting of universal characteristics of human experience (1998, xxiii). Furthermore despite the fact that alternative interpretations are always possible, "some interpretations may be said to fit the patient's experience more than others" (1998, 165).

It is the analyst's participation that is key to Hoffman's epistemology, whereas Schafer's argument is based on the selectiveness of knowledge. They come to a similar conclusion for different reasons. Hoffman would say he has no way to judge that he has "accurately perceived reality," in fact, no such judgment is possible in his view. His use of "fitting the patient's experience" is a backdoor way to bring in a truth criterion without admitting he is so doing.

LK: *But perhaps what is going on then is the construction of a new reality for the patient negating the past that hasn't worked adaptively in any case.*

FS: I think that in any well-conducted psychotherapeutic process the outcome is a newly constructed reality that works better for the patient than the ways of being in the world that have been their historical patterns (Summers 1999, 2001). To be able to work in a way that is satisfactory for the patient cannot mean higher status or more money because many people are more fulfilled in lower status or less remunerative employment. For others, the ability to advance professionally in conventional terms brings emotional gratification of which they never thought themselves capable. The common denominator is that the fulfilling work feels like an authentic expression of who the person is. To find love or work (or any other activity) in a way that is a genuine expression of self is what gives life meaning. I cannot speak for Hoffman on this point, but I am sure he would agree that when analysis works a new reality has been constructed.

Relativism, while attempting to avoid the pitfalls of objectivism, also has significant weaknesses. First, this epistemological position is as self-contradictory as the objectivism it seeks to replace because the claim of relativism is not treated as relative to any perspective, but rather as an absolute, a form of knowledge the relativist claims does not exist. Consequently, relativism cannot be stated without contradicting itself. To claim that the analyst has a perspective on truth assumes knowledge of a reality on which one has a perspective, in contradiction to the perspectivist view of truth.

LK: *Not necessarily, isn't it possible to take a perspectivist position that would say that there is truth and that any particular perspective on it may just be that a bit or piece of a broader whole?*

FS: Perspectivism, by definition, means that one cannot know the truth, but only a perspective on it. If the perspective is meaningful, it must be recognized as the part of a whole. If not, it is blind. To say this another way, how can I have a perspective without knowing what I have a perspective on? If I say, "I can only see one side of this rock," I have knowledge of the rock. If I know nothing of the rock, I cannot have a perspective on it.

If, according to relativism, all interpretations are relative, there is no basis for judging any interpretation as more valid than any other, and psychoanalysis is reduced to solipsism in which all interpretations are equal and psychoanalysis has no claim to knowledge of any sort (Orange 1992).

In order to avoid this problem, each version of relativism offers criteria by which to adjudicate among interpretations. For example, Schafer (1983) contends that all interpretations and constructions are subject to the constraints of coherence, consistency, verification, and completeness. The problem with this solution is that once verification is set forth as a criterion of interpretive validity, a reality beyond the narrative is assumed as a criterion against which the narrative is evaluated. Either the more coherent interpretation of the data is nearer the truth, or there is no basis for this preference. Furthermore, Schafer's criteria of coherence, consistency, verification, and completeness are treated as transcendent, universal criteria of knowledge in direct contradiction to his relativistic epistemology. To avoid lapsing into a solipsistic "anything goes" position, relativism invokes absolute principles it claims do not exist.

Spence's solution was to justify analytic understanding on the basis of aesthetic appeal and rhetorical persuasiveness. The problem here is that many patients enter treatment with accounts of their problems and lives that possess considerable aesthetic appeal and persuasiveness, but maintain symptoms and character pathology. According to Spence's criteria, no analysis should be necessary because the patient already possesses the narrative truth that analysis provides.

LK: *And sometimes they are right but they are not in command of the emotional capacity to do anything about the correctness of their self-report.*

FS: Your comment shows that the analyst cannot live by narrative alone. A coherent account does not necessarily change anything. Furthermore, charismatic leaders and demagogues simplify life into organizing principles and create their own truth, the two sources of efficacy Spence attributes to psychoanalytic interpretation. Therefore, Spence's early position cannot differentiate psychoanalysis from other forms of rhetorical persuasion, such as religious or political zealotry.

Spence's more recent viewpoint is subject to the same pitfalls as any objectivist position (see pp. 2–4). In his first position, he found psychoanalysis valid despite its lack of evidential truth, whereas in his later view, he found the flaws in psychoanalytic knowledge neither acceptable nor necessary.

Hoffman states with the absolute certainty of objectivism that objectivity is not possible because the analyst's participation in the process "skews" her perceptions. To call any perception "skewed" is to assume knowledge of a reality that is distorted. Furthermore, he claims to know objectively that interpersonal experience is ambiguous, and the analyst is a continual participant in the process. According to Hoffman's epistemology, participation in the human process should render objective knowledge impossible. But, he does not treat his view of human experience or the analyst's knowledge as uncertain; he sees it as objective knowledge. If there is some basis for his claim, then he has undermined his paradigm and if there is no basis for it, then he is wrong.

Indeed, Hoffman attempts to separate himself from "radical relativism" by asserting objective knowledge of a "preexisting world" independent of social consensus and universal characteristics of human experience. By laying claim to objective knowledge, Hoffman is adopting a stance of certainty outside his own participation, the impossibility of which is the basis for his constructivism. In the same way, Hoffman insists that some interpretations fit the patient's experience better than others, but the analyst's "skewed participation" should render such a judgment impossible. Hoffman's "qualified relativism" does not work because once he acknowledges that objective knowledge is possible, the stance of standing outside of participation and knowing reality is assumed as a possibility in direct opposition to constructivism.

As Richard Moore (1999) notes, to define an external reality at all departs from the constructivist paradigm, and Hoffman's claim of universality implies that he is able to see beyond the influence of culture and history, a claim that belies his constructivism. Moore's solution is to develop a pure constructivism as the standard for mental health "a fairly explicit, but as yet undeveloped, notion of an optimal process of construction" (1999, 155). Clinical technique, then, would be focused on facilitating a joint construction of reality "rather than correcting a deficient process of adherence to an objective reality" (1999, 155). As soon as Moore begins to apply his "pure constructivist" model, he makes judgments that belie the principles of that model. The judgments "optimal construction" and "deficient constructive process" imply a stance outside the constructive process. Any criteria Moore would apply to make such judgments can have no place in his "pure constructivist model."

One may conclude, then, that relativism sees the error of importing methods from the natural sciences to psychology, but assumes that the only alternative is a relative view of reality. The relativists share with objectivism the assumption that the only valid knowledge is objective knowledge. The difference is that the objectivist believes that such knowledge is attainable in the analytic process, whereas the relativist does not.

LK: *Not only that but that objective knowledge is once and for all, a kind of Platonic inheritance without buying into Plato's epistemology.*

FS: It is certainly assumed to be a "final truth," the difference between the objectivists and Plato is that for the latter final truths are ideas, or forms, and not to be found in the material world. For the objectivists, ultimate truth is discovered only in the material world; ideas are not relevant. Plato would have opposed completely the objectivist position because it is a materialistic reductionism that mistakes reality for materialistic illusion. In Plato's view, objectivism is the position of the poor souls in the cave who see shadows and think they are seeing reality.

After adopting a relativist epistemology, relativists are faced with having to conclude that all interpretations are equally valid; to avoid this problem, each relativist proposes criteria for interpretive validity that contradict relativist epistemology.

LK: *I wonder if part of the real problem is the lack of recognition that something can be real and not universal, that is, for all time.*

FS: That is part of the problem because the assumption of both positions is that any so-called truth that is not universal is simply "true" relative to the observer or the situation and therefore an opinion, not real. The idea of a human reality that is just as real as the physical world, but different, is missing from both views, and, therefore, they see knowledge of others as a chimera, an opinion, but lacking the reality of the physical world. This objectivist assumption runs rampant in the Anglo-Saxon world.

LK: *How does someone objectify the experience of an orgasm and who would deny the experience?*

FS: Objectivists don't deny the experience of orgasm; their claim is that knowledge of it is only scientific to the extent that it is measurable. However, if one does not adopt this positivist view of knowledge and the nature of science, then the failure of objectivism to provide a foundation for psychoanalysis does not vitiate psychoanalytic claims to knowledge. A path is opened for the possibility of another type of knowledge, a uniquely psychoanalytic way of knowing that has its own validity.

LK: *So again the problem may lie to some extent to a silly view of objectivism in relationship to reality.*

FS: Since neither objectivism nor relativism solve our problems, I would like to propose to you a "third path," a human science model. The analyst's task is not to find relationships among observables, as in the natural sciences, but to discover meaning and motivation through the medium of speech. Ricouer (1970) shows that this distinction is absolutely fundamental because without a firm grasp of the unique nature of the psychoanalytic field of investigation psychoanalysis cannot be grounded as a legitimate scientific endeavor.

LK: *Another cultural problem here is the identification of science and knowledge. What can and should be said about that? After all science itself may be ideological as practiced at least certainly in the changing nature of medicine.*

FS: The issue is the legitimacy of psychoanalytic truth, whether one calls that science or knowledge. Then one has the question of what forms of knowledge are science and which forms are not. It is not relevant to the subject of this chapter if one calls psychoanalytic knowledge "science" as long as its knowledge is accorded legitimacy. Today some would say the appellation "scientific" is irrelevant to knowledge. Here we confront the question of how to categorize different forms of knowledge, which is less important than whether psychoanalysis has a legitimate claim to knowledge however it is categorized.

Psychoanalytic inquiry, as an effort to uncover meaning and motivation of an experiencing subject, is a human science in Dilthey's (1883/1979) sense. The *Geisteswissenschaften*, or sciences of the mind, have as their subject matter human experience as opposed to nature, which is the subject of the *Naturwaschten*, or natural sciences. As Dilthey demonstrated, human sciences are interpretive, or hermeneutic. A hermeneutic science is a discipline that attempts to find the meaning or motive in a class of human events, such as the historian's investigation of historical documents to ascertain the motive of historical actors. In Dilthey's view, a hermeneutic discipline is a science the methods of which involve rules of interpretation, not observations or their manipulation.

LK: *Hermeneutics may be relatively young as a term in history but is certainly ancient in application. Exegesis is Talmudic. Jurisprudence, particularly when theocratically based has always had to deal with changing circumstances and thus changing interpretive strategies. I think also that we have probably lost much in the way of interpretive theory when compared with medieval scholasticism.*

FS: Psychoanalysis is a paradigmatic hermeneutic science because its target is the meaning of experience to the patient as expressed through speech. The basis for the hermeneutic is the fact that people do understand each other in ordinary human discourse without recourse to the methods of objective science. Psychoanalytic method and logic of justification involve the rules for interpreting from what is said to what is meant. It follows that the criteria for interpretive validity come from the rules for interpreting speech.

LK: *I disagree here not with the major thrust of what you point to but with the reliance on rules, which themselves invariably require more interpretation. So in law the problem frequently goes to two problems; (1) in any complex more than one rule may be potentially operative, and (2) there may not be an objective application of a rule even if it fits to a particular case. If this is right, how can a hermeneutics itself not be open to problems of relativism and constructionism? Do you have in mind some rules or sets of rules that would ground clinical interpretation? I think almost anyone stated would be open to challenge by an opposing rule. The problem here goes back to a Pyrrhoism.*

FS: I am referring here to rules of intelligibility and inference based on coherence, what underlying motivation and meaning makes the symptom understandable. So, if the patient is unable to be self-assertive, allowing others to control and manipulate him, the analyst looks for what unseen motivation would make sense of the pattern of compliance. Then if the patient recalls that he was abused by his father whenever he disagreed or opposed him, the analyst begins to formulate the proposition that the patient is anxious about retaliation for any use of aggression. Such an idea applies the rule of the best fit. The principle of "fearing responses to aggression" ties together the past and present and makes sense of what otherwise seems like unintelligible self-defeating behavior. The patient's behavior becomes intelligible if one can provide reasons that are not conscious. The application of this rule transcends the particular circumstance of the patient. This is not a rule constructed between this patient and this analyst, but a hermeneutic rule that is used for understanding the other, both in and out of psychoanalysis. The rule is not relative to any circumstance, but rather transcendent, that is, a rule that makes possible any understanding between people.

Husserl's (1925/1977) phenomenological investigations showed that we are able to understand others because we have had similar experiences ourselves. When I tell anyone, even a positivist, I had a dream, the other understands what I said. This understanding is won not by objectivist methods, but by using one's own dreaming experience. We can understand the other because we have enough similar experience to put ourselves imaginatively into the other's experience and know what the other means.

What must be emphasized here is that although our understanding of others is not objective, neither is it relative. When a child says, "I'm going to play baseball," the meaning of the statement is not relative to each listener. It means the child is leaving with the intent of playing a certain game with certain rules. Anyone who thought that the child was going to sit on a park bench would not understand the meaning of the statement. The fact that such meanings are grasped without objective evidence indicates the existence of subjective knowledge, knowledge of meanings and motivation understood in their immediacy, what Dilthey calls *verstehen*.

LK: *Polyani later talks about tacit understanding. How do we define knowledge? Is something knowledge or known only if it can be appropriated for practice? Isn't it possible further that something can be real known and only a one time occurrence and therefore not replicable? Along this line a patient may have a truly unique experience that the analyst comes to know at least approximately and that is significant in only the one time and one place.*

FS: Knowledge is whatever is understood on the basis of principle. To the extent that one has a principle from which the particular fact is derived, to that extent the other is understood. In my view, knowledge is different from empathic resonance, which is a powerful affective experience, and what might even call it an "emotional understanding" but I'm not sure I would call it "knowledge." It is this emotional resonance that I think is at issue in those moments that are grasped uniquely, that may never happen again. This is empathy, a meeting of minds, an affective engagement that is certainly a powerful therapeutic experience but I do not believe it qualifies as knowledge. Often people are helped greatly by affective resonance but that does not mean that we should call that experience "knowledge." I would guess that what I call affective resonance might be what Polanyi means by "tacit understanding." To have knowledge of how someone operates is different from being able to engage a moment empathically, and both are necessary for the analytic process to be successful.

The subject of meaning takes us beyond the dichotomy of objectivism and relativism. As the hermeneutic philosopher Hans-Georg Gadamer (1976) has argued, meaning is neither objective, subject to observable proof, nor relative to the viewpoint of the observer. The meaning of speech inheres in the speech act, but it requires listening to be brought to completion (Gadamer 1976; Bernstein 1981). In this way, Gadamer intends to show us that meaning lies beyond the relativist-objectivist distinction. Any act of speech, like any historical document or other form of human behavior, is not interpretable in any way the observer or listener chooses, but the other is required to complete the act of meaning.

LK: *In fact, sometimes the other is necessary for the speaker to understand himself or herself. In this sense, aren't there times that intention is a field more than directed by an intrapsychic intentionality?*

FS: I think that one could argue that self-understanding can occur only through the other, but the dialectic is that the other gains impressions only through the behavior of the speaker. If the listener is off base, self-misunderstanding can occur. But, without seeing oneself through the other how can one see oneself at all? Although interpretation is limited by the meaning that inheres in the speech, there is no objective way to arrive at meaning.

LK: *So there are no rules of interpretation*?

FS: There are indeed rules of interpretation, and they allow one to arrive at subjective meaning. In fact, to grasp subjective meaning requires the application of rules of inference. All knowledge gained through the psychoanalytic process is subjective meaning; one does not need analysis for objective meaning. Rules are applied to gain subjective knowledge; without them, there is no knowledge of the other, but only moment-to-moment affective resonance.

LK: *I am not so sure of this assertion. Whenever I have seen a claim for rules as defining any kind of assurance I have seen significant problems of application and relevance of competing rules. Perhaps some clinical examples might pin this down to a degree.*

FS: The rules constitute the unique form of psychoanalytic reasoning. The analyst begins where common sense leaves off. So, when something does not make sense according to our usual reasoning process, analytic exploration begins. The psychoanalytic inference is always: Under what conditions would that be true? If the patient goes to great pains to express sorrow over the loss of a competitor and continues to assert his bad feeling, this expression of mourning does not fit, does not make sense from a conscious perspective. So, the analyst wonders why the patient is going to such lengths to avow a bad feeling. The first and most likely possibility is guilt over a feeling of pleasure. This does not mean that this conclusion should be immediately drawn; it means that this speculation is the first guess. Such an analyst is applying a psychoanalytic inference but not a conclusion because she is looking for confirmation in the patient's response. If the inquiry moves the patient to become aware of a guilty feeling of pleasure or a fear that he is feeling joyful underneath the expression of sorrow, that would be confirming evidence in the hermeneutic sense. Nonetheless, this understanding may not be final because subsequent clinical events may alter the understanding of the moment. The difference between this hermeneutic approach and relativism is the latter would claim that no truth is revealed here; and the difference from the objectivist is the latter would draw the conclusion on the basis of reasoning and consider the inquiry over.

Knowledge of meaning is subjective knowledge, but Husserl (1913) and Gadamer make the critical distinction between the subjective and the subjective-relative. Borrowing a phrase from his teacher Heidegger, Gadamer has suggested that the myth of objectivism is that we can somehow get beyond our "world horizon," the context of our experience. However, this impossibility does not mean that we cannot understand another world horizon. Understanding the other is what Gadamer calls a "fusion of horizons." Gadamer illuminated the fact that the nature of language is to be open and continually changing with new experience so that the very essence of language involves a transcendent function. It is this openness of language that allows for the fusion of horizons, the understanding of the other. Gadamer fully recognized that because the self possesses inherent multiplicity, any particular experiential moment can be viewed from different vantage points, but this fact does not render any such viewpoint relative.

LK: *Does this lead to a claim about free choice for a patient to choose among many directions and thereby create many different kinds of self? Or does this mean that the self can take different directions but still be unified in a certain continuity? To put it differently, is there a true self or is the self self-fashioning and if it is what limits on placticity? What position if any do these theorists take on the giveness of the self? Does one need a claim about the fundamental nature of the self to practice clinically? Wouldn't that model limit what could be seen?*

FS: Applying Gadamer to psychoanalysis would imply that the aim of the analytic process is to make possible free choice as to what direction the self may take within the limits of inborn capacities. Winnicott contended that there is an inborn maturational process that cannot be changed, but can either be facilitated with good enough mothering or arrested through impingement. I would say a view of this type is the most commonly accepted object relations view of self-development. This view does not mean there is only one way a self can go; it means that there are only so many possibilities for each of us, and the purpose of the analytic process is to facilitate the realization of at least some of these possibilities. One can put this in Nietzschean terms by saying that analysis helps the patient become who she is. There is a true self in that some ways of being fit our experience and affective direction, and others are performed to avoid anxiety, as in my example earlier of the patient whose compliance is driven by fear of retaliation for self-assertion.

The fact that a variety of truth statements may be made about the self at any given moment does not make knowledge of the self relative any more than the fact that many different correct statements may be made about a chair means that the perception of the chair is relative. The fact that several different interpretations accurately reflect an aspect of self-experience does not vitiate the truth of any particular interpretation.

LK: *But some interpretations are more therapeutic, aren't they? What then would constitute a therapeutic intervention? Also, how does the nature of time fit here, that is, isn't it likely that even a good interpretation will not be therapeutic if it is poorly timed?*

There is a crucial distinction between knowledge and therapeutic effectiveness. The analyst adjudicates among different interpretations that may have some validity according to the affective experience of the patient. The analyst is not looking for anything that might be true, but for what will most likely have the most impact at the moment, and the criterion for that is the accessibility of affect. A poorly timed interpretation is neither "good" nor "therapeutic;" it misses the patient's affective experience.

If we cannot find any corresponding experience of our own with which to grasp what we are being told, we are mystified, and if we wish to understand, we ask questions about the meaning in an effort to find something that we can grasp from our own experience. If our dialogical inquiry is successful in establishing this common ground, we then build by a series of inferences to arrive at understanding. This process consists of a kind of reasoning, an empathic reasoning that is the basis for psychoanalytic reasoning. As Strenger puts it, "analytic reasoning can be broken down into thought processes, which every human being performs constantly. The analyst's implicit knowledge about the appropriateness of behavior, thought, and emotion is not that far removed from the ordinary sensibilities of mature members of a culture" (1991, 93). Psychoanalytic understanding, then, is no more and no less than a highly sensitized refinement of the empathy used in everyday communication.

LK: *Is it possible that some experiences are sufficiently different as to be outside the understanding of the therapist and maybe even the patient? Is there a difference then between empathy and understanding?*

FS: I believe I answered this with my distinction between empathic resonance and knowledge. One can justifiably say that the former is a type of understanding, but it is not knowledge. So, I may be able to grasp your experience, have a feel for what you are going through, and in that sense I understand you, but not have any knowledge of why you feel that way. The difference between the analyst and everyday interaction is that the analyst uses the unique methods of psychoanalytic inquiry to ascertain motivation and meaning.

LK: *What parts cognition and what parts affect in this formulation of empathy, or do we have to have a more organic sense of cognition to include affect following Aristotle?*

FS: Empathy in the psychoanalytic sense is both affective and cognitive because either alone would not allow the analyst to grasp the patient's experience. Affect without cognition is not understanding at all, and cognition without affect is empty of the patient's experience.

The dependence of the observed on the observer so emphasized by relativists such as Hoffman has its foundation not in any paradigm of the natural sciences, but in the intentionality of consciousness (Husserl 1913/1931). Because consciousness intends an object and the world can be known only via our experience of it, neither the world nor consciousness can be conceived of without the other.

LK: *Indeed, isn't this what Heidegger means when he talks about "being-in-the world?"*

FS: This is exactly what Heidegger meant and this is where he got it. This mutual dependence does not mean that perception is "relative," but that the world and consciousness are "co-given" (Husserl 1931/1969). Indeed, Husserl who illuminated this connection, showed that our concepts of reality and fantasy, as well as all other acts of consciousness, are built into the structure of consciousness. We decide that something is real rather than fantasied when the experience fits a rational coherence.

LK: *But isn't this the problem with the delusional? They can construct a model of coherence that does seem whole and rational. This may be a bigger theoretical problem than in most clinical cases. The* Manchurian Candidate *in its newest version plays of this to some degree.*

FS: The problem with the delusional system is that the patient's "coherence" does not fit with experience in the world, both their own and others. The patient who hides in hallways because people are following him is delusional if there are no people giving chase. The patient's system relies on others behaving in certain ways, and if they do not fit the patient's coherence, the patient is delusional. The fact that we know the world only through our experience of it, rather than somehow "relativizing" our knowledge of the world, is the very structure of any such knowledge.

LK: *Not sure about the reach of what you mean by experience here. No one has experienced an atom in some sense. It is an abstraction based on perception to be sure but also on imaginative construction.*

FS: The atom is a construction based on the behavior of what can be seen; if the concept atom fits the behavior that is observed, then atom is a viable concept. So, the atom is experienced in the experiment even if it is not seen. If the concept of the atom does not accord with experience, such as an experiment, the concept must be modified or the experience questioned. So, experience is the criterion; when experiments do not conform to expectation, experience is turning up some new finding.

LK: *How does the therapist deal with countertransference issues and also with ego ideal about reinforcing or negating a patient's field of possibility? How are constraints set and when? How does the therapist differentiate from acting out and authenticate action? Is there such an animal as authenticate action?*

FS: The therapist's responsibility is to be self-aware. It is assumed in virtually every school of psychoanalysis that reactions to the patient are inevitable, perhaps ubiquitous. Countertransference consists of those reactions that interfered with the therapist's ability to maintain an observational stance. When the therapist finds herself losing the ability to look at what is going on in the analytic relationship, she notes a countertransference reaction that must be subject to self-scrutiny. No such dictum is foolproof, of course. The analyst may have blind spots influencing her understanding. The check is to find the evidence: Does the interpretation fit what is known? Any time it does not something is amiss. The therapist's job is to scrutinize herself in those instances to look for a countertransference problem that may be influencing her judgment, but of course that is only one possibility. The bottom line is that the subsequent material, the patient's affective response and the development of the analyst's associations and behavior, provide the best criterion for the validity of the interpretation. The analyst who keeps running into the patient's experience as a continued obstacle is called on to reevaluate her understanding. But, of course, if the analyst is blind to the evidence, the treatment will suffer and probably fall to ruin. Authenticity does exist, but defining it is difficult. It is probably best thought of as action motivated by affects and not anxiety. The degree of anxiety motivation is the extent of defensive protection as opposed to authentic experience.

LK: *From Weber on we have recognized the gap between theory and practice. One way to understand deconstruction is that all theory is not only flawed but maybe not relevant to everyday practice. The only aspect that makes a practice valid is pragmatic on this view. This is Rorty, following Dewey inter alia. So what does theory mean for clinical practice?*

FS: In psychoanalysis, theory does affect how the therapist listens, what is highlighted, noted, and how the material is formulated, but what the analyst hears can also change theory. A good example of this dialectic is Kohut's case of Miss F, the case attributed to his theoretical change from classical to self psychology. His classical approach was not working, so he abandoned his drive/defense assumptions and allowed Miss F to control the evolution of the process. Once he changed theories, he listened to the same material differently. The question is not whether theory is relevant; it is unavoidable. Which theory the analyst uses defines clinical posture. To give one example: if the patient idealizes the analyst, the self psychologist does not interpret nor attempt to affect this transference, but Kernberg or any Kleinian would interpret the idealization as a defense.

As a science of understanding the other, psychoanalysis uses the method that befits its object of investigation. Thus, Kohut's definition of the psychoanalytic method as one of empathy and introspection does not make psychoanalysis a "soft science" of imprecise methods as charged by objectivists, such as Holzman (1985), but defines the very nature of the science of the other and fits the method to the subject of investigation.

LK: *Again, what is at stake in saying that psychoanalysis is not science or soft science? Does this mean that if it were true all practitioners would be malpracticing?*

FS: If psychoanalysis is a "soft science," its claim to knowledge is vitiated. One may say then that the analyst is a writer of fiction, a novelist, who may spin fascinating tales, but has no claim to knowledge of people. The argument is hotly contested precisely because analysts are fighting to legitimate their truth claims. If psychoanalytic truth claims are not regarded as science, but a domain of knowledge, then "science" does not matter. Whether "soft science" constitutes malpractice is a question for lawyers, but I think psychoanalysts are more concerned with being regarded as having knowledge-based practices, rather than people who may have good individual intuition with little knowledge base.

LK: *We are here back on ancient terrain. Plato wanted the poets expelled from the Republic even though he himself was very much a poet. The truth telling of the poet, dramatist, and novelist may well capture reality in its manifold expression at least as reliably as the scientist.*

FS: When we ascribe a motive or meaning to the other, we judge its validity by what it explains, how it fits, the other's actions and behavior. The subject matter of psychoanalysis is precisely the meaning of experience. It follows that coherence is the criterion for interpretive accuracy, and interpretations that do not fit the self experience are inaccurate, just as in ordinary discourse to call a generous person "selfish" does not fit and is, therefore, inaccurate.

For example, the analyst who hypothesizes that her patient sabotages her successes out of excessive guilt due to an oedipal victory derives from the patient's experience of mother, father, and guilt, an explanatory principle by which she hopes to understand the meaning of success and positive experience, their connection with guilt, and related phenomena. The interpretation is accurate to the extent that it makes intelligible the patient's pattern of ruining successful experiences. If analytic material arises that does not fit this explanation, it must be abandoned or modified.

The basis of the principle of coherence is that rational fitting together is the criterion for reality (Husserl 1913/1931). *Narrative* coherence is an insufficient criterion for interpretive validity, as we saw earlier. An interpretation does not simply make a story; it makes intelligible that which had been unexplained. The interpretation that best fits is preferred due to its approximation to the reality of the patient's experience and ability to articulate the previously nonverbalized aspects of it. Other interpretations are always possible, but their validity must be argued on the basis of their ability to account for the unintelligible.

LK: *This may be a prejudice as well, perhaps a fair prejudice, but it may well be and likely sometimes is that the particular problem, symptom, etc., is unintelligible and only the authority of the analyst and the good will of the analyst makes for apparent acceptance.*

FS: It may sometimes be the case that the authority of the analyst carries the day, but that is not legitimate psychoanalysis; it is poorly practiced analysis. There is a legitimacy to analysis to the degree that interpretations are open to argument on the basis of preestablished criteria. This is not to say that one must assume the analyst has some "ultimate truth" about the meaning of the symptom. To make something intelligible means to make sense of it, to bring to light some degree of its meaning. The criteria for that judgment are coherence and the amount of material that can now be understood, that is to say, not authority, but evidence and reason.

Objectivists see the importance of criteria for truth, but they have the wrong ones. Relativists have the right criteria, but do not see that their standards for interpretive validity are transcendent, defining reality and contradicting the relativist position. The human science model adopts the relativist's way of understanding analytic material, but recognizes these principles as criteria for truth. Interpretations of motives are valid to the extent that they can account for actions, that is, the extent to which behavior and actions fit the motive.

LK: *What if motives are inchoate, or multiplied determined, confused, etc.?*

FS: Motives often are, and if they remain that way, analysis has not brought anything to light. If, however, an interpretation clarifies motives, it has served a purpose. Motives cannot always be clarified.

LK: *What if there are just not appropriate or any necessary psychological structures developed by the patient or arrested, etc.? May not the criterion of coherence be limiting and itself wrong?*

FS: There are, of course, many patients with severe character pathology who have very limited and fragile psychological organization; that does not mean they cannot be understood. There are other patients with no psychological organization at all, such people are chaotic and disorganized, but the treatment is not to understand them, but to help them create a structure; interpretation does not apply. Whether one calls such intervention psychoanalytic therapy is another matter. The subject of this chapter is psychoanalytic knowledge, not every form of intervention one might make. Insofar as interpretation is the intervention, the criterion of coherence applies.

This human science criterion does not mean that psychoanalysis can avoid the responsibility of demonstrating its effectiveness compared to other therapeutic methods. However, any such investigation must be kept clearly separate from the validity of: (1) psychoanalysis as an interpretive science; and (2) any particular interpretation.

Psychoanalysis adds to the ordinary experiencing of others the finding that others' motives, although in principle understandable, are often hidden from view. The uncovering of meaning and motives outside of awareness, but accessible in principle, is the unique subject of psychoanalytic investigation. To achieve this end, psychoanalysis uses a specialized method of depth exploration: free association, dream interpretation, and transference. Thus, the science of psychoanalysis may be defined as the systematic, depth investigation of previously unknown meaning and motivation in the other resulting from a unique method of inquiry. These uniquely psychoanalytic tools are technical refinements of ordinary empathy and introspection.

The human science model provides the means for psychoanalysis to pursue its unique search for truth freed from blind faith in objective truth, a faith that has fostered a reductionistic self-misunderstanding and created obstacles to analytic progress. The abandonment of this faith does not damage the analyst's claim to knowledge, as implied in both relativism and the objectivism; rather, it allows the analyst to appreciate the uniqueness of the understanding of others and to conduct psychoanalytic inquiry with criteria befitting its object of investigation. Only by recognizing the distinctive character of human science can the rich uniqueness of psychoanalytic understanding be appreciated and the full potential of psychoanalytic science be realized.

REFERENCES

Ayer, A. J. 1946. *Language, Truth, and Logic*. London: Gollancz.

Bachrach, H. 1989. "On Specifying the Scientific Methodology of Psychoanalysis." *Psychoanalytic Inquiry* 9:282–303.

Bernstein, R. J. 1981. *Beyond Objectivism and Relativism*. Philadelphia: University of Pennsylvania Press.

Blight, J. G. 1981. "Must Psychoanalysis Retreat to Hermeneutics? Psychoanalytic Theory in the Light of Popper's Evolutionary Epistemology." *Psychoanalysis and Contemporary Thought* 4:147–205.

Dilthey, W. 1979. *Introduction to the Human Sciences: An Attempt to Lay a Foundation for the Study of Society and History*. Detroit: Wayne State Press. (Orig. pub. 1883.)

Eagle, M. N. 1980. "A Critical Examination of Motivational Explanation in Psychoanalysis." *Psychoanalysis and Contemporary Thought* 3:324–380.

Gadamer, H-G. 1976. *Philosophical Hermeneutics*. Berkeley: University of California Press.

Grunbaum, A. 1984. *The Foundations of Psychoanalysis: A Philosophical Critique*. Berkeley: University of California Press.

Hoffman, I. Z. 1991. "Discussion: Toward a Social Constructivist View of the Psychoanalytic Situation." *Psychoanalytic Dialogues* 1:74–103.
———. 1992. "Some Practical Implications of a Social-Constructivist View of the Psychoanalytic Situation." *Psychoanalytic Dialogues* 2:287–304.
———. 1998. *Ritual and Spontaneity in the Psychoanalytic Process: A Dialectical-Constructivist View*. Hillsdale, NJ: The Analytic Press.
Holt, R. 1972. "Freud's Mechanistic and Humanistic Images of Man." *Psychoanalysis and Contemporary Science* 1:3–24.
Holzman, P. 1985. "Psychoanalysis: Is the Therapy Destroying the Science?" *Journal of the American Psychoanalytic Association* 33:725–70.
Husserl, E. 1931. *Ideas. General Introduction to Pure Phenomenology*. Trans. W. R. Boyce Gibson. London: Allen and Unwen. (Orig. pub. 1913.)
———. 1969. *Cartesian Meditations: An Introduction to Phenomenology*. Trans. D. Cairns. Den Haag: Martinus Nijhoff. (Orig. pub. 1931.)
———. 1970. *The Crisis of European Sciences and Transcendental Phenomenology*. Trans. D. Carr. Evanston, IL: Northwestern University Press. (Orig. pub. 1936.)
———. 1977. *Phenomenological Psychology*. Trans. J. Scanlon. The Hague: Martinus Nijhoff.
Kohut, H. 1959. "Introspection, Empathy, and Psychoanalysis: An Examination of the Relationship between Mode of Observation and Theory." *Journal of the American Psychoanalytic Association* 7:459–83.
———. 1983. *How Does Analysis Cure?* Chicago: The University of Chicago Press.
Mitchell, S. 1988. *Relational Concepts in Psychoanalysis*. Cambridge, MA: Harvard University Press.
———. 1993. *Hope and Dread in Psychoanalysis*. New York: Basic Books.
Moore, R. 1999. *The Creation of Reality in Psychoanalysis: A View of the Contributions of Donald Spence, Roy Schafer, Robert Stolorow, Irwin Z. Hoffman and Beyond*. Hillsdale, NJ: The Analytic Press.
Nagel, E. 1964. "Methodological Issues in Psychoanalytic Theory." In *Psychoanalysis, Scientific Method, and Philosophy*, ed. S. Hook. New York: New York University Press. (Orig. pub. 1959.)
Nissam-Sabat, M. 1989. "Kohut and Husserl: the Empathic Bond." In *Self Psychology: Comparisons and Contrasts*, ed. D. Detrick and D. W. Detrick. Hillsdale, NJ: The Analytic Press, 151–73.
———. 1991. "The Crisis in Psychoanalysis: Resolution through Husserlian Phenomenology and Feminism." In *Human Studies* 14:33–66.
Orange, D. 1992. "Commentary on Irwin Hoffman's 'Discussion: Toward a Social-Constructivist View of the Psychoanalytic Situation.'" *Psychoanalytic Dialogues* 2:561–66.
Popper, K. 1962. *Conjectures and Refutations: The Growth of Scientific Knowledge*. New York: Basic Books.
Ricouer, P. 1970. *Freud and Philosophy: An Essay on Interpretation*. Trans. D. Savage. New Haven: Yale University Press.
———. 1986. "The Self in Psychoanalysis and in Phenomenological philosophy." Trans. D. Pellauer. *Psychoanalytic Inquiry* 6: 437–58.
Rorty, R. 1980. "Pragmatism, Relativism, and Irrationalism." *Proceedings and Addresses of the American Philosophical Association* 53: 719–38.

———. 1981. "Nineteenth Century Idealism and Twentieth Century Textualism." *Monist* 64:155–74.

Rubenstein, B. 1976. "On the Possibility of a Strictly Clinical Psychoanalytic Theory: An essay on the Philosophy of Psychoanalysis." In *Psychology Versus Metapsychology: Psychoanalytic Essays in Memory of George Klein*, ed. P. S. Holzman and M. M. Gill. New York: International Universities Press.

———. 1980. "The Problem of Confirmation in Clinical Psychoanalysis." *Journal of the American Psychoanalytic Association* 28 397–417.

Schafer, R. 1983. *The Analytic Attitude*. New York: Basic Books.

———. 1992. *Retelling a Life*. New York: Basic Books.

Scriven, M. 1959. "The Experimental Investigation of Psychoanalysis." In *Psychoanalysis, Scientifice Method, and Philosophy*, ed. S. Hook. New York: New York University Press.

Spence, D. P. 1982. *Narrative Truth versus Historical Truth: Meaning and Interpretation in Psychoanalysis*. New York: W. W. Norton & Co.

———. 1993. "The Hermeneutic Turn: Soft Science or Loyal Opposition?" *Psychoanalytic Dialogues* 3:1–10.

———. 1994. *The Rhetorical Voice of Psychoanalysis: Displacement of evidence by theory*. Cambridge, MA: Harvard University Press.

Stolorow, R. D., B. Brandchaft, and G. Atwood. 1987. *Psychoanalytic Treatment: An Intersubjective Approach*. Hillsdale, NJ: The Analytic Press.

Strenger, C. 1991. *Between Hermeneutics and Science*: An Essay on the Epistemology of Psychoanalysis. *Psychological Issues*, Monograph 39. Madison, CT: International Universities Press.

Toulmin, S. 1986. "Self Psychology as a Postmodern Science." *Psychoanalytic Inquiry* 6:459–77.

Waelder, R. 1962. "Psychoanalysis, Scientific Method, and Philosophy." *Journal of the American Psychoanalytic Association* 10:617–37.

Wallerstein, R. 1986. "Psychoanalysis as science: A response to the New Challenge." *Psychoanalytic Quarterly* 55:414-51.

———. 1988. "Psychoanalysis, Psychoanalytic Science, and Psychoanalytic Research." *Journal of the American Psychoanalyic Association* 36:3-31.

II

COMPLEX, CREATIVE, AND SITUATED INTERROGATIONS: SCIENCE IN ACTION

9

Science and Human Well-Being

Toward a New Way of Structuring Scientific Activity

Hugh Lacey

INTRODUCTION

Boaventura de Sousa Santos begins *A Discourse on the Sciences* by posing Rousseau's question: "Is there a relationship between science and virtue?" (Santos 1992, 10). He elaborates: "we must ask about the contribution of all the accumulated scientific knowledge to enriching or impoverishing our lives. In other words, we must ask if science has contributed positively or negatively to our happiness" (Santos 1992, 11–12).

In this chapter, I address the related question: "How should science be pursued in order to enhance human well-being?" Variations of this question have been suggested recently by philosophers of science. Here are two of them:

> the most important topic in the philosophy of science [is] the relation of science to human values. What contributions can or should science make to human well-being? . . . If there is one conclusion of overriding importance to be drawn from the increasing realization in recent times that science is a human product, it is that, like other human products, the only way it can ultimately be evaluated is in terms of whether it contributes to the thriving of sentient beings in the universe. (Dupré 1993, 244, 264)

> Reflective people . . . want to know whether research in various areas is skewed by the values of particular groups and, at the broadest level, how science bears on human flourishing. . . . It has been obvious for about half a century that research yielding epistemic benefits may have damaging consequences for the individual or even the whole species. Philosophical stories about science have been narrowly focused on the epistemic. Faced with lines of research that have

> the capacity to alter the environment in radical ways, to transform our self-understanding, and to interact with a variety of social institutions and social prejudices to affect human lives, there is a much larger problem of understanding just how the sciences bear on human flourishing. (Kitcher 1998, 46)[1]

Elsewhere (Lacey 2004a, xi), with these quotations in mind, I have affirmed that there are rich dialectical interactions among the questions: "How to conduct scientific research?" "How to structure society?" and "How to further human well-being?" Science must be appraised for the cognitive (epistemic) value of its theoretical products and—when positively appraised cognitively—it can (and should) also be appraised for its contribution to social justice and human well-being. In this chapter I briefly explain why this is so.

SCIENCE AS VALUE FREE

Throughout its 400-year history, there has been a strong tradition of the interpretation of modern science that maintains that the question, "How should science be pursued in order to enhance human well-being?" is inappropriate; even posing it (or Rousseau's question) displays misunderstanding of the nature of science and threatens its integrity.

Impartiality, Autonomy, Neutrality

According to this interpretation, there is no role for values[2] at the core of science: they should not be among the criteria for making scientific judgments, and they have no fundamental role in the practices of gaining and appraising scientific understanding: the broad characteristics of scientific methodology should be responsive only to the interest of gaining deepened understanding of phenomena of the world, and the priorities and direction of research should not be shaped systematically by particular values. I call these two views, respectively, "*impartiality* of scientific judgment" and "*autonomy* of methodology."[3] They are held as ideals, values of scientific practices, sometimes not well manifested in actual fact.

Usually they are held together with *neutrality*, that science gives privilege no particular social values—that scientific theories should exhibit *cognitive neutrality*: social value judgments are not among their logical implications; and *applied neutrality*: on application, in principle they can inform evenhandedly interests of a wide range of values. To a first approximation, these three views together sum up what is meant by "*science is value-free*."[4] The integrity, legitimation, prestige and alleged universal value of science have often been tied to *science is value free* being highly manifested in the practices of science for, it tends to be taken for granted, it is science conducted in such practices that has produced the technological applications that have so transformed the world in recent times.

Three Moments of Scientific Activity

We may distinguish three moments (logically, not necessarily temporally distinct), or decision-making points, of scientific activity (Lacey 2004b): M_1, where priorities and direction for research and appropriate methodologies are determined; M_2, where theories (hypotheses) are appraised; and M_3, where scientific knowledge is applied. On the traditional interpretation, M_1 and M_2 are the core of science. M_3 follows them and presupposes positive appraisal at M_2 of the theories to be applied. At M_3 values have not only a legitimate but also an indispensable role; but at M_1 and M_2 there is no proper (or, at least, indispensable) role for them. As Baptista (2002, 96)—an exemplary proponent of the view of science being sketched in this section—wrote in criticising Santos (1992): "How could scientific knowledge have any moral coloring?" Even at M_3, according to *applied neutrality*, science does not serve any particular values; as Baptista (2002, 96) observes: "The results of science are ethically and morally neutral." Scientific knowledge is available, in principle, to inform projects of interest to any values at all; it may be used for good or evil, but when it is used for evil that reflects not on scientific knowledge but on those who applied it. What happens at the core of science cannot be held responsible for abusive uses to which scientific knowledge may be put. Nevertheless, those who share the traditional interpretation insist that scientific knowledge has normally been, and will continue to be, used for a great variety of good ends, because it has been the source of so many "new possibilities and potentialities . . . that have profoundly changed the conditions and even the quality of life of a large part of humanities" (Baptista 2002, 83, 95; my translation). It is a small step from these assertions to the claim that science itself should be considered a universal value.

I recognize the attraction of this interpretation of science and I do not think that it is wholly mistaken. In order to get at its limitations and, at the same time, to begin to explain why it has dominated the tradition of modern science, I will offer an alternative picture of scientific activity.[5]

SCIENTIFIC ACTIVITY

Scientific Understanding

The *objective of science* may be put as follows: To gain, in systematic ways, soundly (rationally) empirically grounded *understanding* of phenomena and things (and of an increasing number and variety of them), where understanding includes accounts of *what* the phenomena (things) are, of *why* that are the way they are, of the *possibilities* (including hitherto unactualized ones) that they allow in virtue of their own underlying powers and the interactions into which they may enter, and of *how* to attempt to actualize these possibilities (Lacey 1999a, ch. 5).[6]

Scientific understanding is articulated in *theories*, each of which offers understanding of a specified domain of phenomena, where the soundness of the understanding articulated should be appraised in the light of available empirical data, of whether the available data suffices for making reliable cognitive judgments, and of cognitive criteria such as empirical adequacy and explanatory power that are rational symptoms of having gained understanding and that do not depend on any value judgments (Lacey 2004a, ch. 3). When appraisals are soundly made solely in accordance with these considerations, I say that a theory (hypothesis) has been accepted in accordance with *impartiality*. The objective of science, then, might be restated as: To generate and consolidate theories, accepted in accordance with *impartiality*, that progressively provide understanding of more and larger domains of phenomena and possibilities.

Impartiality is built into the quest for scientific understanding. *Neutrality* is not—neither *cognitive* nor *applied neutrality* (section 1.1). *Impartiality* does not imply *neutrality* (Lacey 2002a; 2004a, chs. 4, 10). When a theory is accepted in accordance with *impartiality*, its acceptance contributes towards furthering the objective of science, but that does not mean that it makes a *significant* contribution. Reference to the objective (as stated above) does not provide direction to research, define what counts as worthwhile or *significant* research (Kitcher 2001), or provide concrete answers to: What questions to pose, what puzzles to resolve, what classes of possibilities to attempt to identify, what kinds of explanations to explore, what categories to deploy both in theories (hypotheses) and observational reports, what phenomena to observe, measure and experiment upon, who are the appropriate participants in research activity, what are their required qualifications, life backgrounds and virtues?

Strategy: Constraints on Theories, Criteria for Selection of Data; What Possibilities to Investigate

None of questions, just posed, can be addressed without the adoption of what I call a *strategy* (Lacey 2002a; 2004a, ch. 5). The principal roles of a strategy are to *constrain* the kinds of theories (hypotheses) that may be entertained in a given domain of inquiry (so as to *enable* investigation) and the categories that they may deploy—and thus *to specify the kinds of possibilities that may be explored in the course of the inquiry*—and to *select* the relevant kinds of empirical data to procure and the appropriate descriptive categories to use for making observational reports.[7]

Different classes of possibilities may require different strategies for their investigation. To borrow an example that I have used elsewhere (Lacey 2000; 2002c; 2003a; 2003b), the possibilities of seeds *qua* components of sustainable agroecosystems are not identical to their possibilities *qua* objects of genetic engineering; both cannot be investigated under the same strategies and both cannot be coactualized in the same agricultural spaces. A multiplicity of competing strategies is required, if research is not to be limited in compass to a narrow (albeit potentially large) class of possibilities.

The scientific knowledge that informs genetic engineering is a product of research conducted under a version (biotechnological) of what I call *materialist strategies* (Lacey 2002a; 2004a, ch. 6; 2004b). Strategies of this kind are adopted *almost exclusively* in modern science. Under them theories are constrained so that they represent phenomena and encapsulate possibilities in terms of their being generable from *underlying structure, process, interaction and law*, dissociating them from any place they may have in relation to social arrangements, human lives and experience, from any link with value, and from whatever social, human and ecological possibilities that may also be open to them. Reciprocally, empirical data are selected, not only to meet the condition of intersubjectivity, but also so that their descriptive categories are generally quantitative, applicable in virtue of measurement, instrumental and experimental operations. Under materialist strategies, intentional and value-laden categories are deliberately excluded from use in the formulation of theories, hypotheses and data, so that where they are deployed there cannot be any value judgments among the formal entailments of theories and hypotheses. Adopting them thus suffices to ensure *cognitive neutrality*. This is a feature of their design.

Adopting materialist strategies does not suffice, however, to ensure *applied neutrality* (Lacey 2001; 2004a, ch. 8–10; see quotes from Dupré and Kitcher at the outset.). The example about seeds provides a useful illustration. When the biotechnological version of materialist strategies is adopted, the possibilities for the transformation of seeds by techniques of genetic engineering can be identified; but their possibilities, when selected for use in productive, sustainable agroecosystems in which biodiversity is protected and community empowerment is furthered, cannot be, for these latter possibilities are not dissociated from social and experiential context. They can, however, be identified by research conducted under *agroecological strategies* (Lacey 2001; 2003a; 2003b).[8]

Mutually Reinforcing Relations between Adopting a Strategy and Holding Values

Why, then, in the institutions of modern science do materialist strategies tend to be adopted virtually exclusively so that, for example, research under biotechnological strategies is strongly supported and that under agroecological strategies downplayed? Why, in these institutions, is it so difficult for research, other than that conducted under versions of materialist strategies, even to be recognized as science? Why is the interpretation of science described in section 1 so dominant in these institutions that normally only choice among versions of materialist strategies is recognized as belonging to the methodological component of M_1? The reasons cannot always be purely cognitive (epistemic), for the interest of gaining understanding (knowledge) by itself does not account for one kind of possibilities of seeds (either their possibilities *qua* dissociated from social and other contexts, or those *qua* ecological and/or social objects) being considered more significant than another.

Not only has adopting materialist strategies enabled research that has been extraordinarily fruitful, but also it underlies great adaptability, ushering in new varieties of materialist strategies as research has unfolded—so that, at various historical moments, when opportune we find strategies that deploy (e.g.) mechanical categories, mathematically formulated laws of various types (deterministic and probabilistic, compatible or not with physicalistic reductionism), use of the mathematics of complexity and computer modelling (see section 2.5). The fruitfulness and resourcefulness of materialist strategies are undoubtedly impressive and rightly establish a central position for them in scientific institutions. But, by itself, this does not warrant exclusivity. What else is involved?

Elsewhere (Lacey 2002a; 2004b, ch. 6) I have argued in detail that mutually reinforcing relationships between adopting materialist strategies and holding *specifically modern* values concerning the control of natural objects are crucial.[9] To give a summary statement, these values concern the scope of control, its centrality in daily life, and that relatively they are not subordinated to other moral and social values—so that, for example, the expansion of technologies into more and more spheres of life and into becoming the means for solving more and more problems is especially highly valued, to such an extent that the kind of ecological and social disruption caused by this can be seen simply as the price of progress.[10] It is acceptance of these specifically modern values about control of natural objects, and their widespread manifestation in actual hegemonic institutions, not cognitive factors alone, that account, for example, for the general perception that research conducted under biotechnological strategies is more significant than that conducted under agroecological ones (Lacey 2000; 2001). In line with this, I have maintained that materialist strategies are adopted virtually exclusively in the institutions of modern science *in important part* because, on application, knowledge gained under them serves interests strongly linked with these values about control. These interests conflict with those of the emancipatory movements (like those that have been meeting at the World Social Forum), among whom it is common to value highly (among other things) agroecological practices that aim simultaneously to further productivity, sustainability, preservation of biodiversity and social empowerment of poor communities (Lacey 2002b; 2003b; Santos 2002). Knowledge gained under materialist strategies serves these latter interests to a much lesser degree (for, by definition, these strategies have nothing to contribute directly to identifying possibilities that are not defined in dissociation from human and social contexts); and, in some cases, it contributes to undermine them.

On application, research conducted under materialist strategies (not in each and every instance, but on the whole when pursued virtually exclusively) especially favors interests linked with the modern values about control (Lacey 2004a, ch. 10). Moreover, I have argued in neo-Baconian vein (Lacey 2004a, ch. 6), that is why materialist strategies are virtually exclusively adopted: the deep embodiment of the values about control in modern hegemonic institutions—where they are reinforced, on the one hand, by the values of capital and the market and, on the other hand, by those of the military—*explains* this exclusivity, and the most compelling *argument* for the exclusivity draws upon (in conjunction with the fruitfulness and resourcefulness of materialist strategies) mutually reinforcing relations that exist between adopting materialist strategies and holding the values about control.[11]

It is the deep grip of these values about control on modern sensibilities that renders the interpretation of science (summarized in section 1) intelligible—and for many irresistible. These values pervade that interpretation so thoroughly, and it tends to be taken for granted that they represent universal values, that their role becomes invisible to the proponents of the interpretation. Then, indeed, there is no question of choosing between some version of materialist strategies and other strategies The methodological strictures of materialist strategies become casually identified with those of science *per se* ("the only explanation of reality," Baptista 2002, 94). Then, applicability that is *neutral among projects valued partly because they embody the values about control* (whether the projects be those of capital, the military or those of medicine—see Santos (1989, 149), but that otherwise are valued differently depending on different personal and institutional values, is misidentified with *applied neutrality* in general.

Where the specifically modern values about control are held, there is heightened interest in the possibilities of things *qua* dissociated from their social and human context, that is, the possibilities that become actualized in modern technologies (e.g., genetic engineering); so much so that, when materialist strategies are adopted virtually exclusively, few barriers are posed to inferring *ceteris paribus* from the confirmation of a theory that identifies a particular technological possibility to the legitimacy of the social actualization of the possibility (Lacey 2004b). The inference, of course, is mediated by positive value judgments about control; and these, I have argued, are effectively invisible where the interpretation of section 1 is dominant. Then, scientific resources cannot easily be brought to bear decisively on the *ceteris paribus* condition (just mentioned), for its satisfaction includes confirming such hypotheses as that there are no unacceptable side effects of the technological implementation, and no better means to produce the goods it is intended to produce. Hypotheses like these cannot be addressed exclusively under materialist strategies; these are matters on which "science" (as understood in the predominant institutions of modern science) is largely silent, but where the opinions of scientists can forever be defended by claiming that there is no "scientific" evidence against them. This is clearly illustrated in the case of genetically engineered seeds: what are the side effects of their widespread use on the environment and social arrangements, and are they better than the agroecological alternatives (Lacey 2000; 2002c; 2003a; 2003b; 2004b)?

In the absence of research conducted under alternative strategies (e.g., agroecological ones), positive assessments of hypotheses about side effects and alternative cannot be made in accordance with *impartiality.* Instead such assessments tend to reflect the value judgments linked with the adoption of the strategy. This leaves untouched that the judgments made about the (potential) *efficacy* of the technology may have been made in accordance with *impartiality.* Research conducted exclusively under materialist strategies can generate hypotheses about efficacy that are accepted in accordance with *impartiality,* but generally not relevant hypotheses that are presuppositions of legitimacy of application. That values may play an indispensable role in the decision to adopt a strategy does not undermine that, in research conducted under the strategy, theories may be accepted in accordance with *impartiality* (cf. Proctor 1991). The lack of *applied neutrality,* which comes with this role of values, is consistent with *impartiality.*

Three Moments of Scientific Activity; Possible Roles for Values at Them

To put some clear order and generality into these remarks consider again the three moments that I introduced in section 1. On the model that I have been sketching, values have legitimate and often indispensable roles at M_1 and M_3, but *no indispensable role at* M_2. At M_3, no one disputes that an application is made because it is intended to serve specific interests, and thus to further the manifestation of specific values, and that judgments of its legitimation depend upon a multiplicity of value judgments. On my model, the key methodological decision made at M_1 is about the adoption of a strategy; and a strategy may (but not always) be adopted—subject, in the long run, to research conducted under it being fruitful in generating theories that become accepted at M_2 in accordance with *impartiality*—in view of mutually reinforcing relations between adopting it and holding certain values, and the interest in furthering those values. Adopting a strategy defines the kinds of possibilities that may be identified in research (in important cases), possibilities that, if identified and actualized, would serve interests cultivated by the values linked with adopting the strategy. Adopting a strategy *per se* does not imply that possibilities of these kinds exist and, if so, concretely what they are; such matters can only be settled at M_2 where *impartiality* is the reigning aspiration.

Theories that are accepted in accordance with *impartiality* at M_2, given that they arise under a strategy, will offer understanding of domains of phenomena and possibilities of limited scope, whose limits can often be clarified by attention to the values implicated in adopting a strategy. Precisely because of this, *applied neutrality* cannot generally be counted on to hold (although there are plenty of well-known instances where a result, confirmed under materialist strategies, serves a great variety of interests). On application, at M_3, theories will tend to serve especially well the values linked with the strategy under which they were developed and confirmed. Thus, the only way in which the activity of science, as a whole, can aspire to *applied neutrality* is to require that research be conducted under a multiplicity of strategies. Then, provided that a sufficient variety is actually well developed, we could expect that there would be scientific knowledge available to inform projects valued from different value-outlooks.

According to my model there is no objection, in principle, to the active development of a multiplicity of strategies, even where each particular strategy has mutually reinforcing relations with a particular value-outlook. This achieves the "dedogmatization of science" that Santos (1989, 27) is striving for and shows that *autonomy* of methodology (section 1.1) cannot be sustained. Furthermore, a significant measure of *applied neutrality* can arise from methodological pluralism (Longino 1990; 2002), while it cannot from the methodological monism that ensures *cognitive neutrality*. Moreover, returning to the agricultural example, key presuppositions of the legitimacy of the use of genetically modified seeds (no adverse side effects, no better alternatives) can only be investigated empirically if multiple strategies are in play (Lacey 2000; 2001; 2002c; 2003a; 2003b). While I do not doubt that attempting to further the ideal of *applied neutrality* will generate difficulties and tension in scientific practice (Lacey 2004a, ch. 10), it is integral to the alternative way of structuring science that my model underlies. It permits us, in relevant fields, to develop multiple strategies with a clear awareness of the way in which a strategy may have links with particular values—so that, in the first place, the values will not play an unrecognized role in the acceptance and rejection of theories; secondly, controversies about values may become part of the discussions of the worldwide community of scientists, and scientists will have the freedom to opt for approaches that will enable them to identify possibilities that may serve interests like those of the movements of the World Social Forum; and thirdly, science will not be not excluded from the domain of democratic discussion.

Comment on the Controversy about A Discourse on the Sciences

Santos (1992, 13) speaks of "a global model of scientific rationality, with some internal variation." I think that commitment to materialist strategies best defines this model since, I have claimed, normally in the institutions of modern science, science, *qua* rational activity, is said to be characterized by the features that I attribute to research conducted under materialist strategies. In these institutions knowledge gained in accordance with this model normally excludes other forms of knowledge from serious consideration and considers them to be "less than scientific." This captures much of what Santos says about the "global model." At times, however, he appears to identify the model with the forms of materialist strategies present in mechanistic conceptions or in classical mechanics, for he considers that the transition from them to the strategies deployed in relativity theory and quantum mechanics begins to break with the "global model" (Santos 1992, 22). In contrast, I interpret these as transitions to more sophisticated versions of materialist strategies, resourceful developments and strengthening of the model that are rationalized by empirical and other cognitive factors, without any special input from values and without bringing into theories, generated under them, categories (like mental or holistic ones) that have no place within materialist strategies (see also Freire 2005). The relationship that I have identified between adopting materialist strategies and and holding the specifically modern values about control remains constant across such transitions.

Santos foresees and wishes to hasten the transition from this "global model" to an "emerging paradigm"—I prefer to anticipate a "new way of structuring of scientific activity" (see section 3 below)—one (as expressed insightfully in this book's subtitle) aiming to generate "prudent knowledges for a decent life." He appears, as I interpret his text within my framework, to discern two necessary, interconnected components in this transition: (1) the gradual internal dissolution of materialist strategies, which he thinks is already well under way, and (2) the recognition of a multiplicity of forms of knowledge and its acquisition, which are reflective of cultural diversity and the needs and interests of emancipatory movements.

I do not expect that (1) will happen; I do not desire it in the physical sciences, and I am not swayed by the arguments Santos proposes for it. Where he sees the "global model" as in internally generated decline, I, especially when I reflect on the character of the strategies that frame relativity theory and quantum mechanics and the enormous growth of the cognitive and brain sciences, see its strength and its expansionary tendencies left unchecked. Proposing (1) has been interpreted as rhetorical and obscurantist (Baptista 2002); and Santos' own label for his desired form of science, "postmodern," appears to add an element of credence to making the same charge against (2), for so-called postmodernist criticisms of science have often defended (or been caricatured as defending) the view that scientific theories are only open to sociological explanation and not to cognitive appraisal. This is irrelevant, however, in the present context. Those "postmodernist" criticisms of science are not made by people close to emancipatory practices, as is Santos. Emancipatory practices cannot afford the luxury of drawing upon forms of understanding that do not meet the most rigorous cognitive standards. (As indicated at the outset, all defensible forms of appraisal of scientific theories, including that of a theory's potential relevance on application to emancipatory interests, presuppose their positive cognitive appraisal.)

The important point is that the intelligibility of (2) and movement towards it are quite independent of (1)—and, in other works of Santos (1989; 1999), (2)'s supposed link with (1) is not so prominent (even if not totally abandoned). In my opinion, research conducted under materialist strategies (preserving its own internal integrity) should remain important, within the kind of restructuring of scientific activity that I will attempt to articulate, as one approach to gaining scientific knowledge among others. What degree of prominence it has should be a matter for ongoing negotiation within strong democratic institutions. This, I believe, coheres with the main lines of Santos' argument.

A NEW WAY OF STRUCTURING SCIENTIFIC ACTIVITY

To pose the question, "How should science be pursued in order to enhance human well-being?" or to pose Rousseau's question, does not show misunderstanding of the nature of science or threaten its integrity. Quite the contrary! Not to pose it is to leave unquestioned the ongoing practice of science, which is predominantly shaped by mutually reinforcing relationships with the specifically modern values about the control of natural objects and the values (connected with capital and the military) that, in turn, reinforce them.[12] This does not fit with the ideals of scientific practice.

In view of the model of scientific activity presented in section 2, we can now restate the question: For the sake of enhancing human well-being, what range of strategies needs to be pursued in scientific research?[13] (We might add: Who should be the participants in the research at each of the key moments of decision making, and what kinds of scientific institutions are needed to pursue it effectively?) Conducting science in response to these questions would point to a new way of structuring scientific activity. I will conclude this chapter by introducing four theses that point to features of this new way. (I do not pretend that I have presented conclusive arguments here, only indicators that need much more extensive development.) I offer these theses as suggestions made in the spirit of those proposed in the final section of Santos (1992): "All natural-scientific knowledge is social-scientific" (Santos 1989, 31); "All knowledge is local and total" (Santos 1992, 37); "All knowledge is self-knowledge" (Santos 1992, 40); "All scientific knowledge aims at becoming common sense"(Santos 1992, 43). I leave it for the reader to determine if my theses qualify, clarify, develop or contest those of *A Discourse on the Sciences.*

Reciprocity of the Natural and Social Sciences

Answering the question, "For the sake of enhancing human well-being, what range of strategies needs to be pursued in scientific research?" requires, not an *a priori* response, but the conduct of systematic empirical investigation that is, in part, social scientific. That is because a fundamental dimension of human well-being is the exercise of cultivated, effective agency. Such agency is exercised when one acts in the important aspects of one's life, informed by one's beliefs, so that desires that are expressions of a broad array of the values one holds are regularly satisfied (Lacey 2002b). Thus, answering the question requires understanding the range of values (personal, moral, social) that people aspire to manifest in their lives, their variation with culture, class and other factors, and what projects people have developed (and aspire to develop) in order to further the exercise of their agency.[14] We cannot discover what range of strategies need to be pursued in scientific research for the sake of enhancing human well-being (or of gaining "prudent knowledge for a decent life"), unless we conduct this kind of social inquiry.

Institutions of the restructured science will need to deal with the complex reciprocal interplay between the natural and social sciences (cf. Santos 1989, 69ff). Only in the course of inquiry in the social sciences can we discover what is the range of value-outlooks that frame conceptions of well-being that are espoused throughout the world. This inquiry will help to identify the limits of *neutrality* of research conducted under materialist strategies, and any other strategy that may be developed. It will also identify value-outlooks that may be interested in possibilities (e.g., those of emancipation from dominant social structures), the exploration of which requires alternative strategies. Once the range of these value-outlooks has been identified, we can explore the set of strategies that may be desirable, in response to my question, for directing research in current social circumstances. The conduct of research, however, requires extensive and often expensive material and social conditions, so that it may not be possible to conduct research to the desirable extent under all of the strategies. It follows that under which of these strategies research actually will take place, and how priorities are set among them, should be a matter for democratic deliberation and enactment (Santos 1989, 86).[15] (It is a major question for emancipatory politics to address the issue of how such democratic decision-making could happen in a reliable and effective way.) At the same time developments and applications of the natural sciences intervene in the social order, opening new possibilities for human life, reorderings of priorities and, sometimes, revised value-outlooks, thus setting new tasks for the social sciences.

> *Thesis 1.* There is a profound unity between the natural and the social sciences based on their reciprocity.

There are also other dimensions to the interplay of the two kinds of sciences (Lacey 1990). Where the values about control and thus the interpretation of section 1 are uncontested, it is common to develop social sciences and especially psychology (Lacey 2003c) under versions of materialist strategies, so that the kind of categories (pertaining to the formulation of laws, etc.) that are proper to the natural sciences developed under these strategies penetrate into the social sciences—with the legitimation that this is how to make these sciences "scientific." I have indicated (note 13) that I think that materialist strategies will retain an important place in the restructured science, with research conducted under it properly maintaining relative autonomy, without the penetration of categories from the human sciences.[16] But it is part of the distinctness of materialist strategies that, in the research conducted under them, phenomena and the possibilities they permit are dissociated from links with lived experience, value and social place. There is no reason, as I pointed out using the examples of "seeds," to hold that all the possibilities of phenomena can be identified when such dissociations are made. Moreover the same possibility identified with a description deploying the dissociated categories of materialist strategies may be identical to a possibility characterized using social categories. For example, seeds that are produced by the techniques of genetic engineering (describable under materialist strategies) are simultaneously commodities; so that if we want to know the full range of consequences of using these seeds, we need to investigate their effects *qua* commodities (social entities) as well as their effects *qua* biological entities (Lacey 2001; 2002c; 2003a). Strategies that are alternatives to materialist strategies, for example, agroecological ones, may be designed in order to avoid these kinds of dissociations and aim to identify possibilities of natural objects in relation to their ecological and social contexts. In such strategies the routine categories of the natural and social sciences will cut across each other in ways that make it impossible to separate the natural (biological) and the social. In these cases, rather than reciprocity, we have thorough interpenetration.

Additional Theses

> *Thesis 2*. Any item of scientific knowledge is produced and confirmed under a particular strategy, which may be adopted because of its mutually reinforcing relations with particular values; but, if it is soundly confirmed, consistency with it is universally (rationally) demanded.

This follows directly from the model of scientific activity of section 2. All strategies have a "local" or "limited" compass; none can aspire to encapsulate all the possibilities of phenomena. Gaining comprehensive understanding of phenomena requires the cultivation of a multiplicity of strategies, where each one (in principle) is able to identify a particular class of possibilities, and where interest in which may derive from valuing that some of those possibilities be realized in (technological) practice. A strategy may draw upon the categories of traditional, local knowledge of a culture, and then any knowledge gained under it will constitute a continuation of the tradition. A considerable amount of knowledge gained under agroecological strategies is of this kind (Altieri 1995.) Alternatively, it may have mutually reinforcing relations with values (e.g., the values about control) that are linked with forces with "global" reach; but global reach is not the same thing as universal assent. Whatever the case may be, items of knowledge are only gained when theories are accepted in accordance with *impartiality*. While it would be irrational to deliberately act in contradiction of knowledge confirmed in accordance with *impartiality* or to reject it, simply because it was investigated under a strategy linked with values that one does not endorse, nevertheless one is not obliged rationally to enter into practices that are informed by such knowledge. As we have seen, *impartiality* does not imply *neutrality*; positive judgments of the cognitive value of a proposition are consistent with negative judgments of actions informed by it.

> *Thesis 3*. Scientific theories are produced and confirmed in the course of intentional activity, and the limits of the knowledge expressed in a theory are dependent on the goals of the agents engaged in that activity; so that a scientist's understanding of what these limits are is well served by his having sound self knowledge.

Scientists, I have claimed, have often misidentified their activity, incorrectly identifying research conducted under materialist strategies with scientific research *per se*. Self-knowledge—knowing well the goals of the activities in which one is engaged and why those goals have been adopted—is a good antidote to the tendency to misidentify the "local" knowledge gained under materialist strategies with knowledge of universal social value.

> *Thesis 4.* Scientific activity should (and can) be for the sake of furthering human well -being; its products thus enter deeply into the lives of everyone, so it should be carried out constantly under democratic guidance and scrutiny.

Concluding Remark

A final observation: Santos describes his objective in *A Discourse on the Sciences* as attempting to discern, with the rich array of sociological and historical methods that he commands, the contours of the emerging new way of structuring science (or, in his words, the "emerging paradigm"). My approach, developed in the philosophy of science, complements Santos' and it can be deepened by ongoing dialogue with his. I have argued, with basis in my model of scientific activity (section 2), for the necessity of engaging in research under a multiplicity of strategies. My four theses have elaborated some of the implications of taking seriously the idea of multiple strategies. I believe that this idea is indispensable for restructuring scientific activities so that they become carried out for the sake of furthering human well-being (cf. Santos 1989, 44). It is also indispensable for science to approach its long-affirmed ideal of *applied neutrality* (as distinct from the *neutrality*, bounded by consensus on the specifically modern values about control of natural objects, that is endorsed by the interpretation of science laid out in section 1). The interests of genuine scientific *neutrality* and human emancipation can live happily together and mutually support each other.

NOTES

1. Philip Kitcher has raised questions like these repeatedly, within a broadly liberal perspective, over the past decade (especially in Kitcher 1997, 2001). Compare Kitcher's formulations with the more radical ones of Paul Feyerabend and Robert Proctor: "being related to each other in lawful ways, [the entities unearthed by science] can be manipulated or predicted by using these laws. There can be new combinations of them and new entities may in this way arise at the phenomenological level. But these entities are important only if the resulting world is pleasant to live in, and if the gains of manipulation more than compensate for the losses entailed by the losses of unscientific layers. The objection that the entities and laws that connect them are 'real' and that we must adapt to them, no matter how dismal the consequences, has no weight" (Feyerabend 1999, 12). "What we need is a political philosophy of science, a philosophy that focuses on the forms of power in and around the sciences. The question for the political philosopher of science is [:] . . . Who benefits from knowledge (or ignorance!) of a certain sort, and who suffers? How might

the practice of science be different? . . . We must also ask *political, ethical, and activist questions*: Why do we know what we know and why don't we know what we don't know? What *should* we know and what shouldn't we know? How might we know differently? (Proctor 1991, 13)

Compare these remarks with others made by Feyerabend (1993) on criticizing science in the light of "humanitarian" and "democratic" considerations and his repeated suggestions (Feyerabend 1993, 1999) that following a single approach to science threatens human well-being (Lacey 2001). Feminist philosophers of science (e.g., Anderson 1995; Longino 1990, 2002) have proposed yet further variations of questions like these.

2. Throughout the chapter by "values" I mean "moral and social values;" similarly "value judgment." There is a basically uncontroversial role for *cognitive values* at the core of science (Lacey 2004, ch. 3), but it is not be discussed here. The claim, which is crucial for my arguments, that cognitive values (or the criteria for appraising the cognitive value of theories) are distinct from moral and social values, has been disputed. For my response, see Lacey (2004a, ch. 9; 2004b).

3. *Autonomy* should be qualified by: "provided that the research practices do not involve morally unacceptable actions or outcomes." Sometimes, then, in certain domains empirical data needed for making theoretical appraisals should not be obtained. Sometimes the value of gaining knowledge is subordinate to other moral values. This does not imply that moral values may function among the criteria of theory appraisal, but only that sometimes we may limit the quest for knowledge on moral grounds. My formulation allows that, at any given time, where the general characteristics of a line of research are settled, cognitive interests leave open a range of problems that are (we might say) 'cognitively equally worthy' of further investigation. Then which of these problems is chosen (e.g., to study one virus rather than another) as a research priority may reflect values. *Autonomy* implies that, *qua* scientific problem, the one chosen has no special primacy, so that scientific institutions should not (as a whole) be dominated by an agenda defined by these values. As described, *autonomy* is compatible with values playing an important part in the choices of research projects of individual scientists (and particular scientific institutions). Arguments for *science is value free* usually draw upon premises such as: there is a sharp separation between fact and value; values express subjective preferences, etc. Then, while it may be a fact that individual scientific choices reflect values, the subjectivity of values is taken to make it likely that choices representing a diversity of values will be made among the community of scientists, thus ensuring that particular values do not come to dominate the research agenda.

4. For extensive analysis and criticism of *science is value free*, and its component notions, see Lacey (2002a; 2004a).

5. For the full details and the arguments, see especially Lacey (2004a). Most of the next section is taken from Lacey (2002b) and draws upon Lacey (2004b).

6. I have stated the objective (albeit contentiously) so as to encompass all inquiries that are called "sciences" (including social sciences) and those that bear close affinities with them (see Lacey 1991, ch. 5). I include under "science" all forms of *systematic empirical inquiry*, because I do not want to rule out by definitional *fiat*, or to assume *a priori*, that forms of knowledge that are in continuity with traditional

forms of knowledge have comparable epistemic (cognitive) status to those of modern science. Also, I want a notion of "science" that is sufficiently inclusive that, in principle, no proposition about the world need be held to be excluded from the realm of scientific inquiry. (That does not mean that scientific inquiry will, in fact, be able to address them all.) In accordance with this usage of "science," any reasonably systematic structure that expresses understanding of some domain of phenomena counts as a theory. (Often it is preferred to limit "theory" to that which has mathematico-deductive structure or that which contains representations of laws, and not merely richly descriptive or narrative structure.) The effect of deploying this wider notion of "science," which—as we will see—underlies the new way of structuring scientific activity (that permits a multiplicity of research strategies) that I propose, is to strengthen the empirical credentials that may be provided for claims made under the authority of science.

7. For the relationship between "strategy" and Thomas Kuhn's notion of "paradigm," see Lacey (2004a, 261). "Strategy" is closely connected with what Ian Hacking calls "form of knowledge" (Hacking 1999).

8. The success of research conducted under agroecological strategies (Altieri 1995; references in Lacey 2002c) shows that accord with *impartiality* can be reached under strategies that are not reducible to materialist ones (Lacey 2001; 2003b).

9. In Lacey (2004a, ch. 5) I consider and reject other arguments for the virtually exclusive adoption of materialist strategies, including arguments based upon materialist metaphysics and Kuhn's philosophy of science. Note that for Santos (1992, 18) "modern science [as a] hegemonic model of rationality" is linked with "its capacity to control and transform [the real]."

10. In Lacey (2002a) I fill out this summary statement by explicitly offering a list of the values, and exploring the presuppositions of rationally holding them.

11. I describe the multiple dimensions of this mutually reinforcing relationship in Lacey (2002a; 2004a, 117–26). I do not claim that all research under materialist strategies is conducted *for the sake of* furthering the values about control. Deploying them exclusively for exploring many phenomena (e.g., celestial, subatomic) is unobjectionable and is not rationalized by reference to the values. (Even so, given the dependence of research in these areas on "high" technological developments, links with the values about control are not absent.) Moreover (see section 2.5) the furtherance of the values about control was not a causal factor (although it was a consequence) in the transition from the strategies of classical mechanics to those of relativity and quantum mechanics.

12. To question the values about control, as members of the World Social Forum do, does not mean that no value is attributed to control of natural objects, but only that the value of control is subordinated to or balanced with other social values (e.g., community empowerment, cultural integrity, preservation of biodiversity). Cf. the following remark, referring to Santos' desire to "destroy the hegemony of modern science without undercutting the expectations that it generates . . . The new configuration of knowledge is, thus, the guarantee of the desire and the desire of the guarantee that technological development would contribute to the deepening of cognitive and communicative competence, and so it becomes transformed into practical knowledge and helps us to provide sense and authenticity to our existence" (Santos 1989, 46; my translation).

13. I believe that materialist strategies will always be in the range, even that it may properly be granted a special place (Lacey 2001; 2004a, ch. 10). One reason for this is that strategies like agroecological ones utilize results of research conducted under materialist strategies in numerous indispensable ways. Such strategies should not be seen as complete alternatives to materialist strategies, but more as an interlocking set of local approaches each of which draws upon the results gained under materialist strategies, where convenient. Each one also qualifies claims that are put forward when materialist strategies are granted exclusivity. I do not know what eventual answers to my restated question might be like. I have explored it in detail only in connection with agricultural science. There I have argued that a multiplicity of strategies (including agroecological ones) is essential (Lacey 2000; 2002c; 2003a; 2003b). I am confident that this is so in other domains too. But, it might turn out to be the case that in some fields, for example, subatomic physics, research under materialist strategies suffices; certainly, regarding this field, I have no alternative speculations to offer.

14. In note 10, I referred to the presuppositions of rationally holding the modern values about control. I have argued that all value-outlooks have presuppositions of a broadly factual character (Lacey 2004a, ch. 2). The social science inquiry, which I am referring to here, should also appraise the evidence that supports such presuppositions, including those of the value-outlooks of emancipatory movements. Not all value-outlooks that are articulated by people are worthy of serious recognition; those that rest upon clearly false presuppositions certainly are not.

15. Certainly there is nothing about the expertise cultivated by professional scientific practitioners under any strategy that especially prepares them for making these choices.

16. Democratic decision-making applies to priorities and allocation of resources at M_1 and it is obviously relevant in all sorts of ways at M_3, but it does not apply at the moment of accepting theories. Maintaining *impartiality* at M_2 is of utmost importance, regardless of the strategy under which the theories have been developed.

REFERENCES

Altieri, Miguel. 1995. *Agroecology: The Science of Sustainable Agriculture*, 2nd ed. Boulder, CO: Westview.

Anderson, Elizabeth S. 1995. "Knowledge, Human Interests, and Objectivity in Feminist Epistemology." *Philosophical Topics* 23:27–58.

Baptista, António M. 2002. *O Discurso Pós-moderno contra a Ciência: Obscurantismo e Irresponsibilidade*. Lisboa: Gradiva.

Dupré, John. 1993. *The Disorder of Things: Metaphysical Foundations of the Disunity of Science*. Cambridge, MA: Harvard University Press.

Feyerabend, Paul K. 1993. *Against Method*, 3rd ed. London: Verso.

———. 1999. *Conquest of Abundance: A Tale of Abstraction versus the Richness of Being*. Chicago: University of Chicago Press.

Freire, Olival, Jr. 2005. "E. P. Wigner and the Foundations of Quantum Physics." Ch. 10 of this book.

Hacking, Ian. 1999. *The Social Construction of What?* Cambridge, MA: Harvard University Press.

Kitcher, Philip. 1997. *The Lives to Come: The Genetic Revolution and Human Possibilities*. New York: A Touchstone Book.

———. 1998. "A Plea for Science Studies." In *A House Built on Sand: Exposing Postmodernist Myths about Science*, ed, Noretta Koertge. New York: Oxford University Press.

———. 2001. *Science, Truth and Democracy*. New York: Oxford University Press.

Lacey, Hugh. 1990. "Interpretation and Theory in the Natural and Human Sciences." *Journal for the Theory of Social Behavior* 20:197–212.

———. 2000. "Seeds and the Knowledge They Embody." *Peace Review* 12:563–69.

———. 2001. "Incommensurability and 'Multicultural Science.'" In *Incommensurability and Related Matters*, ed. Paul Hoyningen-Huene and Howard Sankey. Dordrecht, Netherlands: Kluwer.

———. 2002a. "The Ways in Which the Sciences Are and Are Not Value Free." Vol. 2 of *In the Scope of Logic, Methodolology and Philosophy of Science*, ed. P. Gardenfors, K. Kijania-Placek and J. Wolenski. Dordrecht, Netherlands: Kluwer.

———. 2002b. "Explanatory Critique and Emancipatory Movements." *Journal of Critical Realism* 1:7–31.

———. 2002c."Assessing the Value of Transgenic Crops." *Ethics in Science and Technology* 8:497–511.

———. 2003a. "Seeds and their Socio-cultural Nexus." In *Philosophical Explorations of Science, Technology and Diversity*, ed. Sandra Harding and Robert Figueroa. New York: Routledge.

———. 2003b. "Alternativas para a Tecnociência e os Valores do Forum Social Mundial." In *O Espírito do Porto Alegre*, ed. Isabel. M. Loureiro, Maria E. Cevasco and José Corrêa Leite. São Paulo: Paz e Terra.

———. 2003c. "The Behavioral Scientist *qua* Scientist Makes Value Judgments." *Behavior and Philosophy* 31:209–23.

———. 2004a. *Is Science Value Free? Values and Scientific Understanding*. Paperback edition. London: Routledge.

———. 2004b. "Is There a Significant Distinction between Cognitive and Social Values?" In *Science, Values and Objectivity*, ed. Peter Machamer and Gereon Wolters. Pittsburgh, PA: Pittsburgh University Press.

Longino, Helen E. 1990. *Science as Social Knowledge*. Princeton, NJ: Princeton University Press.

———. 2002. *The Fate of Knowledge*. Princeton, NJ: Princeton University Press.

Proctor, Robert N. 1991. *Value-free Science? Purity and Power in Modern Knowledge*. Cambridge, MA: Harvard University Press.

Santos, Boaventura de Sousa. 1987. *Um Discurso sobre as Ciências*. Porto, Portugal: Afrontamento.

———. 1989. *Introdução a uma Ciência Pós-moderna*. Porto, Portugal: Afrontamento.

———. 1992. "A Discourse on the Sciences." *Review* 15 (1):9–47. [English translation of Santos (1987).]

———. 1999. "On Oppositional Postmodernism." In *Critical Development Theory: Contributions to a New Paradigm*, ed. R. Munck and D. O'Hearn. London: Zed Books.

———. 2002. *Reinventing Social Emancipation*, http://www.ces.fe.uc.pt/emancipa/en/index.html (accessed July 31, 2004).

10

Orthodoxy and Heterodoxy in the Research on the Foundations of Quantum Physics:

E. P. Wigner's Case

Olival Freire Jr.

Dealing with Eugene Wigner's ideas on the measurement procedure in quantum physics and unearthing the controversy that pitted him against supporters of the interpretation of complementarity, I will show how Wigner and his followers contributed to the defeat of a seemingly unshakeable consensus. In fact, although he intended to defend what seemed to him to be orthodoxy, he himself became a heterodox. I suggest that Wigner's conjectures on the role of consciousness in physical phenomena were not fruitful and were discarded, being today part of the *history* of physics rather than physics proper. However, his ideas and actions left an indelible mark on the physics of the second half of the twentieth century. Namely, he formulated his ideas in opposition to the "Copenhagen monocracy," which held a stronghold on the interpretation of quantum physics until the late 1960s; he stressed the unsolved status of the measurement problem; and he effectively defended his ideas and supported those who were willing to investigate the foundations of quantum physics. He thus contributed to the creation of a new field of physical research, that of the foundations of quantum physics, which attributed a higher scientific status to the old controversy on the interpretations and foundations of this theory. This new field has had to deal with important theoretical, experimental, and philosophical issues, and significant repercussions have arisen in the last decades.

The intrinsic historic worth of Wigner's case should be enough to justify its inclusion in a book organized to criticize the hubris of the contemporary scientism and to suggest, instead, the role of *prudent knowledge* for a decent life. However, Wigner's case is of general interest as this topic involves a contradiction: some contemporary physicists are inclined to deprecate

Wigner's insight into the role of the mind in the measurement problem of quantum theory,[1] yet these same individuals read Wigner's work neglecting to take into account the role he played in the context of the 1960s. Judging any historically significant work by contemporary standards constitutes an error of anachronism, and preventing anachronistic narratives of past science may foster collaboration between science and society and counter irrationalism more effectively than distorted images of an idealized science.

THE STAGE OF A NEW SCIENTIFIC THEORY AND ITS PROBLEMATIC FOUNDATIONS

To understand the role Wigner played in the 1960s, one should take into account quantum theory's intellectual and historical context at that time. This theory emerged between 1925 and 1927 as an ensemble of three different mathematical formalisms plus an interpretation of them offered by the Danish physicist Niels Bohr, which was christened the "complementarity view," later known as the "Copenhagen interpretation," the "orthodox interpretation," and even the "usual interpretation," besides simply "complementarity."[2] Bohr's interpretation had a philosophical flavor, which was unproblematic to physicists such as Pauli, Born, Heisenberg, but it became a hindrance to physicists from, for instance, the American and British milieux. These physicists retained rules of formalism but did not incorporate its philosophy (Heilbron 2001). It is beyond the scope of the current work to define complementarity,[3] I will, however, demonstrate its relevance to the solution of the measurement problem, which was the central focus of the divergence between Bohr's views and those of John von Neumann and Eugene Wigner. As an exemplar of this philosophical flavor, one may cite the way in which Bohr solved the so-called measurement problem; which is related to the reason why the superposition of quantum states—the quantum identity—did not appear in the macroscopic world of the measurement devices. Bohr suggested that such devices had to be treated within the framework established by classical physics, not because one could not treat them from a quantum point of view, but because they need to be treated classically in order to communicate one's measurement results to other researchers. As communication is a requirement to attain objectivity, and communication requires ordinary language refined by concepts from classical physics (e.g., concepts indicated by classical terms such as "work," "force," etc.), the classical treatment of measurement devices is a condition for preserving objectivity in scientific research. As it happens, Bohr's solution was not accepted by John von Neumann, as we shall see below. He rejected it as a consequence of his own mathematical work, in which, for the first time in the history of physics, he axiomatized quantum theory, identifying Hilbert spaces as the mathematical structure implied in those different mathematical formalisms (von Neumann 1932). Von Neumann's position received continued support from Wigner.

In the 1930s, complementarity also suffered strong criticisms from Albert Einstein and Erwin Schrödinger. However legitimate those criticisms might have been, at that time they were not seen as such by the vast majority of physicists. Of course, they are now central to the field of research we call foundations of physics. These criticisms were not necessarily shared by those who excluded complementarity from their view of quantum theory, nor did these researchers necessarily foresee their significance. In fact, before World War II, most physicists accepted that Bohr, Heisenberg, Pauli, and others had solved the foundational problems of quantum mechanics, even if they did not consciously share that solution. Max Jammer (1974, 247–51) depicted this state of affairs as "the almost unchallenged monocracy of the Copenhagen school in the philosophy of quantum mechanics." This diagnosis, however, does not reveal the existence of a true labor division among physicists. Monocracy of Copenhagen school meant two types of physicists. A few of them involved with foundational problems, extension of quantum mechanics to new phenomena and its applications to old and new problems, and the others involved with extension and applications, but believing that the foundational problems were solved by the founding fathers of quantum mechanics.

Beginning the early 1950s, however, this scenario slowly began to change. Besides Einstein and Schrödinger's renewed criticisms, other characters arrived on the scene. These quantum mavericks included physicists like David Bohm, Imre Fényes, Friedrich Bopp, Louis de Broglie, Henry Margenau, Alfred Landé, D. Blokhintsev, I. P. Terletski, Hugh Everett, Jean-Pierre Vigier, and philosophers such as Karl Popper and M. Omel'ianovskii. The new main challenge to complementarity came from the causal, or "hidden variables" interpretation, suggested by David Bohm, in 1952. In fact, he developed an approach that was able to reproduce known quantum predictions in the non-relativistic domain, but his approach was embedded in a causal framework, not a probabilistic one in the manner of complementarity. But no matter how interesting the causal interpretation might be, it was not accepted by physicists throughout the 1950s, in such a way that the Copenhagen monocracy was shaken, but not disabled (Freire 2005).

ENTER WIGNER

If the causal interpretation presented the greatest challenge to complementarity coming from outside the circle of the founding fathers of quantum mechanics, another major challenge came from within. E. P. Wigner was born in 1902, in Budapest, where he received his chemical engineer diploma. Early, at the Lutheran high school, he met John von Neumann and became his friend and admirer forever.[4] Together, they immigrated to the United States in order to jointly develop a mathematical physics program at Princeton in the 1930s. In the early 1960s, Wigner's prestige was approaching its zenith. He was recognized early on for his use of the theory of groups in quantum mechanics, and his recognition increased with his contributions to nuclear physics, including his participation in the Manhattan Project (Mehra 1993).

Since the late thirties, Wigner began to play a role beyond physics proper, motivated by the military implications of recent discoveries in nuclear physics. He and his Hungarian colleague L. Szilard suggested that Albert Einstein write the famous letter to President Roosevelt calling for the development of the U.S.'s nuclear program (Doncel, Michel, and Six 1984). For the many roles he played in the Manhattan Project, he was consequently awarded the title of "the founder of nuclear engineering" (Weinberg 2002). After the war, Wigner's involvement with defense matters did not wane. In the late fifties, he was one of the "Princeton three," along with John Archibald Wheeler and Oskar Morgenstern, who pushed the American government to build an enormous national laboratory dedicated to defense research, an initiative that failed but eventually led to JASON, a group of academic physicists who advised the U.S. Department of Defense on defense matters (Aaserud 1995). Wigner committed himself to advertising the role of civil defense in the Cold War context, and even built a nuclear fallout shelter in his own home.[5]

In the early 1960s, Wigner decided to intensify his public involvement beyond physics, publishing papers on the philosophy of science, and dealing with the measurement problem of quantum mechanics. This central issue in the foundations of quantum physics would be of interest not only to physicists but to other audiences as well, such as philosophers. As we shall see, Wigner believed that the measurement problem was part of the philosophy of physics, which in turn he saw as an integral part of physics itself, a view that many of his colleagues did not share. Before going to the 1960s, it should be noted that since the 1930s, working with von Neumann, Wigner was interested in and contributed to measurements issues in quantum mechanics (Shimony 1997). In the early 1950s he resumed the subject showing how quantum formalism exhibits limitations of measurability (Wigner 1952). In the same year, with G. Wick and A. Wightman, they further extended such limitations introducing the concept of a superselection rule. None of those papers, however, were as influential as his 1960s' works.[6]

Between 1961 and 1963, Wigner published two papers that would become the centerpiece of his views on the foundations of quantum mechanics. He revisited a distinction first emphasized by von Neumann between two kinds of evolution of quantum states. The first one, continuous and causal, is governed by the Schrödinger equation. The second one, discontinuous and erratic, occurs during the measurement processes. Additionally, but still following von Neumann, he treated measuring devices quantum mechanically, instead of treating them classically as suggested by Bohr. The latter choice leads to the propagation of the singular superposition of quantum states from the system under scrutiny to the ensemble: system plus the measuring apparatuses. In mathematical terms, this propagation is represented by the inner product between the two Hilbert vectors, one related to the system and the other related to the devices. As nobody has ever seen such a bizarre superposition in our macroscopic world, one needs to answer how, where, and when this superposition becomes a vector with just one component. After all, what we get after measurements is related to vectors and probabilities rather than to superposition of vectors. Wigner emphasized this point and arrived at the next conclusion: in order to eliminate this superposition one needs to admit that measurement leads eventually to the role of the observer's introspection, that is, when the information enters the mind of the observer.

Conjecturing that mind plays a not eliminable role in the description of quantum measurements was one of Wigner distinctive feature when approaching the measurement problem. According to him (Wigner 1961), "when the province of physical theory was extended to encompass microscopic phenomena, through the creation of quantum mechanics, the concept of consciousness came to the fore again: it was not possible to formulate the laws of quantum mechanics in a fully consistent way without reference to the consciousness." He presented his arguments following two steps. The first, and less incisive, was that the quantum state changes every time the observer gets new information from observations. While in classical mechanics you also need observation to get the initial conditions and establish the classical state, when you get them and solve the equations of motion, new information is no more relevant to change the state. In the second step, he strengthened his case, presenting an idealized experiment in order to demonstrate the difference between quantum descriptions of measurements with and without human observers. Nowadays the argument related to Wigner's idealized experiment is known as "Wigner's friend" (Wigner 1961). Wigner suggests you observe, helped by a friend, an object quantum described by a linear combination of two states. Your friend observes the object hence, to him it is in one of the two states and no more in a linear combination of the two. Before he tells you the result of the observation, there will be a conflict between your description of the object (linear combination of the two states) and that given by your friend (one of the two states). Since the right quantum description is yours, you must admit that your "friend was in a state of suspended animation before he answered" your question, which is not a reasonable conclusion, and allowed Wigner to arrive to the conclusion that quantum theory is not enough to deal with measurements. So, if quantum theory is to encompass not only inanimate bodies, but also life and mind, it needs to be changed, and Wigner suggested explicitly to look for a non linear equation of motion.[7] In fact, he was suggesting a true research program: to solve the measurement problem, including, from Wigner's point of view, the inclusion of life and mind in the scope of physical theories.[8]

Furthermore, Wigner's arguments entailed a more sociological and historical question: to define the orthodox interpretation of quantum mechanics, and to identify its protagonists. He wrote: "The standard view is an outgrowth of Heisenberg's paper in which the uncertainty relation was first formulated. The far-reaching implications of the consequences of Heisenberg's ideas were first fully appreciated, I believe, by von Neumann, but many others arrived independently at conclusions similar to this. There is a very nice little book, by London and Bauer, which summarizes quite completely what I shall call the orthodox view." (Wigner 1963). Bohr's paper on complementarity is only referred in a footnote. In Wigner's account, therefore, Bohr and complementarity go to the backstage corridor of the quantum story, and Heisenberg and von Neumann become its chief protagonists. I interpret this excerpt as a dispute over the intellectual heritage of the founder fathers of quantum mechanics. It was not by chance that Wigner wrote this text after von Neumann's and Bohr's deaths, and while scientists and historians in the United States were involved in one of the largest projects ever to collect and store records significant in the creation and evolution of a scientific theory, which would come to be known as the *Archives for the History of Quantum Physics* (Kuhn, Heilbron, and Forman 1967).

Wigner's papers drew both support and opposition. Abner Shimony was very impressed by it: "I found your paper on the mind-body problem extremely stimulating. It is one of the few treatments of the problem which considers the mind-body relationship to be a legitimate subject for scientific investigation, without achieving this scientific status for the problem by reducing it to behavioristic or materialistic considerations."[9] M. Satosi Watanabe also reacted very favorably to Wigner's suggestion about the role of consciousness in physical processes, and we find in their correspondences a true anticipation of subsequent opposition to Wigner's ideas from Rosenfeld, as well as Rosenfeld's Marxist motivation to take such a position.[10]

It was up to Rosenfeld to oppose Wigner in defense of complementarity. Rosenfeld was Bohr's former assistant since the 1930s, and a physicist very sensitive to epistemological matters. Rosenfeld and Wigner had, however, quite different profiles on a number of issues. Politically, Wigner was very conservative—he was a follower of the Republican Party and was supportive of U.S. foreign policy to the point of receiving a telegram from President Richard Nixon in thanks for his support of the Vietnam War effort.[11] In contrast, Rosenfeld was engaged in Marxist philosophy since the thirties. Rosenfeld's Marxism was closer to Western Marxism than it was to Soviet Marxism, to use terms introduced by Perry Anderson (1979) in order to make sense of Marxist trends in the twentieth century.[12] Rosenfeld participated actively in the late 1940s and 1950s in organizations and movements, such as the "World Federation of Scientific Workers," "Science for Peace," and the "Manchester University Socialist Society." His political records led him to doubt about applying a visa to visit the United States, in the Cold War times.[13] Intending to preserve what seemed to him to be a dialectical feature of complementarity, Rosenfeld criticized both the Soviet and Marxist physicists, like D. Blokhintsev and D. Bohm, who were themselves critics of complementarity, and physicists like Heisenberg, who leaned towards idealism (Rosenfeld 1953). So, for a number of reasons—political, ideological, and philosophical—Rosenfeld could not accept a view like Wigner's, which assigned a central role to the mind in physical phenomena.

Wigner and Rosenfeld also displayed significant differences in their approach to the measurement problem. We could also speak of different scientific styles. For Wigner, following von Neumann, dissecting the mathematical formalism of quantum physics in order to exhibit its axiomatic structure was a necessary step in grasping the theory's full implications. That is not to say that for Wigner axiomatic theories were necessary for all research in physics since in other fields, nuclear physics for instance, his approach was phenomenological.[14] But for Rosenfeld, following Bohr, a phenomenological insight into a physical theory was the best way to understand it, and he always emphasized his distrust of the reach of axiomatic treatment of physical theories. Even before his dispute with Wigner, he wrote: "the 'axiomatizers' do not realize that every physical theory must necessarily make use of concepts which *cannot*, in principle, be further analyzed, since they describe the relationship between the physical systems which is the object of study and the means of observation by which we study it: these concepts are those by which we give information about the experimental arrangement, enabling anyone (in principle) to repeat the experiment. It is clear that *in the last resort* we must here appeal to *common experience* as a basis for common understanding."[15] Last but not least, Rosenfeld maintained that complementarity was the great epistemological lesson of quantum theory, and for this reason, he could not accept Wigner's position, according to which, Bohr's complementarity played no role in the orthodox interpretation of quantum theory.

THE HEATED DISPUTE: WIGNER VERSUS ROSENFELD AND THE ITALIANS

Rosenfeld's strategy for criticizing Wigner's view was to give strong praise to a certain work, by writing, "these misunderstandings [i.e., that the translation of Bohr's argument into the formal language of the theory should present unrecognized difficulties], which go back to the deficiencies in von Neumann's axiomatic treatment, have only recently been completely removed by the very thorough and elegant discussion of the measuring process in quantum mechanics carried out by Daneri, Loinger and Prosperi" (Rosenfeld 1965). These Italian physicists had used the ergodic theorem to explain quantum measurements as a thermodynamic amplification of a signal, triggered by the interaction between quantum systems and measurement devices (Daneri, Loinger, and Prosperi 1962). Indeed, the Italian physicists had quantum mechanically treated both the system and the interaction between the system and the measurement device, but, after the interaction ended, they considered the measurement device evolving according classical statistical physics, which was compatible with Bohr's requirement that the measurement devices should be considered classical bodies. It's clear that if Rosenfeld's point of view about the reach of the Italian work were accepted, Wigner's claims would be considered ungrounded. The dispute lasted throughout the second half of the 1960s, and it was marked by bitter arguments, even though it dealt with rather technical content, that is, to determine whether the Italian work was a rigorous solution or just an approximation. Wigner was specially struck by the next paper of the Italian physicists (Daneri, Loinger, and Prosperi 1966), in which they criticized Wigner, Shimony, Moldauer, Yanase, and Jauch's analyses of the measurement problem, and wrote that "none of them gives new substantial contributions to the subject." He wrote to J. M. Jauch suggesting a reply, and admitting that he was specially irritated not by the attack on him but by its significance for young researchers like Abner Shimony and Michael Yanase, his ex-students.[16] The letter to Jauch was a typical maneuver for recruiting allies, as Wigner was not in complete accord with Jauch, notwithstanding the fact that the latter was trying to refine von Neumann's mathematical treatment. Jauch did not agree with Wigner's conjecture on the role of mind in the measurement process and believed that the changes he himself had introduced in von Neumann's treatment had transformed the difference between the two kinds of evolution of the state vector into a pseudo-problem. Wigner suggested to Jauch that they write a common response, together with Yanase, Wigner's former student. Looking for allies, Wigner went so far as to suggest that Jauch include in their joint paper a favorable citation of a paper by David Bohm and Jeffrey Bub. However, Jauch could not accept this last proposal since, in his own work, he was trying to reinforce von Neumann's proof against hidden variables, while Bohm and Bub's work was in line with that last approach.[17] Afterwards, Rosenfeld (1968) and Loinger (1968) replied to Wigner, Jauch and Yanase's paper, but the next round did not take place in papers published in journals. Instead, Wigner waited for the Varenna school to have a live debate, as we shall see. First, however, consider that Rosenfeld was not in an easy position during this dispute. In 1952, he labeled David Bohm, the leader of the first round of dissent, as a "tourist," or a "dilettante," in the field of the foundations of quantum mechanics,[18] but he could not deal in the same way with the 1963 Nobel Prize winner, Eugene Wigner. His manner of arguing likely had something to do with Rosenfeld not being invited to talk at that summer school by Bernard d'Espagnat, its director. Instead, for Rosenfeld's side of the dispute, d'Espagnat invited Prosperi.[19] Let us say that the very existence of this school evidenced the changing mood among physicists concerning the status of research on foundations of quantum theory (Freire 2003).[20]

The dispute eventually ended at the Varenna summer school, which Wigner succeeded in transforming into an agreement on the need for a research program on quantum measurement processes. Wigner gave the main talk at Varenna (d'Espagnat 1971, 1–19), and Prosperi spoke about "macroscopic physics and the problem of measurement in quantum mechanics," in a section dedicated to "Measurement and basic concepts" (d'Espagnat 1971, 97–122). An informal discussion between Wigner and Prosperi ensued from these lectures. Assisted by Shimony and d'Espagnat,[21] Wigner reconstructed his arguments and had them published (d'Espagnat 1971, 122–26). It's worth taking a detailed look at Wigner's conclusions because we can discern in them two distinctive features of Wigner's approach to foundations of quantum physics: his diplomatic and open-minded attitudes, and his consideration of the philosophy of physics as part of physics. He divided them into two items, the first one related to "the philosophical problem," and the second about "questions of physical theory." He explained that the main question at the center of both Prosperi's and of his own concerns was related to the knowledge of the "reason for the statistical, that is probabilistic, nature of the laws of quantum-mechanical theory." In other words, how can one understand that quantum predictions are not "uniquely given by the inputs" even though equations of quantum and classical physics are deterministic? He suggested one might answer this question in different ways, and cleverly framed Prosperi's and his own responses into the same side. This type of answer implies that "the possible reason for the probabilistic nature of quantum theory's conclusions concerning the outcomes of measurements is that the theory cannot completely describe the process of measurement, that some part of the process is not subject to the equations of quantum mechanics." The difference between Wigner's and Prosperi's views resided in "the *area* to which quantum mechanics is inapplicable." For Prosperi, probability is necessary for the translation of the quantum-mechanical description into classical description because this translation is not unique. To Wigner, as for von Neumann, either quantum mechanics does not "apply to the functioning of the mind" or "the conscious content of the mind is not uniquely given by its state vector." Finally, arguing on more scientific grounds, Wigner remarked that Prosperi and collaborators were using phrases such as "macroscopic variables" and "macroscopic objects" without giving a precise definition of these terms. He remembered examples of phenomena with macroscopic bodies but which exhibit quantum features, such as permanent currents in superconductors and spontaneous magnetization in different directions, besides the observable difference between dextro and levorotatory sugar, which is based on a quantum relation of microscopic phases. In his conclusion, Wigner once more looked for areas of agreement between the two physicists and presented a proposal for a genuine research program. He concluded that since Prosperi's premises could not be rigorously formulated (at least not at that time), and their formulation would entail a significant modification of the theory current at that time, the convergence resided in the conclusion, common to Prosperi and Wigner's views, namely "the inapplicability of quantum mechanics to some part of the measurement process has to be postulated or admitted." Surely, Wigner had been postulating this for some time, and he was asking Prosperi to admit it. If it were Rosenfeld at the Varenna school, one might suppose that he would not admit it because for him, according to Bohr's views, concepts such as "macroscopic variables" or "macroscopic bodies" should be admitted without previous definitions since one gains nothing when trying to axiomatize, or to define, all theoretical terms. As close as his views were to Bohr's, Prosperi thought differently and conceded to Wigner. He did not publish any additional reports on his own arguments, and returned from Varenna to Milan convinced that the measurement problem was still unsolved.[22]

It is important to consider how Wigner's contemporaries interpreted his dispute with Rosenfeld and the Italian physicists. Otto Frisch, in a colloquium held in 1968, said: "I understand that at present there exists a controversy, roughly speaking between a group of people which includes Wigner as the best known person and another group centered on Milan in Italy, and that these two have different views on how this reduction happens." (Frisch 1971, 14). For the first time in the literature, the name "Princeton school" was used to differentiate Wigner's views from the Copenhagen school. According to Leslie Ballentine (1970, 360), there were "several versions of the Copenhagen interpretation" and, "although both claim orthodoxy, there now seems to be a difference of upholds between what may be called the Copenhagen school represented by Rosenfeld, and the Princeton school represented by Wigner." Since then, labeling Copenhagen and Princeton schools has become current in the literature (Home and Whitaker 1992). The monocracy was thus broken, from inside, as was the fate of many other monocracies of the twentieth century. I do not want to say that Wigner was the sole driving force to break that monocracy. In the same second half of the 1960s, the Irish physicist John Bell also made major contributions to it. He demonstrated flaws in von Neumann's mathematical proof against "hidden variables" in quantum mechanics, and revealed that quantum formalism leads to the strange physical property of entanglement between systems that are separated in the space-time.[23] Other factors also contributed to change the physicists' attitudes related to the research in foundations of quantum mechanics, but I will not discuss them here (Freire 2004, 2006; Bromberg 2006). What I am saying is that Wigner made a major contribution in this direction, which is not always recognized today.

WIGNER'S STYLE OF INTELLECTUAL LEADERSHIP

The portrait of Wigner as just a disputant in the creation of the field of foundations of quantum mechanics is not completely fair to him. He engaged in a variety of activities and assumed a kind of nondogmatic but highly influential leadership. He formed a group of students to work on the subject, such as Abner Shimony, a philosopher who held a Ph.D. in Physics with a dissertation on the foundations of statistical mechanics, and Michael Yanase, a Jesuit priest whose dissertation treated the measurement problem. We have already seen how he mobilized Yanase to join the debate with the Italians. Shimony (1993, xii) gave us a very impressive testimony about the role Wigner played in his career: "I am most deeply grateful to Eugene Wigner, [who] encouraged my later work on foundations of quantum mechanics. The preponderance of the physics community at that time accepted some variant of the Copenhagen interpretation of quantum mechanics and believed that satisfactory solutions had already been given to the measurement problem, the problem of Einstein-Podolsky-Rosen, and other conceptual difficulties. My decision to devote much research effort to these problems would have been emotionally more difficult without Wigner's authority as one of the great pioneers and masters of quantum mechanics." The bulk of the correspondence exchanged between Shimony and Wigner on philosophical matters suggests that Wigner also benefited from that intellectual relationship because, in fact, Shimony acted informally as Wigner's assistant on philosophical matters. Wigner was also supportive of entrants and nonentrants in the field, such as Bernard d'Espagnat, Henry Margenau, and John Archibald Wheeler. Bernard d'Espagnat was already a senior-level high-energy physicist when he decided, in the 1960s, to resume a project begun in his younger days, to philosophize on the problems of contemporary physics.[24] He found Wigner a dialoguer, even if he did not completely share Wigner's view, and Rosenfeld an ironic and bitter critic, when he accepted some of Wigner's positions.[25] Unlike d'Espagnat, Margenau was a seasoned veteran in the field of foundations of physics as he had been criticizing the complementarity view since the 1930s. In the early 1960s, he welcomed Wigner's analysis of quantum measurement, which motivated him to resume his own ideas and to present them in a clearer and more concise way.[26] They engaged in a published debate with Hillary Putnam over the conceptual structure of quantum mechanics and planned to write a book together, but this project was never seriously initiated.[27] In the late 1960s, Wigner accepted Margenau's invitation to be a member of the editorial board of a new journal, *Foundations of Physics*, aimed at fostering research of "disciplined speculations suggestive of new basic approaches in physics,"[28] including those concerning the foundations of quantum mechanics. Wigner not only accepted this invitation but assumed editorial responsibilities of the journal, suggesting papers and influencing the choice of editor who would replace Margenau upon retirement.[29]

A final example, and perhaps the most significant example of Wigner's influence on certain contemporaries, is John Archibald Wheeler, a physicist well known for his insights, both sound and speculative, in fields as diverse as cosmology and quantum physics. The two men were very close not only in science, but also in political matters and in defense-related research. As we have seen, Wigner and Wheeler were two of the "Princeton three" who were involved at the outset of the JASON project. In foundations of quantum mechanics, however, Wheeler's views were very close to Bohr's, but Wigner was so influential that, by the mid-70s, he began to doubt what Bohr's real opinion was on the role of consciousness in the quantum measurement process. Haunted by this doubt, he wrote the following to Niels Bohr's son, Aage Bohr: "I have the impression, perhaps mistaken, that your father at one time thought that for the making of an observation it only took in the end an irreversible account of amplification; but that later on he changed his position to something closer to the idea that no observation is an observation unless and until it enters the consciousness. However, I am not able to find anything to document this supposed change of view and my understanding of the history may be quite wrong." Without a response from the younger Bohr, he tried the unusual procedure of asking a friend who was in Copenhagen, John Hopfield, to answer a questionnaire, after consulting Bohr, which included the following item: "Niels Bohr did change position from (a) 'Measurement requires irreversible act of amplification' to (b) something closer to Wigner's 'a measurement is not a measurement until the result has entered the consciousness' YES ___; NO ___; QUESTION ILL DEFINED ___." This time, however, it did not take long for

Aage Bohr to send a reply corroborating the accuracy of Rosenfeld's interpretation of Niels Bohr's views. Aage Bohr wrote: "our reactions can be deduced from the answers to the questionnaire which you have formulated so cleverly that no evasion is possible. Let me just add that it is quite true that my father strongly emphasized that for an unambiguous description it is essential to include the detection device in the definition of a quantum phenomenon and even advocated that one reserved the word 'phenomenon' for processes that are 'closed' in this sense. However, I do not think he meant this to imply that the act of observation need have any effect on the processes which generated the phenomenon in question."[30]

It is worthwhile to conclude these comments about Wigner's style of intellectual influence with a remark about one of his characteristics that the reader has surely noted, namely, the nondogmatic manner in which he dealt with the subjects related to the foundations of quantum mechanics. Shimony was well situated to attest to this characteristic because his point of view on the role of the mind in quantum mechanics, different as it was from Wigner's, did not impinge on their close collaboration.[31] According to Shimony (2002), one of the salient features of Wigner's contribution to the measurement problems in quantum mechanics was "[f]reedom from dogmatism, open-mindedness towards new ideas, . . . and in general an exploratory attitude regarding the frontiers of physics, other sciences, and of philosophy." Still, according to Shimony, "consequently, it is a historical error and a misunderstanding of his work, to speak of 'The Wigner solution to the measurement problem' without attention to his exploratory attitude." One last example is related to the reaction to the approach suggested by H. Dieter Zeh (1970). This approach was critical both of Wigner's and of the Italian physicists' approach because both admitted the validity of Schrödinger's equation to describe the measurement devices, and according to Zeh measurement devices are not closed systems to which such an equation could be applied. Wigner received a preprint version of Zeh's paper, supported its publication in the first volume of *Foundations of Physics*, and opened his Varenna talk with six possible solutions to the measurement problem, Zeh's solution being the last. According to Zeh, a preliminary version (in German) of this paper had been rejected by several journals in 1967, "the usual answer being that 'quantum theory does not apply to macroscopic objects,'" a kind of answer based on Bohr's and Rosenfeld's point of view.[32]

CONCLUSION—ORTHODOXY BECOMES HETERODOXY

Let us conclude with three remarks: on Wigner's self-awareness of the role he played in the foundations of quantum mechanics; on the success of his ideas and action; and on a very different question, anachronism in the history of science, which is not strange to the aim of this book. Shimony had the insight to record Wigner's feelings about the attitudinal changes he underwent. These changes may also help us understand changes in the *Zeitgeist* of physicists of the '60s and '70s with respect to the foundations of quantum mechanics. Intending to defend what he considered to be the "quantum orthodoxy," he in fact helped to legitimize heterodoxy on this subject, and he himself became a dissenter. In Shimony's (1997, 412) words: "Wigner recognized with some relish a similarity between the 'heterodox' view that quantum mechanics is only approximate in the physical world and the 'orthodox' view that a reduction of the wave packet occurs only when there is a registration upon the consciousness of an observer." Shimony concluded, citing Wigner: "Both points of view come to the conclusion that the validity of quantum mechanics' linear laws is limited."

During the 1970s, the community working on the foundations of quantum mechanics was mainly occupied with another subject, Bell's inequalities and their experimental tests. Wigner was not as interested in this subject as in measurement problems, but he continued to play an active role until his intellectual vigor began to fade. However, physicists continued to work on the measurement problem research program and, in the 1980s and 1990s, it matured into the decoherence approach with its first experimental results in 1996. Where Wigner saw a role for the mind in quantum measurements, the contemporary trend looks for an exchange of information between the experimental devices and the environment (Zurek 1991; Haroche 1998). Today, Wigner's conjecture about the role of the mind in the quantum measurement process is no longer part of physics, but rather part of the *history* of physics. Nevertheless, the question persists, and from time to time, physicists devote some time to building technical arguments against it (Brandt 2002). In contrast, Wigner's research program—to understand from a physical point of view what quantum measurement is—has flourished and *is* part of physics. Furthermore, in order to create this subfield of physics, foundations of quantum physics, it was necessary to break what Jammer called the "Copenhagen monocracy." As stated by the French physicist Alain Aspect (2004), who is a leader in the field of foundations of quantum physics, and is not at all a critic of the complementarity view, "questioning the 'orthodox' views, including the famous Copenhagen interpretation, might lead to an improved understanding of the quantum mechanics formalism, even though that formalism remained impeccably accurate."[33] Wigner made major contributions in achieving this goal.

Finally, we can conclude that deprecating Wigner's contribution to the foundations of quantum mechanics constitutes an anachronistic reading of those events. Anachronism does not facilitate our understanding of how science really works because it yields to a distortion of real science. The history of science, as a historical discipline, rectifies anachronistic readings of science because, according to Lucien Febvre (1982), historians should prevent the sin of all sins—the unforgivable sin, anachronism. Historians know, however, the tension implied in such prevention, since Marc Bloch and Lucien Febvre, the creators of new perspectives to the historical disciplines, advocated that historians should ask questions of the past, and that these questions could be provoked by contemporary questions. This tension is inherent in a discipline on its way to becoming a science, a science still in its infancy, as remarked by Bloch (1997).

There is further significance in quoting Marc Bloch at the end of this chapter, which is included in a book edited by Boaventura de Sousa Santos. There are some parallelisms between Bloch's and Santos's intellectual *démarches*. In "A Discourse on the Sciences," Boaventura de Sousa Santos (2001) took into account what seemed to him to be lessons from the natural sciences in order to reflect on the changing paradigms of social sciences. Marc Bloch (1997), in his beautiful but unfinished essay about the historian's craft, written in Nazi prisons sometime before being killed on June 16, 1944, by order of Klaus Barbie, affirmed that our mental environment was not the same anymore. Quantum physics, relativity theory, and the kinetic theory of gases soundly changed the idea we had formed about science, making it more flexible, he wrote. Bloch added that we were then better prepared to admit that knowledge, like the historical, even without Euclidian proofs or immutable laws of repetition, could nevertheless aim to be named scientific.

ACKNOWLEDGMENTS

I am thankful to CNPq (Grant 303967/2002-1), Dibner Institute for the History of Science and Technology, American Institute of Physics, and the American Philosophical Society for the grants, which supported this research, and to Abner Shimony, Anja Jacobsen, David Kaiser, Joan Bromberg, Michel Paty, and Sam Schweber, for their comments about previous versions of this chapter. Thanks to James Giangola for providing feedback and editorial suggestions on matters of language. I also thank the following institutions and their staffs for facilitating consultation and for permission to quote from their archives: Henry Margenau Papers [*MP*], Manuscripts and Archives, Yale University Library; John A. Wheeler Papers [*WheP*] and Archives for the History of Quantum Physics [*AHQP*], American Philosophical Society Library, Philadelphia; Eugene P. Wigner Papers [*WP*], Manuscripts Division, Department of Rare Books and Special Collections, Princeton University Library; Léon Rosenfeld Papers [*RP*], Niels Bohr Archive, Copenhagen; Center for the History of Physics—American Institute of Physics [*CHP-AIP*], College Park, MD.

NOTES

1. Gell-Mann (1994, 155), 1969 physics Nobel Prize, wrote: "many sensible, even brilliant commentators have written about the alleged importance of human consciousness in the measurement process. Is it really so important?" Omnès (2002, 174), in a recent analysis, tried to put the mind-body thesis into context, but in previous papers he used derogatory phrases to describe such ideas. Baptista (2002, 63–74) asked where it was that a serious physicist said consciousness plays a role in measurements. He also wrote that Wigner speculated outside the boundaries of the natural sciences.

2. For a technical distinction between formalism and interpretation, see (Jammer 1974).

3. For the purposes of this text, I just need to emphasize that Bohr's complementarity treats measurement devices according to classical physics, not according to quantum physics. This view stands in opposition to Wigner's, whereby measurement devices should instead be treated quantum mechanically. For a standard, comprehensive description of complementarity, see Bohr's (1982) report of his discussions with Einstein.

4. Wigner considered von Neumann's mathematical work (1932) on the foundations of quantum mechanics "more important than any of these inventions [computing machine and implosion bomb]. See E. P. Wigner, interviewed by W. Aspray, April 12, 1984. American Institute of Physics.

5. *Trenton Evening Times*, November 6, 1961: "Princeton Scientist Who Did Work on Atom Bomb Has Own Shelter." See *WP*, Box 97, Folder 1. For detailed information about my archival sources, see the end of this chapter.

6. For a technical presentation of Wigner's papers on quantum measurements, see (Shimony 1997), and G. Emch, "Annotation," in (Wigner 1995, 1–28). Wigner's emphasis on the inability of quantum theory to deal with measurement was not independent of these early works on the limit of measurability implied by the mathematical formalism of the quantum theory, but I do not discuss this issue in this chapter.

7. In his 1961 paper, he wrote one section under the heading "Non-linearity of Equations as Indicators of Life." Later (Wigner 1973), he kept the same stance, "it seems unlikely . . . that the superposition principle applies in full force to beings with consciousness. If it does not, or if the linearity of the equations of motion should be invalid for systems in which life plays a significant role, the determinants of such systems may play the role which proponents of the hidden variable theories attribute to such variables. All proofs of the unreasonable nature of hidden variables are based on the linearity of the equations."

8. Wigner's conjecture about the role of mind in quantum physics was strongly intertwined with his metaphysical and epistemological beliefs. He kept a dualistic view about mind and matter and maintained the former was primary. He criticized mechanistic approaches to the question of life since, for him, the phenomenon of consciousness entreats us to admit the existence of biotonic laws, that is, laws of nature not contained in the laws of physics (Wigner 1972, 1997a, 1997b). I will not, however, extend here my analysis of his broader philosophical views. To a discus-

sion on such issue, see (Esfeld 1999). Thanks to Ron Anderson for bringing this paper to my attention.

9. Letter from Abner Shimony to Wigner, May 1, 1961. *WP*, Box 94, Folder 1.

10. Apparently, Wigner underestimated the ideological background of the quantum controversy. He wrote to Watanabe on August 30, 1961, "Do you know of any political background that has come into the open in these discussions? I am under the happy impression that we can keep the discussion on these subjects free from politics and am not aware of anyone having brought in any doctrine into the argument." Watanabe's reply of December 15, 1961, is premonitory of the Wigner-Rosenfeld's dispute: "I have indeed had quite a few experiences myself of being exposed to shameless attacks by Marxists in Japan for what they call my bourgeois idealism. In spite of the fact that Marxism is not a mechanical materialism, they are dead against giving any kind of independent reality to consciousness. There are Marxists who are quite broad-minded (like Prof. Rosenfeld) in many respects, but they usually become quite emotional when the topic touches upon their basic dogmas. (I was rather disappointed by the partisan emotion which tainted Prof. Rosenfeld's paper on Statistical Mechanics which was published in Poland. Even a broad-minded Marxist like Prof. Rosenfeld acts like this.") *WP*, Box 63, Folder 12, and box 71, Folder 1, respectively.

11. R. Nixon to Wigner, June 22, 1970: "Encouragement is always gratifying, but I particularly appreciated your very thoughtful letter and I want you to know how pleased I was to hear from you. Your support for our policies toward Southeast Asia means a great deal to America's fighting men, and needless to say, it means a great deal to me." *WP*, Box 97, Folder 3.

12. Anderson's distinction is driven to label those Marxist intellectuals, such as Lukacs, Korsch, Bloch, and Adorno, who took distance from the Soviet Marxism and the Western Communist parties related to it. The distinction is not trivial since Antonio Gramsci, the leader of Italian Communism, is considered, for his works, part of the Eastern Marxism. Roughly used, however, it is useful to understand Rosenfeld's Marxism.

13. See letters to L. Rosenfeld, from J. A. Wheeler (March 27, 1952); R. E. Marshak (September 24, 1954); and A. Roberts (December 22, 1955). *RP*.

14. I am thankful to Sam Schweber for discussion on this issue. Commenting on the founder fathers of quantum mechanics, Schweber (1996) wrote: "Wigner stands out by being, on the one hand, the theorist who had perhaps the greatest affinity to pure mathematics and, on the other, probably the most phenomenologically inclined among them."

15. L. Rosenfeld to Saul M. Bergmann, December 21, 1959. *RP*. The subject of the letter concerns Everett's approach to quantum physics.

16. "I just finished reading the article of Daneri Loinger and Prosperi in the July issue of Nuovo Cimento and am really a bit irritated by it. First of all, it is not good taste to say about a set of articles that they do not make substantial contributions to a subject. Needless to say, I am less concerned about myself than about other people who are much younger than I am and whose future careers such statements may hurt." Letter from E. Wigner to J. M. Jauch, September 6, 1966. *WP*, Box 94, Folder 7.

17. The suggestion to quote Bohm and Bub's paper came from Shimony (Letter to Wigner, January 1, 1967). This letter records Shimony's first reaction to what we now call "Bell's inequalities." *WP*, Box 83, Folder 7.

18. (Rosenfeld 1953, 56), and letter from L. Rosenfeld to N. Bohr, January 14, 1957. *AHQP*, Bohr Scientific Correspondence, reel 31.

19. Bernard d'Espagnat's, interviewed by the author, October 26, 2001. Deposited at *CHP-AIP*.

20. Varenna's courses, organized by the Italian Society of Physics, had been regularly carried out since 1953, in the summer, in Varenna, on Como Lake. The 1970 course was dedicated to the theme "Foundations of Quantum Mechanics," following a suggestion from Franco Selleri supported by Toraldo di Francia, then the president of that society. It had eighty-four participants, and its proceedings (d'Espagnat 1971) reveals a diversified spectrum of subjects, such as measurement, hidden variables and nonlocality, interpretations, and people, such as Wigner, Jauch, Shimony, d'Espagnat, Bell, de Broglie, Selleri, and Bohm. D'Espagnat, who was its head, suggested in an invitation letter some diplomatic rules to be followed in order to have a pacific and creative coexistence in which to discuss scientific controversies.

21. "It would indeed be a service to people interested in foundations of quantum mechanics for you to reconstruct your discussion with Prosperi." Letter from Shimony to Wigner, (w/d 1971), *WP*, Box 72, Folder 2.

22. G. Prosperi, July 3, 2003, interviewed by the author.

23. Comparing the two physicists, Schweber (1996) wrote: "Except for John Bell, no one addressed these foundational issues as critically as Wigner did."

24. d'Espagnat, quoted interview.

25. As early as February 18, 1964, d'Espagnat criticized Jauch for his idea that mixture and pure state are in the same "equivalence class," and supported Wigner: "This is a matter into which I always took a great interest and I found your article in AJP very illuminating." (*WP*, Box 94, Folder 1). Later, Rosenfeld praised d'Espagnat's book (1965), but not the paper in which d'Espagnat (1966) suggested a generalization of Wigner's point of view, according to which "the framework of the orthodox theory of (ideal) measurements" means that these cannot as a rule be described by means of linear quantum mechanical laws. d'Espagnat wrote to Rosenfeld (February 26, 1966), "*Je tiens à vous remercier pour . . . l'approbation que vous avez la gentilesse d'y exprimer à l'égard de mon livre.*" Four months later, Rosenfeld wrote (July 8, 1966), "*votre dernier travail* 'Two Remarks on the Theory of Measurement" *semble indiquer que vous avez besoin de vous retremper dans l'air pur de Copenhague* (d'Espagnat had been there in 1954, and had received, in January 1966, an invitation to return from Rosenfeld). *Il n'y a rien de tel comme cure de cette wignérite dont vous paraissez subir une atteinte, que j'espère légère.*" The letters are in the *RP*.

26. "I have read your illuminating paper in the American Journal of Physics. This, together with thoughts about other materials from your pen and further recent publications, has prompted me to put together what I consider a simple and consistent theory of measurement. . . . I believe I have not made sufficiently clear in the past what I regard as important, for I really think that my basic concepts do not differ from your version." Margenau to Wigner, January 21, 1963. *MP*, Box 1, Folder 12.

27. See Margenau and Wigner (1962). For the book they planned, as suggested by Wigner, see the letter from Margenau to Wigner, October 4, 1974 (*WP*, Box 56,

folder 13); idem, December 26, 1974 (*WP*, Box 72, Folder 3); and the letter from Wigner to Margenau, December 28, 1974 (*MP*, Box 1, Folder 12). Later, however, Wigner apparently did not follow Margenau's admission of extrasensory perception and remained skeptical about Margenau's essays on blending science and religion. See letter from Margenau to Wigner, May 27, 1988; and Wigner's to Margenau, June 30, 1988 (*WP*, Box 56, Folder 13). For Margenau's views on extrasensory perception, see the material in *WP* (Box 56, Folder 13), and *MP* (Box 1, Folder 6).

28. See *Foundations of Physics*, 1970, 1, "editorial preface."

29. Letter from Wigner to Robert Ubell, September 24, 1974, *WP*, Box 72, Folder 3.

30. Wheeler to Aage Bohr, February 25, 1977; Wheeler to John Hopfield, May 2, 1977; Aage Bohr to Wheeler, May 16, 1977. The letters are at the *WheP*, Series V, Notebook October 1976–April 1977.

31. In his first published paper on the foundations of quantum mechanics, Shimony (1963) analyzed these two proposals for interpreting quantum state evolution during measurements: von Neumann's and Bohr's approaches. His point of view on the former was that "although this interpretation appears to be free from inconsistencies, it is not supported by psychological evidence and it is difficult to reconcile with the inter-subjective agreement of several independent observers."

32. Wigner to Margenau (editor of *Foundations of Physics*), March 31, 1970, "I am really very glad that Zeh's paper was accepted." *MP*, Box 1, Folder 12. Zeh (1970) thanks Wigner for his support of his paper. Information about the refusal of this paper was obtained at www.rzuser.uni-heidelberg.de/~as3/, on May 14, 2007.

33. We have an impressive piece of evidence for the low status of research in foundations of quantum physics; just before the Wigner-Rosenfeld dispute, in J. S. Bell's remarks, first published in 1966: "The minority view is as old as quantum mechanics itself, so the new theory may be a long time coming. . . . We emphasize not only that our view is that of a minority, but also that current interest in such questions is small. The typical physicist feels that they have long been answered, and that he will fully understand just how if ever he can spare twenty minutes to think about it" (Bell and Nauenberg 1966). I thank Osvaldo Pessoa for bringing this interesting excerpt to my attention.

REFERENCES

Aaserud, Finn. 1995. "Sputnik and the 'Princeton Three': The National Security Laboratory that Was Not to Be." *Historical Studies in the Physical and Biological Sciences* 25 (2):185–239.

Anderson, P. 1979. *Considerations on Western Marxism*. London: Verso.

Aspect, A. 2004. "Introduction." In *Speakable and Unspeakable in Quantum Mechanics*, 2nd edition, ed. J. S. Bell. Cambridge University Press.

Ballentine, L. E. 1970. "The Statistical Interpretation of Quantum Mechanics." *Review of Modern Physics* 42 (4):358–81.

Baptista, Antonio M. 2002. *O discurso pós-moderno contra a ciência - obscurantismo e irresponsabilidade*. 2nd ed. Lisbon: Gradiva.

Bell, John S., and M. Nauenberg. 1966. "The Moral Aspect of Quantum Mechanics." In *Preludes in Theoretical Physics*, ed. A. De Shalit, H. Feshbach, and L. Van Hove. Amsterdam: North Holland.

Bloch, Marc. 1997. *Apologie pour l'histoire, ou le métier d'historien*. Paris: Armand Colin.

Bohr, Niels. 1982. "Discussion with Einstein on Epistemological Problems in Atomic Physics." In *Albert Einstein: Philosopher-Scientist*, 3rd edition, ed. Paul A. Schilpp. La Salle and London: Open Court/Cambridge University Press.

Brandt, Howard E. 2002. "Deconstructing Wigner's Density Matrix Concerning the Mind-Body Question." *Foundations of Physics Letters* 15 (3):287–92.

Bromberg, Joan L. 2006. "Device Physics vis-à-vis Fundamental Physics in Cold War America: The Case of Quantum Optics." *ISIS* 97 (2):237–59.

Daneri, A., A. Loinger, and G. M. Prosperi. 1962. "Quantum Theory of Measurement and Ergodicity Conditions." *Nuclear Physics* 33:297–319.

———. 1966. "Further Remarks on the Relation Between Statistical Mechanics and Quantum Theory of Measurement." *Nuovo Cimento* 44B (1):119–28.

d'Espagnat, Bernard. 1965. *Conceptions de la physique contemporaine—Les interprétations de la mécanique quantique et de la mesure*. Paris: Hermann.

———. 1966. "Two Remarks on the Theory of Measurement." *Suppl al Nuovo Cimento* 1:828–38.

d'Espagnat, Bernard, ed. 1971. *Foundations of Quantum Mechanics—Proceedings of the International School of Physics "Enrico Fermi."* London: Academic Press.

Doncel, Manuel G., Louis Michel, and Jules Six. 1984. "Interview de Eugene P. Wigner sur sa vie scientifique." *Archives Internationales d'Histoire des Sciences* 34 (112):177–217.

Esfeld, Michael. 1999. "Wigner's View of Physical Reality." *Studies in History and Philosophy of Modern Physics* 30B:145–54.

Febvre, Lucien. 1982. *The Problem of Unbelief in the Sixteenth Century, the Religion of Rabelais*. Trans. Beatrice Gottlieb. Cambridge, MA: Harvard University Press.

Freire, Olival, Jr. 2003. "A Story Without an Ending: The Quantum Physics Controversy 1950–1970." *Science & Education* 12:573–86.

———. 2004. "The Historical Roots of 'Foundations of Physics' as a Field of Research (1950–1970)." *Foundations of Physics* 34 (11):1741–60.

———. 2005. "Science and Exile—David Bohm, the Cold War, and a New Interpretation of Quantum Mechanics." *Historical Studies in the Physical and Biological Sciences* 36 (1):1–34.

———. 2006. "Philosophy Enters the Optics Laboratory: Bell's Theorem and Its First Experimental Tests (1965–1982)." *Studies in History and Philosophy of Modern Physics* 37 (4):577–616.

Frisch, O. R. 1971. "The Conceptual Problem of Quantum Theory from the Experimentalist's Point of View." In *Quantum Theory and Beyond–Essays and Discussions Arising from a Colloquium*, ed. Ted Bastin. London: Cambridge University Press.

Gell-Mann, Murray. 1994. *The Quark and the Jaguar: Adventures in the Simple and the Complex*. New York: W. H. Freeman and Company.

Haroche, Serge. 1998. "Entanglement, Decoherence and the Quantum/Classical Boundary." *Physics Today* 51 (7):36–42.

Heilbron, John. 2001. "The Earliest Missionaries of the Copenhagen Spirit". In *Science and Society—The History of Modern Physical Science in the Twentieth Century*, vol. 4, *Quantum Histories*, ed. P. Galison. M. Gordin, and D. Kaiser. New York: Routledge.

Home, D., and M. A. B. Whitaker. 1992. "Ensemble Interpretations of Quantum Mechanics. A Modern Perspective." *Physics Reports* 210 (4):223–317.

Jammer, Max. 1974. *The Philosophy of Quantum Mechanics—The Interpretations of Quantum Mechanics in Historical Perspective.* New York: John Wiley & Sons.

Kuhn, Thomas S., John L. Heilbron, and Paul Forman. 1967. *Sources for History of Quantum Physics. An Inventory and Report.* Philadelphia: American Philosophical Society.

Margenau, Henry, and Eugene P. Wigner. 1962. "Discussion: Comments on Professor Putnam's Comments." *Phil. Sci.* 29:292–93. Reprinted in Wigner 1995.

Mehra, Jagdish. 1993. "Eugene Paul Wigner—A Biographical Sketch." In *The Collected Works of Eugene Paul Wigner*, part A, vol. 1. Berlin: Springer, 3–14.

Omnès, Roland. 2002. "Edmond Bauer et la conscience de l'observateur en mécanique quantique" *Épistémologiques* 2 (1–2):171–76.

Rosenfeld, Léon. 1953. "L'évidence de la complementarité." In *Louis de Broglie—Physicien et Penseur*, ed. André George. Paris: Albin Michel.

———. 1965. "The Measuring Process in Quantum Mechanics." *Progress of Theoretical Physics* Supplement: 222–35.

———. 1968. "Questions of Method in the Consistency Problem of Quantum Mechanics." *Nuclear Physics* A108:241–44.

Santos, Boaventura S. 2001. *Um discurso sobre as ciências*. 12th ed. Porto, Portugal: Afrontamento.

Schweber, Silvan S. 1996. "Wigner's Vast Output: Four of the Eight Planned Volumes." *Physics Today* October:65–66.

Shimony, Abner. 1963. "Role of the Observer in Quantum Theory." *American Journal of Physics* 31:755–73.

———. 1993. *Search for a Naturalistic World View*, vol. 1. New York: Cambridge University Press.

———. 1997. "Wigner on Foundations of Quantum Mechanics". In *The Collected Works of Eugene Paul Wigner*, part A, vol. 3, *Particles and Fields; Foundations of Quantum Mechanics.* Berlin: Springer.

———. 2002. "Wigner's Contributions to the Quantum Theory of Measurement," *Proceedings of the Wigner Centennial Conference*, Pécs [Hungary], Paper 51.

von Neumann, J. 1932. *Mathematische Grundlagen der Quantenmechanik*. Berlin: Springer.

Weinberg, Alvin M. 2002. "Eugene Wigner, Nuclear Engineer." *Physics Today* 55 (10):42–46.

Wigner, Eugene P. 1952. "Die Messung quantenmechanischer Operatoren." *Zeitschrift für Physik* 133:101–8. Reprinted in Wigner 1995.

———. 1961. "Remarks on the Mind-Body Question." In *The Scientist Speculates*, ed. I. J. Good. London: W. Heinemann. Reprinted in Wigner 1995.

———. 1963. "The Problem of Measurement." *American Journal of Physics* 31:6–15. Reprinted in Wigner 1995.

——. 1972. "The Place of Consciousness in Modern Physics." In *Consciousness and Reality*, ed. C. Muses and A. M. Young. New York: Outerbridge and Lazard. Reprinted in Wigner 1995.

——. 1973. "Epistemological Perspective on Quantum Theory." In *Contemporary Research in the Foundations and Philosophy of Quantum Theory*, ed. C. A. Hooker. Dordrecht: Reidel. Reprinted in Wigner 1995.

——. 1995. *The Collected Works of Eugene Paul Wigner*, vol 6, part B. *In Philosophical Reflections and Syntheses*. Berlin, Germany: Springer.

——. 1997a. "The Probability of the Existence of a Self-Reproducing Unit." In *The Collected Works of E. P. Wigner*, vol. 3, part A. Berlin: Springer.

——. 1997b. "Are We Machines?" *Idem*.

Zeh, H. D. 1970. "On the Interpretations of Measurement in Quantum Theory." *Foundations of Physics* 1:69–76.

Zurek, Wojciech H. 1991. "Decoherence and the transition from quantum to classical". *Physics Today* 44 (10):36–44.

11

The Rhetoric of Science in Darwin's *On the Origins of Species*

*Anna Carolina Regner**

INTRODUCTION

The *science wars* that were going on in the 1990s have become silent among scientists on duty, but they left deep "wounds" asking for an epistemological treatment. A welcome new edition of Boaventura de Sousa Santos' *Um discurso sobre a ciência* (2002) reminds us of what was (and still is) at stake. Among other things, we are faced with the question about the rationality of science.

Traditionally, the meaning of rationality is associated to our ability to discern properties, establish relationships and build arguments to present and defend our beliefs. On the one hand, rationality is the exercise of a cognitive faculty—we do call it "reason." On the other, it is the property that pervades all products of this exercise—let us call them "reasons." Among these products, there is the generation of reason's own regulatory principles and criteria to evaluate reliable and objectively valid knowledge, be it considered as solidly founded, or as critically fallible knowledge. Would the traditional view of scientific rationality be able to deal with the questions raised by recent analyses of science? Would the bases for the traditional view of rationality be strong enough to support its own criteria for the demarcation of "scientific rationality"?

An extensive literature has examined the conditions under which contemporary science is determined or made possible. These conditions include the impact of the new technological approaches on concepts such as "theory," "experience," "evidence," "cognitive process," "cognitive faculties," "subject" and "object," the formation of new disciplinary and interdisciplinary charts, and the institutional, cultural, political and material aspects of doing science. In short, the new conditions pointed out by the recent analyses of science break with our traditional way of thinking about science. From these analyses, we derive a negative answer to our first question above. Facing the new scenario, we cannot appeal to a tacit understanding of what "scientific rationality" means, and it seems that we must choose between recognizing science as an "irrational activity" and investing on a new view of "scientific rationality." Both alternatives demand the examination of the meaning of "rationality," and there is no way to step out of "rationality" in order to examine it from a irrational point of view. In order to tell what "rationality" may mean, we must think of it as "rationality in action," rationality "doing something," in "some way," and, then, as "rationality in context." Contextuality does not conflict with the traditional requirement that reason's principles and operations should be "universal" or "universally valid," since their universality is measured against the "particular" they determine and thus demands us to look at the "universal-particular" relationship, which is always given in a context.

Therefore, in order to analyze the bases for the traditional view of "scientific rationality," we will look at its original context, which is found in Aristotle's texts. In order to see how this view could (or not) succeed in dealing with modern scientific argumentation, we will compare the traditional view with one landmark of our scientific tradition, Charles Darwin's *Origin of Species*[1]. To make our examination still more convincing, we focus on a basic distinction found in the traditional view: the distinction between "scientific" and "rhetorical" arguments, which are at the bases of other several important epistemological distinctions, such as "subjective / objective," "verisimilar/true."

ARISTOTLE'S RHETORIC AND DARWIN'S SCIENCE

At the very bottom of the traditional view, we find Plato's *Teetetos* on science and *Gorgias* on rhetoric. These texts show us, on the one hand, rhetoric as the art of the eloquence and persuasion based on the speaker's subjective intentions of manipulating the dispositions and decisions of the listener through language. On the other hand, they show science as opposed to rhetoric and as based on the objectivity of what is, at the service of knowledge and virtue. Aristotle maintained some of the Platonic distinctions. He distinguished between science and opinion, and between science and rhetoric, but he brought important changes in the approach to the latter.

For Aristotle, genuine knowledge (science) is the one obtained by demonstration starting from premises that express necessary truths and looking for the causes or necessary connections that determine the way things are. What is verisimilar belongs to the realm of opinion and concerns what ordinarily happens but does not do so in an absolute way. This means that what is verisimilar can be otherwise, and then can be true or false. Rhetoric (and dialectics) does not deal with the way things are, but with the way some cognitive faculties of seeking for reasons are determined by our faculty to discern what can be believed in each circumstance, what is verisimilar (*Rhetoric*, Book I, chapter 2).

According to Aristotle and differently from Plato, rhetoric is the faculty to discern what is verisimilar which has an object of its own and constitutes a legitimate field of investigation (*Rhetoric*, Book I, chapter 1). Its object is not persuading, but the examination of the persuasive ways proper to each case. Rhetoric (and dialectics) is concerned with things that do not belong to any science in particular, but that are somehow known and used by everybody, since everyone attempts to inquire about or resist to some reason, defends his/her own ideas and accuses his/her opponents (*Rhetoric*, Book I, chapter 1).

Aristotle's fundamental difference between "rhetorical" and "scientific" arguments is epistemic rather than formal. It is proper to scientific syllogisms to start from premises stating "what cannot be otherwise," while it is proper to rhetorical syllogisms to start from premises stating "what can be otherwise." Nevertheless, they have important points in common, which opens the way to a less canonical interpretation of the Aristotelian distinction. Both of them are syllogisms, although the rhetorical ones are presented as enthymemes. In both cases it is necessary, in the first place, to have the respective premises and things whose occurrence is acceptable (*Rhetoric*, Book I, chapters 3 and 4). Starting from necessary truths is not exclusive to the scientific syllogisms, although these syllogisms always start from this kind of premises, and few are the premises about necessary things from which the rhetorical ones start (*Rhetoric*, Book I, chapter 2). Both kinds of arguments can be constructed starting from topics or common notions, as well as from specific notions, although the rhetorical syllogisms are more commonly obtained from common notions, and the scientific syllogisms depart start from specific notions (*Rhetoric*, Book I, chapter 2).

Besides, according to Aristotle, the understanding of what is true and of what resembles it belongs to the same faculty (*Rhetoric*, Book I, chapter 1). Demonstrating and persuading are very often intertwined. In the *Poetics*, under the topic "reason and language," Aristotle says: "I deal with reason more fully in my *Art of Rhetoric*. The word "reason" includes all the effects that can be produced by the language: proof or refutation of an argument, the arousal of feelings like pity, terror, anger, and the others, the capacity to exaggerate and understate" (*Poetics*, chapter 19).

"Demonstrating" and "persuading" are equally included in this list. Rhetoric is also concerned with facts, but in order to produce the desired dispositions in the listener by means of the speaker's use of language.

Clarity, certainty and conviction of truth are also dispositions of the mind. Demonstrating is also a way of persuading. And Aristotle calls attention to the fact that

> This is so even if one is addressing a single person and urging him to do or not to do something, as when we scold a man for his conduct or try to change his views: the single person is as much your "judge" as if he were one of many; we may say, without qualification, that any one is your judge whom you have to persuade. Nor does it matter whether we are arguing against an actual opponent or against a mere proposition; in the latter case we still have to use speech and overthrow the opposing arguments, and we attack these as we should attack an actual opponent (*Rhetoric*, Book II, chapter 18).

Rhetoric is revealed as necessary and valid not only to undo the reasons of those who try convince us of reproachable things, but also to persuade certain audiences when it is not easy to do so by means of science:

> Still, the whole business of rhetoric being concerned with appearances, we must pay attention to the subject of delivery, unworthy though it is, because we cannot do without it. The right thing in speaking really is that we should be satisfied not to annoy our hearers, without trying to delight them: we ought in fairness to fight our case with no help beyond the bare facts: nothing, therefore, should matter except the proof of those facts. Still, as has been already said, other things affect the result considerably, owing to the defects of our hearers. The arts of language cannot help having a small but real importance, whatever it is we have to expound to others: the way in which a thing is said does affect its intelligibility" (*Rhetoric*, Book III, chapter 1).

Scientific discourse is concerned with instruction (*Rhetoric*, Book I, chapter 1), and the rhetorical art is necessary for all teaching (*Rhetoric*, Book III, chapter 1).

Aristotle's view does not restrict rationality to science, nor does it exclude rationality from rhetoric, but Aristotle restricts necessary and absolutely true knowledge to science. The Aristotelian view will suffer the impact of more contemporary criticisms based on the conjectural character of scientific knowledge, which, for this reason, belongs in the realm of "what can be otherwise." However, most of these criticisms would like to preserve a distinction between "scientific" and "rhetorical" arguments and to maintain the authority of the first. But, as this distinction could not be based on the Aristotelian epistemic distinction, any other attempt would have to start from or lead to a comparison between procedures supposedly characteristic of each kind of those arguments. Nevertheless, when we compare the procedures traditionally labeled "rhetorical" with those that characterize an exemplary case of argumentative science, namely, the "one long argument" of Charles Darwin's *Origin of the Species*, that distinction collapses.

From the very beginning of the *Origin of the Species*, we are surprised at the use Darwin makes of arguments and explanatory patterns traditionally called "rhetorical" to elaborate his "one long argument." He was not a professional epistemologist concerned with defining "explanation" before using this term. The only passage in which he offers us something closer to a definition is in his Autobiography: to "understand" or "explain" what one observes means "to group all facts under some general laws" (Darwin 1958, 55). When we build a Darwinian meaning of explanation from the analyses of the occurrences of "explanation" and its cognates in the text of the *Origin of Species*, we find a multifaceted semantic horizon of innovative connotations of "explanation," like the use he makes of imagination, metaphors, and the appeal to our ignorance and to the psychological and sociological values of scientific investigation. And the more usual ones—like observation and experiment, subsumption of facts under a rule, study of exemplary cases, use of diagrams, illustrations, discussions, comparisons, analogies and calculations—acquire a particular meaning in the light of Darwin's argumentative strategies.

In the context familiar to Darwin, it was common to learn from John Herschel's, William Whewell's, and Stuart Mill's analyses of science that "explanation" is a rational activity of understanding the meaning of facts and clarifying concepts, giving reasons to support our beliefs and expectations, building an argument, looking for causes (and, more than that, looking for a *vera causa*, proceeding according to patterns accepted by the relevant community, and aiming at the Victorian ideals of social and moral improvement of mankind.

The Darwinian understanding of the meaning of facts is unfolded in the usual "what," "how," and "why"—questions about different levels and the nature of facts, but also in "knowing what to ask," generating "patterns of questions and answers," and "establishing conjectures." "Explaining" encompasses "elucidating concepts," "treating difficulties and objections," and accounting for "procedures" (as those used in Classification). "Explaining" means to give reasons, but without restricting them to the traditional logical and empirical reasons; among other things, "beliefs" or "mental habits" can also be "reasons." "Explaining" means to build an argument, but the traditional meaning of "deductive" or "inductive" arguments receives new connotations from the Darwinian explanatory effort. Instead of inductive generalizations, Darwin undertakes the study of "exemplary cases." Quite often, Darwin's deductive arguments do not meet strict deductive patterns: their premises are dependent on each other and even on the conclusion. Although "causal explanation" is the explanation Darwin looks for, the Darwinian causality is not limited to the traditional Humean and Kantian patterns, and makes use of causality in terms of functions, goals, purposes, and of the causal analysis in terms of causal networks.

Darwin's argumentative strategies will be a central element in his complex explanatory effort—to show how new species are originated in Nature—and the impossibility of supporting his thesis—species originated in Nature by means of Natural Selection—on immediate and conclusive empiric evidence.

DARWIN'S ARGUMENT

The Darwinian argumentative strategies operate on different levels, under the general strategies which structure the "one long argument," such as *the particular whole-part movement* designed to assemble Darwin's argument, his appeal to *the explanatory power as a whole* to deal with the evidence, *the comparison of his view with those of his opponents* to emphasize his superior explanatory power, *the treatment of difficulties / objections / exceptions* as part of the entire explanatory power of his theory, *the interplay of the actual and the possible* in order to work out the evidence and the difficulties, and *the weight of reasons* of both sides. In order to analyze the nature of these strategies and their importance to rethink the traditional distinctions related to the traditional view of "scientific rationality," we will compare the "rhetorical" procedures recommended by Aristotle and the "scientific" ones adopted by Darwin.

AS TO THE GENERAL STRUCTURING OF THE WORK

Aristotle

One should begin by establishing one's own arguments and then attack the reasons raised by the opponent (*Rhetoric*, Book III, chapters 17 and 19).

Darwin

Darwin begins with the establishment of the logical-conceptual frame that structures his theory (chapters I–V). Immediately afterwards, Darwin treats of the difficulties rose his theory has to face (chapters VI–IX) and makes this treatment culminate in the transformation of the key unfavorable evidence—the geological records—into a favorable one (chapter X). This treatment prepares the reader to plainly accept the cases clearly favorable to Darwin's theory (chapters XI–XIV) and maximizes its explanatory force. In chapter XV, in one single breath, Darwin recapitulates all the arguments and facts raised in each step of the general argument and evaluates the explanatory power of his argument as a whole.

Aristotle

Preamble

The speaker should start by praising, criticizing, persuading and dissuading in order to indicate the tone of *his / her* speech (*Rhetoric*, Book III, chapter 14).

Darwin

In his Historical Sketch, Darwin evaluates the historical path of the investigation on the origin of species, and places his own theory at the culmination of that process. In his Introduction, he presents his objectives, praises the "wonderful" facts to be explained, points out the new demands related to the problem of the origin of species, and criticizes the attitude of believing in the explanatory power of evolution without showing how it happens.

Aristotle

In the *narration*, the themes should be composed of related and truth resembling parts (*Poetics*, chapter 24)

Darwin

A particular whole-part movement structures the Darwinian narrative. The whole, the entire book, the "one long argument," sustains the parts, the chapters, the short arguments, and, at the same time, the whole is sustained by its parts through a continuous projective and retrospective movement from one chapter to another chapter. This very general strategy embraces the other strategies. Together, they act on the Darwinian explanatory procedures and give them an innovative character. As an example, let's look at Darwin's reference to "experiment." The main line of Darwin's argumentation does not rely on empirically proving his theory by means of immediate and conclusive evidence. The "empirical evidence" is not just collected by direct observation, but integrates findings from diversified sources of information (scientific literature, farmers, gardeners, an old Chinese encyclopedia, classic Roman authors, the biblical book of Genesis), as he does in the central case of the origin of the races of domestic pigeons. When examining the support provided to his theory by the available evidence, instead of measuring the theory against singular events, he emphasizes its explanatory power as a whole. He appeals to the interplay of what is actually given and what is logically and/or factually possible and skillfully compares this conjoined possibility with the impossibility and/or actual inexistence of contrary evidence.

Aristotle

Epilogue

Briefly, the speaker should tell what the demonstration served to, say that the demonstration fulfilled what was promised, confront his/her opponent, and compare his/her own reasons with the ones raised by his/her opponent one by one, or face the objections according to the order "natural" to the refutation (*Rhetoric*, Book III, chapter 19).

Darwin

In his epilogue (chapter XV), Darwin recapitulates and evaluates his "one long argument" in one single breath, but still apparently much longer than Aristotle's recommendation to be brief. However, considering the extension of Darwin's argument constructed throughout the book, his conclusive chapter presents the demonstration in a concise way and emphasizes the execution of what was promised in the Introduction. He showed "how the innumerable species inhabiting this world have been modified, so as to acquire that perfection of structure and coadaptation which justly excites our admiration" (Darwin 1875, 2), answering the main question: "how are new species originated in Nature?" And he does so by confronting his opponents, comparing the reasons given by each side, and facing them in the "order" natural to the refutation.

AS TO THE GENERAL ARGUMENTATIVE STRATEGIES

Aristotle

Pondering all the reasons and circumstances that may be involved in the narrative is the most appropriate form to demonstrative speeches (*Rhetoric*, Book I, chapter 9).

Darwin

In his Introduction as well as in his Conclusion, Darwin asks the reader to judge his theory "as one long argument" in the light of its explanatory power as a whole after considering all the facts and reasons on both sides of the subject:

"I am well aware that scarcely a single point is discussed in this volume on which facts cannot be adduced, often apparently leading to conclusions directly opposite to those at which I have arrived. A fair result can be obtained only by fully stating and balancing the facts and arguments on both sides of each question; and this is here impossible" (Darwin 1875, 3).

As this whole volume is one long argument, it may be convenient to the reader to have the leading facts and inferences briefly recapitulated. (Darwin 1875, 404)

Aristotle

Whenever two things are very similar, one should examine them from the point of view of their consequences (*Topics*, Book III, chapter 2).

Darwin

By comparing the explanatory power of the hypotheses in dispute, the nature and number of facts that they explain, Darwin finds a solid reason in favor of his theory against its opponents: "It can hardly be supposed that a false theory would explain, in so satisfactory a manner as does the theory of natural selection, the several large classes of facts above specified" (Darwin 1875, 421).

Aristotle

Make an opportune use of the interrogations, leading the opponent to contradict him/herself or to concede the point we want to defend (*Rhetoric*, Book III, chapter 19).

Darwin

Darwin makes an innovative use of "asking questions," addressing them to elucidate what should be asked, searching for the meaning of what seems to be "curious" or difficult to understand, bringing into regularity what seems to be strange, generating "questions-answers patterns," and leaving to the opponent the burden of proof: "We see in these facts some deep organic bond, throughout space and time, over the same areas of land and water, independently of physical conditions. The naturalist must be dull who is not led to enquire what this bond is" (Darwin 1875, 318).

The answer to the question about what seems "curious" or difficult to understand can be found in the significance that some facts spell out to us, as in the case of the "rudimentary organs" that "plainly declare their origin and meaning in various ways" (Darwin 1875, 397). Or the answer can be found in looking at the fact in the light of the intelligible pattern to which it belongs, as in the case of the similarity of organs by virtue of their conformity to a given type (Darwin 1875, 382). The pattern configures not only a set of questions and answers, but also what should be said, even concerning intricate matters, such as the one on the instinct of bees (Darwin 1875, 133).

Aristotle

Start by examining contrary things or predicates, and proceed comparatively (*Rhetoric*, Book II, chapter 23).

Darwin

All Darwinian explanatory efforts are made through the comparison of opposed theoretical views. Since there is no one single fact that could not be seen in one or another way (Darwin 1875, 2), we must compare the explanatory power of the opposed views. In fact, even the qualification of the empiric evidence depends on its examination in the light of that comparison. As a result of this procedure, the Darwinian theory shows itself as a possible and then as the best alternative, and, finally, as the only rational "explanation" (the only one accessible by "rational means," according to our cognitive condition as "natural" beings who investigate "natural" objects and processes).

Aristotle

Start from the examination of "the possible/the impossible": if the similar is possible, that which is similar to the first is possible too; if the most difficult is possible, the easiest is also possible; if the parts are possible, the whole they are parts of is possible, too; if a thing is possible for those who are inferior, or weaker, or less intelligent, it will be still more so for those whose qualities are the opposite (*Rhetoric*, Book II, chapter 19).

Darwin

The interplay of the actual and the possible is one of the most innovative aspects of the Darwinian explanation. At crucial moments of his argumentation, Darwin starts from what is / what is not actually given, and from what is logically or factually possible, and conjugates the possibility of the favorable evidence with the impossibility nonexistence of contrary evidence. The Darwinian interplay of possibilities / impossibilities enlarges the field of investigation[2]. Himmelfarb refers to this game as to a new "logic of the possibility":

> His essential method was neither observing nor the more prosaic mode of scientific reasoning, but a peculiarly imaginative, inventive mode of argument. . . . What Darwin was doing, in effect, was creating a "logic of possibility." Unlike conventional logic, where the compound of possibilities results not in a greater possibility, or probability, but in a lesser one, the logic of the *Origin* was one in which possibilities were assumed to add up to probability. (Gertrude Himmelfarb, 1968, 333–34)

Aristotle

Prefer what is more useful and demands less effort (*Rhetoric*, Book I, chapter 3–10), and pay attention to the characteristics of the audience and to what facilitates learning (*Rhetoric*, Book I, chapter 1, and Book III, chapters 1 and 10).

Darwin

A substantial part of the Darwinian theory incorporates argumentative resources traditionally taken as rhetorical, such as the appeal to familiarity, to the simplification of the systematist's work routine, to mental habits (in order to account for the well-known naturalists' rejection of Darwin's theory), to the progressive mind of those from whom Darwin expects adherence to his theory (Darwin 1875, 422 and 423; Francis Darwin 1888: Vol. II, 245—letter to Asa Gray of December 21, 1859, p. 355—letter to Huxley of December 2, 1860), and to the revolutionary spirit of his theory. Darwin anticipates the revolution in Natural History to be caused by his theory (Darwin 1875, 425ff.) by demanding the restructuring of disciplinary fields and the creation of new ones, and thus contributing to the advancement of knowledge. Darwin refers to the occurrence of revolutions in decisive moments of human knowledge in such a way that we are led to think he considered revolutions as a sign of knowledge advancement (Darwin 1875, 226, 285).

Aristotle

Make comparisons with famous people (*Rhetoric*, Book I, chapter 9, and Book II, chapter 25).

Darwin

Throughout his work, Darwin appeals to renowned scientific authorities in order to support the results that directly or indirectly favor his theory.

Aristotle

If one has no enthymemes, one must employ actual or invented examples as demonstrative proofs, for conviction is produced by these; but if one has them, examples must be used as evidence and as a kind of epilogue to the enthymemes because they resemble evidence, and a witness is in every case likely to induce belief (*Rhetoric*, Book II, chapter 20).

Darwin

Darwin makes frequent use of exemplary cases and imaginary illustrations. The study of exemplary cases—as in the case of the origin of domestic pigeons (Darwin 1875, 15)—is a fundamental Darwinian strategy. Such a procedure allows him to meticulously analyze the question and fairly scrutinize the available evidence. Requirements for precision and style are met at the same time (Darwin 1875, 321). If Darwin founded the credibility of his statements just on the amount of examined facts, he would be always exposed to criticism based on the fact that the list of the requested evidence was imperfect. The use of "exemplary cases" by Darwin meets the persuasive condition that Aristotle attributes to them in rhetorical arguments, because they contain the universal in an implicit way. As examples of a universal condition, they cannot be blamed for being at the service of just inductive generalizations.

In general, the role of illustrations is to confirm Darwin's theory (Darwin 1875, 360), and to account for the existence and nature of the exceptions to the rules according to his theory (Darwin 1875, 391, 392). In the use of illustrations, we can see the powerful use Darwin makes of imagination as an explanatory resource that goes beyond the mere portrait of what is empirically given. The use of illustrations is combined with the use of other strategies, and all of them are framed within the main perspective of evaluation of the explanatory power of the theory by means of weighing pro and con reasons.

Aristotle

Knowing how to use metaphors requires accurate perception of the relationships of similarity and meets the requirement for a clear and precise elocution: neither trivial, nor obscure (*Poetics*, chapters 21 and 22). Metaphors name what does not have a name (*Rhetoric*, Book III, chapter 2). Metaphors are images (*Rhetoric*, Book III, chapter 4). The proportionate metaphors bring us knowledge (*Rhetoric*, Book III, chapter 10). The more appreciated metaphors are those based on analogy (*Rhetoric*, Book III, chapter 10).

Darwin

In his use of metaphor and analogy Darwin goes much beyond what the Aristotelian remarks above suggest. Darwin not only introduces his central issue, Natural Selection, on the basis of its analogy to selection by man, but he also enlarges its scope by means of several other analogies. For instance, analogies sustain the general laws concerning the processes of modification and extinction of species in the animal kingdom (Darwin 1875, 299), the principles that regulate the relationships between inhabitants of different neighboring places (Darwin 1875, 357), or the universal rule on the endemic productions of oceanic islands (Darwin 1875, 354).

Along Darwin's intellectual trajectory toward his main work, *Origin of Species*, the use of metaphor provides key images that articulate his theory. They do not only serve to "name," as Aristotle would say (*Rhetoric*, Book III, chapter 2), or to introduce fundamental concepts with the due clarity, elegance and brevity (*Rhetoric*, Book III, chapters 10 and 11), but they also provide corroboration to theory, for instance when Darwin explains the absence of the transitional organic forms through "Lyell's metaphor," which accounts for the imperfection of the geological records (Darwin 1875, 289).

Aristotle

Making objects sensible or putting them "before the eyes" is to give their meaning "in action" (*Rhetoric*, Book III, chapter 11).

Darwin

The use of *diagrams* plays a central role in the Darwinian explanation. *Diagram* is an important tool to fit the conceptual synthesis into a sensible scheme, and thus to lead to the application of the theory of natural selection to questions about classificatory and philogenetic issues, and to facilitate their precise resolution (Darwin 1875, 369, 379–80).

AS TO THE TREATMENT OF DIFFICULTIES AND OBJECTIONS

The treatment of difficulties and objections is a Darwinian strategy that closely resembles the Aristotelian recommendations regarding rebutting accusations (*Rhetoric*, Book III, chapters 15–19). Being able to respond to difficulties, exceptions and objections strengthens Darwin's positive argument. According to Aristotle, refutation is part of the positive argumentation, and the comparison of reasons results in the reinforcement of the view that is being defended. In the *Origin of the Species*, from chapter VI to chapter IX Darwin aims at corroborating his theory and its key principle through the treatment of the most difficult and complex cases against it, by answering the objections or showing their impropriety, or by diminishing their apparent weight and dismissing their apparently decisive character. As a final stage of this process, Darwin converts seemingly "unfavorable" converts seemingly "unfavorable" evidence to his theory into "favorable" evidence (chapter X). Next, he exploits the clearly favorable cases for the explanatory superiority of the Darwinian theory in relation to its rivals (from chapter XI to chapter XIV).

Aristotle

Answer the opponent's questions by establishing distinctions (*Rhetoric*, Book III, chapter 19).

Darwin

Darwin always examines thoroughly the nature of the objections raised against his theory. Part of his effort encompasses a usual procedure in any explanatory task, namely, to achieve a proper classification. In Darwin, however, the classificatory task acquires a particular meaning and projects the definitory job onto a new dimension, breaking with the Aristotelian requirement that the definition must contain all that belongs to the defined and only those properties which are separately necessary and together sufficient to characterize the defined and only the defined (*Posterior Analytics*, Book II, chapter 13). When the classificatory procedures usually adopted by naturalists are intertwined with the Darwinian procedures, they form an explanatory network that strengthens Darwin's theory as a whole and provide a "real" (genealogical) foundation for classification. The opponents' theories cannot provide it.

Aristotle

Start by examining and undermining the bases on which the suspicion is founded (*Rhetoric*, Book III, chapter 15).

Darwin

He begins his examination of the difficulties (chapter VI) by facing those he initially took as being very serious. At the end, very few difficulties remain and none of them is fatal to his theory. These serious difficulties concern four great topics: absence or rarity of evidence for transitional forms; the modification process required to produce structures and habits very different from those of the ancestral forms, and to produce so complex organs like the human eyes, or so trifle ones like the giraffe's tail; the acquisition and modification of instincts (chapter VIII); the sterility among species and the fertility among varieties of the same species (chapter IX).

When dealing with the rarity or absence of transitional forms, Darwin not only shows that there are no reasons to expect to find them in the degree and number supposed to have existed according to his theory. He also shows the consistence of such an absence with the conditions established by his theory, according to which the intermediate forms, those inhabiting intermediate areas and being less capable and fewer expanded than their neighbors, would be supplanted by the most capable. About these intermediate forms there would be just fossil registrations, excepting except for the case of aberrant forms living at isolated areas. However, the very fossilization conditions, in accordance with the principles of modern (Lyellian) Geology, would not be favored by those conditions which would in turn be favorable to the arising of transitional forms.

As to the case of the formation of peculiar structures and life habits, as well as of organs of extreme complexity (chapter VII), the objections point to the inability of natural selection to account for the incipient stages of useful structures. In dealing with these difficulties, most of them raised by George Mivart, Darwin looks at the difficulty or objection from the point of view of the theory in its integrity. He also attacks the bases from where the objections emerge. Thus, he criticizes Mivart's ideas about a "force or intrinsic tendency to change" that the organic forms would possess and about the occurrence of "abrupt changes." Darwin shows their inconsistencies. Such an attack strengthens Darwin's defense of his own ideas and is in accordance with the Aristotelian remark that refutation is part of the positive argument and that the comparison of reasons results in a reinforcement of the position one wants to defend (*Rhetoric*, Book III, chapter 13).

Aristotle

Allege that the imputed fact does not exist (*Rhetoric*, Book III, chapter 15).

Darwin

As in the case of the formation of the human eye by means of the gradual action of natural selection, Darwin alleges that there is no difficulty to accept it if one accepts the general argument of the theory as a whole. If one accepts the general argument that there can be small variations useful to each step of the process and they can be preserved, and if the logical and factual possibilities are guaranteed given the available evidence, then the particular difficulties for the acceptance of Darwin's argument are avoidable. The psychological difficulty in imagining gradations does not determine their logical or factual impossibility when several other analogies (logical and empirical reasons) are considered.

Aristotle

Allege that the fact is not harmful (*Rhetoric*, Book III, chapter 15).

Darwin

Darwin alleges that even when the difficulties cannot be all settled, they do not undermine his theory, as in the case of the formation of the instincts (chapter VIII) and of hybridism (chapter IX). After weakening the possible initial impact caused by the objection, if it is not possible to move the difficulty completely away, Darwin redirects the difficulty to another target. After careful discussions, the remaining difficulties are revealed to be such that any theory would have to face them. In other words, the difficulties become cases that should be appraised within the larger explanatory scope of each theory. And when brought to this larger ground, the explanatory power of the Darwinian theory and principle prevails over its opponents. The examination then accomplished brings to the open an advanced understanding of the nature and scope of the Darwinian principle.

Aristotle

Allege that the fact is not unjust (*Rhetoric*, Book III, chapter 15).

Darwin

Through meticulous discussions and by making use of the entire network of his explanatory procedures and argumentative strategies, Darwin alleges that the due evidence was appropriately examined.

Aristotle

Allege that the fact is not so harmful; that the fact is not so unjust (*Rhetoric*, Book III, chapter 15).

Darwin

Darwin diminishes the weight of the difficulty or objection.

Aristotle

Allege that the fact is not shameful (*Rhetoric*, Book III, chapter 15).

Darwin

Darwin alleges that the evidence does not disqualify the theory. On the one hand, the difficult cases are not fatal to it, and, on the other, it is capable of explaining facts that are "strange" or inexplicable by the opposing theory. When dealing with the case of the formation of the instincts by natural selection, the general structure of the argument in favor of that thesis consists of the following. First, it starts from the same argumentative structure established for the acceptance of the production of new organic forms according to the principle of natural selection, namely: that there are light variations in Nature; that such variations are of the largest importance for the survival of organic beings (for the most part, they will be either useful or injurious, although there are also neutral "variations"); thus, there is no difficulty to admit the action of the natural selection in preserving the useful variations and destroying the injurious ones in the light of the "conditions of existence." Second, the action of the use/disuse of parts and habits combined with the action of natural selection is admitted (chapters VI and V). Third, it is shown why the difficult cases are not fatal to the theory: they are avoidable or revealed as being difficulties to any theory. Fourth, the argument emphasizes those conditions implied in the examination of the instincts that are favorable to an explanation in terms of the principle of natural selection, namely: the instincts are not perfect, therefore, they are able of being improved; instincts are never produced for the good of a being other than their possessor (although other beings can take advantage of that); the great canon in Natural History, "Natura non facit saltum," is applicable to instincts as it is to corporal structures. Finally, the argument emphasizes those cases that seem to corroborate the explanatory power of the Principle of Natural Selection, and, in particular, those that, otherwise remain "strange" or unexplained by Darwin's greatest opponent, Creationism.

Aristotle

Allege that the injustice was committed to compensate an injury; allege that, if the action was harmful, it was, however, honorable; allege that, if the action provoked sadness, it was, however, profitable (*Rhetoric*, Book III, chapter 15).

Darwin

Darwin alleges that the favorable and unfavorable evidence should be examined in the light of the integrity of the explanatory context. He begins his conclusive chapter by saying: "As this whole volume is one long argument, it may be convenient to the reader to have the leading facts and inferences briefly recapitulated" (Darwin 1875, 404.).

In its Conclusion, the argumentative structure of the *Origin* is to be evaluated as a whole, and it is in relation to this whole that a fair balance of the partial arguments should take place. These partial arguments, in turn, weave the fabric of this seamless network. Therefore, to conclude the work means to recapitulate the facts and reasons that are intertwined in making the "long" argument and bring it to the unit of a long narrative. Here are the subtitles covered in Darwin's last chapter (Darwin 1875, chapter XV): "Recapitulation of the objections to the theory of Natural Selection; Recapitulation of the general and special circumstances in its favor; Causes of the general belief in the immutability of species; How far the theory of Natural Selection may be extended; Effects of its adoption on the study of Natural History; Concluding remarks."

In the final balance carefully prepared along the book, the explanatory superiority of the Darwinian theory prevails.

Aristotle

Allege that the fact is not important (*Rhetoric*, Book III, chapter 15).

Darwin

Darwin alleges that the contrary evidence does not affect any theory in particular, or that, in any case, the contrary evidence is not solidly founded, as in the case of the sterility among the species taken as evidence against his theory. Darwin reminds us that apparently the elimination of sterility results from a gradual adaptation to frequent changes in the life conditions. When species exposed to the same life conditions during a long time are subjected to great changes, they perish or become sterile. In a similar way, hybrids of two distinct species, being composed of two different organizations, suffer a great change in the life conditions. Darwin then lightens the difficulty by saying: "He who will explain in a definite manner why, for instance, an elephant or a fox will not breed under confinement in its native country, whilst the domestic pig or dog will breed freely under the most diversified conditions, will at the same time be able to give a definite answer to the question why two distinct species, when crossed, as well as their hybrid offspring, are generally rendered more or less sterile, whilst two domesticated varieties when crossed and their mongrel offspring are perfectly fertile" (Darwin 1875, 406).

Aristotle

Accuse the accuser by discrediting his/her reasons (*Rhetoric*, Book III, chapter 15).

Darwin

Darwin disqualifies his opponents' reasons and the explanatory power of the opposing theory (Creationism) for not being able to account for several facts that are explained by Darwin's theory, especially in the fields of geographical distribution, paleontology, morphology and classification (Darwin 1875, 305, 359, 384–85, 402, 416, 420). Creationism does not meet the scientific requirements then accepted:

> Nothing can be more hopeless than to attempt to explain this similarity of pattern in members of the same class, by utility or by the doctrine of final causes. The hopelessness of the attempt has been expressly admitted by Owen in his most interesting work on the Nature of Limbs. On the ordinary view of the independent creation of each being, we can only say that so it is,—that it has pleased the Creator to construct all the animals and plants in each great class on a uniform plan; but this is not a scientific explanation" (Darwin 1875, 383). In works on natural history, rudimentary organs are generally said to have been created "for the sake of symmetry" or in order "to complete the scheme of nature." But this is not an explanation, merely a restatement of the fact. (Darwin 1875, 400)

Aristotle

> If the written law tells against our case, clearly we must appeal to the universal law, and insist on its greater equity and justice. We must argue that the juror's oath "I will give my verdict according to my honest opinion" means that one will not simply follow the letter of the written law. We must urge that the principles of equity are permanent and changeless, and that the universal law does not change either, for it is the law of nature, whereas written laws often do change. (. . .) Or if a law is ambiguous we shall turn it about and consider which construction best fits the interests of justice or utility, and then follow that way of looking at it. If however the written law supports our case, we must urge that the oath "to give my verdict according to my honest opinion" not meant to make the judges give a verdict that is contrary to the law, but to save them from the guilt of perjury if they misunderstand what the law really means. (. . .) Or that not to use the laws is as has to have no laws at all. (. . .) Or that trying to be cleverer than the laws is just what is forbidden by those codes of law that are accounted best. (*Rhetoric*, Book I, chapter 15)

Darwin

When it is favorable to his argument, Darwin appeals to the possible interference of factors like the use/disuse of parts, or the laws of variation and growth with the action of the Principle of Natural Selection. Sometimes this interference helps the action of Natural Selection, sometimes it excludes the action of Natural Selection. Depending on the circumstances, Darwin may appeal to the wisdom of the scientific community and its maxims, such as *Natura non facit saltum*, over common sense, or to the wisdom of common sense. In relation to the geological records, he appeals to their imperfection when they oppose his theory, but they are considered reliable when they corroborate it.

Aristotle

Blame the accuser for slandering (*Rhetoric,* Book III, chapter 15).

Darwin

In a passage added to the third edition of the *Origin* (Darwin 1875, 63), Darwin alleges that his use of 'natural selection" was misinterpreted by several authors. Darwin refers to objections that assumed "natural selection" to be committed to the ideas of "conscious choice," "Divinity's power" and "personification of Nature." Darwin calls attention to the metaphorical character of such expressions. He compares the function of the expression "natural selection" in his theory to the function of other expressions in scientifically accepted theories, such as the "elective affinities" of chemical elements and the "gravitational attraction" in the regulation of the planetary movements. Darwin admits that a meaning can be legitimate metaphorically speaking, and false if literally taken. But he says that "every one knows what is meant and is implied by such metaphorical expressions: and they are almost necessary for brevity."

Aristotle

Allege that the action performed is not harmful to the opponent (*Rhetoric,* Book III, chapter 15).

Darwin

Darwin alleges that his theory does not hurt any religious feeling. His allegation responded to the most popular Creationist criticism of his views and was addressed to the public in general, namely, the attack on religion supposedly contained in Darwin's theory. In favor of his view, Darwin cites the support that a renowned theologian (without publishing his name) gives to his theory (Darwin 1875, 422).

Aristotle

Exalt the accused prolixly, and immediately censure him/her strongly and concisely (*Rhetoric,* Book III, chapter 15).

Darwin

Darwin carefully examines the seemingly successful aspects of the opposing theory and then immediately points out their deficiencies, as he does in the treatment of the most serious objection to his theory: the absence of fossils of intermediate forms in the degree and number requested. Initially, Darwin admits the seriousness of the difficulty. The general answer to be given is based on the imperfection of the geological records as condensed in Lyeel's metaphor of the history of the world written in a mutant dialect, of which we possess only the last volume relatively to two or three regions, and of this volume we possess only a short chapter preserved here and there, and from each page only a few lines (Darwin 1875, 289). He concisely says: "He who rejects this view of the imperfection of the geological record, will rightly reject the whole theory" (Darwin 1875, 313).

Aristotle

Point to the worst and best motives for the performed action (*Rhetoric*, Book III, chapter 15).

Darwin

Darwin disqualifies the unfavorable evidence and emphasizes the evidence that is favorable to his theory in the balance of all facts and reasons presented in his Conclusion. The recapitulation of evidence allows Darwin to show the superiority of his theory. The theory of natural selection makes it possible to explain: (a) questions about concepts; (b) facts, empiric regularities, laws, tendencies, principles and maxims for the scientific practice; (c) methodological-conceptual questions related to biological investigations (like the arrangement of organic beings in a "natural" system); (d) cases of presence or absence of beauty in Nature (to which Darwin brings the concurrence of "sexual selection"); (e) criterion for relative perfection ("natural selection" acts by means of competition, adapting and improving the inhabitants of each area only in relation to their coinhabitants), and for absence of perfection of organic forms; (f) cases that initially seem to be difficulties to the Darwinian theory and that are reverted into cases for the explanatory superiority of the theory (like the facts revealed by the geological records); (g) the compatibility of the acceptance of the Principle of Natural Selection with the principles of human rationality, in general, and scientific rationality, in particular; (h) the compatibility of the Principle of Natural Selection with any religious feeling.

Additionally, it is credited to the set of reasons that are favorable to the theory of natural selection the fact that Darwin's theory (i) is supported by progressive minds and minds capable of fair judgment; (j) conforms to philosophical axioms ; (k) is the only rational explanatory alternative for the theme of the origin of species; (l) renders the study of Natural History much more interesting; (m) advances the scientific investigation by demanding the creation of new fields, instruments and research tasks; (n) arouses an optimistic "view" concerning the future (the survivors will be the fittest ones).

Aristotle

Appeal to "such is our nature" (*Rhetoric*, Book III, chapter 16).

Aristotle recommends the appeal to "such is our nature" in those cases in which there are no reasons to give, because we suspect that the action was performed with a good intention and not just for convenience

Darwin

Appeal to "our ignorance."

In Darwin, the appeal to "ignorance" also takes place in limiting situations based on an "epistemological" good faith and plays a central and innovative role in the Darwinian argument. According to Aristotle, ignorance is lack of knowledge or a positive state or position of mind produced by an inference (*Analytical Posterior*, Book I, chapter 16). In Darwin, the appeal to ignorance involves a process that is much more complex and closer to the limiting condition "such is our nature." In different passages and after careful argumentation, Darwin emphasizes that the most important objections are related to questions about which we are admittedly ignorant and thus are not sufficient to abandon the theory of common descent with modification (Darwin 1875, 410). The appeal to our ignorance and the recognition that there are limits to our knowledge that may or may not be overcome, become part of Darwin's explanatory effort. In this way Darwin transforms our ignorance, which results from the imperfection of the geological records, and the impossibility of necessarily determining which organic forms are transitional without rebuilding the entire chain into a solid circle of ignorance which, in its turn, is explained by Darwin's theory in accordance with the new and well-accepted (Lyellian) Geology.

CONCLUSION

On the one hand, the comparison between Aristotle's rhetorical recommendations and Darwin's argumentation emphasizes a vein of the Aristotelian rhetoric that has been little explored by the tradition. In spite of having maintained the difference between "science" and "rhetoric," demonstration and persuasion many times penetrate each other in Aristotle's text. Knowing how to appropriately use language and persuasive resources is not indifferent to necessary demonstration. Rhetoric is legitimately within the scope of rationality. The tradition that followed Aristotle became trapped by a stiff reading of the Master's text and accentuated that distinction in such a way that rhetoric was for the most part restricted to treatise on style lacking its epistemological dimension. On the other hand, the examination of the elaboration and defense of a theory plainly accepted as scientific, namely, Darwin's theory of Natural Selection, shows us procedures called "rhetorical" as being constitutive of scientific argumentation. The bases to find a rigid distinction between "scientific" and "rhetorical" arguments vanishes, be it in terms of the nature of the premises, be it in terms of the specific procedures of both of them. Should we then refuse to call the Darwinian theory a "scientific" theory? Very few would take this option instead of choosing to rethink that traditional distinction and the concept of rationality linked to it. This second option is more in line with the idea of science as conjectural and fallible knowledge rather than knowledge based on necessary truths.

Thus the view of science in terms of knowledge obtained from irrefutable truths (that is, the knowledge of "what cannot be otherwise") succumbs. And different regulatory ideas of science will lead us to different analyses and projects. We come again to Boaventura's reflections: why privilege a given view of science? The judgment that will guide us will be a judgment of value. And judgments of value are not isolated pieces of a social, cultural and natural context. It is within this context that values are created, changed, and become concrete guides for us while being integrated with cognitions and actions. The very dynamic character of "rationality," which is always known as "in action," is framed within that context. This dynamic condition also affects the other distinctions involved in the distinction between "scientific" and "rhetorical" rationality, such as "objectivity / subjectivity," "true / probable," "universality / contextuality." The epistemic impact that may be caused by challenging these distinctions does not seem so dramatic if we exploit the possibility—already present in the Aristotelian text—of considering the demonstrative syllogism from necessary truths as a modality within the wider spectrum of the argumentative and demonstrative field. Demonstrating may be thought of as a privileged means of persuading.

In its actual practice, the rationality of science turns out to be much more flexible than its rigid caricature. Its rigid caricature ensures the "cultural authority" of science, but is detrimental to the view of science as promoting "innovation." Innovation, as Boaventura says, "consists of inventing persuasive contexts to lead to the application of methods out of their 'natural' habitat" (Santos 2002, 49; my translation).

It is undeniable that, in any case, we have to deal with questions related to the use of language. Allegations frequently used to establish the distinction between "scientific" and "rhetorical" on the basis of the distinctiveness of the "true discourse," where "the facts speak for themselves," may be equally seen as rhetorical. Frequently, the maintenance of the distinction becomes a rhetorical "resource," in the old sense of the word, to leave aside the opponents' arguments and even to depreciate them while strengthening one's own arguments. We dare to say that our opponents are making use of "merely" rhetorical arguments! (Soyland 1994, 2–3). However, the "scientific" and "rhetorical" distinction, rather than being immediately given, cannot be thought of and built outside of its thematic and historical contextuality.

The examination of the relationships between "rhetorical" and "scientific" rationality illuminates the more general question about "rationality." In this sense, the case of the *Origin of Species* reveals that the self-confidence and artifices of reason go much beyond the strict deductive or inductive limits in constructing and supporting our explanations. In its origins, rhetoric was the art of public and spoken speech; later it bloomed within the realm of the written word. Nowadays, the acknowledged presence of rhetorical arguments in scientific discourse together with the materialities of scientific communication open the doors to a new blooming of rhetoric beyond the limits of the spoken and written discourse and to its playing in new language games. Rationality thus flexibilized approximates scientific and literary discourse. Rather than vulgarizing "rationality," such a flexibility gets closer to its original meaning.

Finally, our analysis allows us to see that, in order to adequately preserve our rationality, we must contextualize it and let it flow in the intrinsic dynamism of "rationality in action." Its intended universality is not denied, but, parodying Boaventura Santos, "every knowledge is local and total." The particular, in its exemplarity, must be determined within the universality of its context, and the universal can only be revealed as such by means of the determination it imposes on the particular. Therefore, particular and universal are necessarily related to each other. The artifices of reason, which the tradition recognized as "rhetorical," break with the limits of traditional rationality and make science move ahead as a rational enterprise. Parodying what Philip Kitcher says regarding rhetoric, rationality is inescapable. It permeates our lives, our historicity and contextuality.

NOTES

*Graduate Program in Philosophy—Universidade do Vale do Rio dos Sinos (UNISINOS), São Leopoldo, RS, Brasil. *E-mail*: aregner@portoweb.com.br

1. The 6th English edition will be used here, which was the last one revised by Darwin.

2. Whewell criticized Darwin's theory because he considered this interplay a non-scientific maneuver: "For it is assumed that the mere possibility of imagining a se-

ries of steps of transition from one condition of organs to another, is to be accepted as a reason for believing that such transition has taken place. And next, that such a possibility being thus imagined, we may assume an unlimited number to take place in, and that this indefinite time may extinguish all doubt that the transitions really have taken place" (Whewell *Astronomy and General Physics*, xvii–xviii, cited in Himmelfarb 1968, 333–34).

REFERENCES

Aristotle. 1981. *Analytica Priora et Posteriora*. Ed. Sir Davis Ross. Oxford: Oxford University Press.

———. 1997. *Poetics*. Trans. Malcon Heat. New York: Penguin Books. Penguin Classics, New Edition.

———. 2004. *Rhetoric*. Trans. W. Rhys Robert. New York: Dover Publications.

Darwin, C. 1875. *On the Origin of Species by Means of Natural Selection or the Preservation of Favored Races in the Struggle for Life* (from the 6th English edition). New York: Appleton.

———. 1958. *The Autobiography of Charles Darwin and Selected Letters*. Ed. Francis Darwin. New York: Dover Publications.

———. 1987. *Charles Darwin's Notebooks, 1836–1844*. Ed. Paul H. Barret et al. Ithaca, NY: Cornell University Press.

Herschel, John F. W. 1966. *A Preliminary Discourse on the Study of Natural Philosophy*. Introduction by Michael Partridge. New York: Johnson Reprint Corporation.

Himmelfarb, Gertrude. 1968. *Darwin and the Darwinian Revolution*. London: W. W. W. Norton Library.

Kitcher, P. 1991. Persuasion. In *Persuading Science: The Art of Scientific Rhetoric*, ed. M. Pera and W. Shea. New York: Science History Publications/USA.

Mill, John Stuart. 1979. *Sistema de lógica dedutiva e indutiva*, 2nd ed. São Paulo: Abril Cultural (Coleção Os pensadores).

Regner, A. C. K. P. 1997. "O papel da metáfora no longo argumento da 'Origem das Espécies'." *VI Seminário Nacional de História da Ciência e tecnologia—Anais* 6:35–40.

———. 1998. "Argumentos retóricos na ciência: re-pensando Aristóteles." *Episteme—Filosofia e História das Ciências em Revista*, 3 (7):64–83.

Santos, Boaventura de Sousa. 2002. *Um discurso sobre as ciências*, 13th ed. Porto, Portugal: Edições Afrontamento.

Whewell, William. 1967. *The Philosophy of the Inductive Sciences Founded upon Their History*. 2 vols. Introduction by John Herivel. New York: Johnson Reprint.

12

Science on the Edge:

Some Reproductive Biology Paradigms

João Ramalho-Santos

For

Jerry Schatten

And in memory of

Stephen Jay Gould, W. G. Sebald and Joe Strummer

A REPRODUCIBLE NARRATIVE

A large round cell, surrounded by a protective layer and by a cumulus-like "cloud" of smaller cells is transported down a flexible tube. Coming towards it, thanks to muscle contraction around the tube and to the movement of a long tail, are several small, aerodynamic, cells. The round cell grew and developed, together with other, similar, cells, in an organ not far away. However, its sisters degenerated, and it was the only one released into the tube. The smaller cells came from a different organ, actually an organ located in another organism altogether. Those that reach the big cell are merely the brave survivors of a long journey. Both cell types have two things in common: they are totally different from other cell types in the body and have only half of the nuclear genetic material (DNA) one normally finds in other cells. The small cells can hardly be classified as cells at all, they are terminally differentiated, have lost most cell components, and are made up of a compact pointed head (that houses the DNA), the tail, and a intermediary piece (the midpiece) linking one to the other, which provides some of the energy required for movement.

Once in close contact, several of the smaller motile cells release components from a vesicle placed on the tip of the head, called the acrosome. The released substances help the small cells penetrate the protective layer that surround the large cell. One of them eventually reaches the surface and fuses with the large cell, slowly sinking into it as if treading quicksand. But let us call these cells by their proper names, shall we? The large oocyte and the small motile sperm (Ramalho-Santos et al. 2000a, 2002; Wassarman et al. 2001; Yanagimachi 1994). Joining to become a zygote. Cued by the sperm, the oocyte triggers a series of changes, mobilizing its reserves of mass and energy that will fuel the earliest stages of embryo development. Sperm structures are usually destroyed by the embryo, with the exception of parental DNA, half the contribution to the nuclear DNA of the embryo (another type of DNA, mitochondrial DNA, is inherited from the maternal side). A notable exception, in primates, is the male-borne centrosome, a structure derived from a centriole located between the sperm head and midpiece, which plays a crucial role in bringing together maternal and paternal genomes (Schatten 1994).

Henceforth, many other events take place. The zygote divides into a multicellular embryo that implants in the uterus. Its cells start to differentiate and specialize in different functions, tissues and organs, as well as accessory structures (the amnion, the placenta) start to form. But this is another tale, for other storytellers. Or scientists. For our purposes here, the elaborate choreographies of sperm and occyte are sufficient. And when we look in the mirror to face one end result of such events, it is difficult to believe the fragility of the many individual steps that led to it. Any researcher working in reproductive or developmental biology faces this epiphany at some point of her/his career. With so many interconnected mechanisms, with so many steps that can go wrong, it is almost unbelievable how millions of new organisms are born every day. And yet, that is exactly what happens.

But not always. Sometimes functional gametes, sperm and oocyte, are not produced in sufficient quantities, sometimes they may not be fully mature or functional, sometimes their coming together may be made difficult for a variety of reasons. Sometimes things don't work, and the system has to be helped.

The visibility of assisted reproductive technologies (ART) in contemporary western society is such that one almost forgets that Louise Brown, the first "test tube" baby, was born in the United Kingdom in 1978 (Steptoe and Edwards 1978). The technology of the *in vitro* fertilization (IVF) process used in this case consisted of joining sperm and oocytes in a controlled laboratory environment, given that they could not do so *in vivo*. Subsequent developments involved the injection of a single sperm cell inside an oocyte (IntraCytoplasmic Sperm Injection, or ICSI), employed when there are few spermatozoa or they are not very mobile (Palermo et al., 1992, 1995). Or the use of immature sperm collected directly from the testicle, if none can be found in the ejaculate (Kupker et al. 2002). Or even the controversial procedure of injecting the contents of one oocyte as a "cellular cure" for another one, from a different woman, that may have decreased functionality (Cohen et al. 1998). Or finally, the use of *somatic* cells (that is, cells that do not belong to the *germ line*, which produces oocytes and sperm) to produce new individuals, by introducing the DNA of those cells into a donor oocyte from which DNA has been removed, and then using several manipulations to make this reconstructed construct behave and develop like a fertilized embryo (Wilmut et al. 2002; Rhind et al. 2003). This is what constitutes somatic cell nuclear transfer, of which reproductive cloning is merely the most publicly visible (and controversial) side. Therapeutic cloning, or the creation of a stem cell line derived from a given individual that may then be used to regenerate some of the same individual's diseased tissues, or the creation of "somatic gametes" in the absence of normal gametogenesis in the testis and ovary, are other possibilities of the technology (Tesarik 2002).

It should be noted that ICSI was developed in 1992 from a technique that involved placing the sperm close to the oocyte (Palermo et al. 1992). It also involves a sperm-oocyte choreography distinct from what takes place *in vivo* or in IVF. For example, as mentioned above, the acrosome is a sperm vesicle that is destroyed and thus releases its contents to help the sperm penetrate the oocyte. On the other hand, during ICSI the acrosome is injected, intact, inside the oocyte together with the sperm, and its presence may influence the first stages of early embryo development (Hewitson et al. 1999; Ramalho-Santos 2000b). This didn't stop the technique from becoming a standard method to treating some forms of human infertility, given the high success rate. As far as somatic cell nuclear transfer/reproductive cloning is concerned, public interest was piqued after the birth of "Dolly the sheep," the first mammal born from an adult cell, via cloning, in 1997 (Wilmut et al. 1997). Although success has been achieved in several species using similar techniques, extension to humans remains controversial and unexplored. Not only because success rates are alarmingly low and the process often results in embryo or fetus malformations (Wilmut et al. 2002; Simerly et al. 2004), and thus application to humans would border on criminal negligence, or result in fraud (Rhind et al. 2003; Schatten et al. 2003); but also because there are many species-specific differences. For example, although cloning a mouse is almost a widespread routine technology in many laboratories, the same cannot be said of the rat, although the casual observer would probably consider the latter a mere larger version of the former. Even using similar procedures, some animals seem much harder to clone from adult cells than others (Wilmut et al. 2002; Rhind et al. 2003). This low success rate implies that a large supply of oocytes as recipients for the adult somatic DNA would be needed, which, in the human, is a limiting, though certainly not insurmountable factor.

The establishment of ART raised many ethical, political and sociological issues. Gamete cryopreservation and donation or surrogate motherhood are two examples. But these procedures (technically not terribly complicated) have broadened our conceptions of conception, both in time and space (Franklin 1998). The number of individuals that can be potentially involved in the birth/raising of a new being has increased, and the way they can be involved, or how to determine the "weight" of different contributions (from the genetic/biological to the familiar/cultural) is the subject of much discussion. At the same time, thanks to sperm and embryo banking (possibly to be followed by the technically more challenging banking of oocytes or ovarian tissue), there is a potentially much wider time frame between gamete production, fertilization and birth. It is possible to imagine an individual conceived via gamete donation, with the embryo implanted in a surrogate mother, and finally raised by a familiar nucleus that is not related to any of the other participants. In the case of sperm cryopreservation, it is not even necessary that the donor be alive by the time his sperm contributes DNA and a centrosome to the embryo.

This hypothetical scenario (already played, in several variations) opens the way for several debates (see Charo 2002). And this is without touching on the very recent findings that open the possibility of producing new oocytes in the adult ovary (Johnson et al. 2004), thus challenging the unavoidability of menopause, and, together with other findings and techniques, also opening new avenues for pregnancy at advanced ages in the female (Powell 2004). But there is a double insight that is useful to mention about ART and its social acceptance. Namely, that the accusations leveled at the time against the physicians responsible for Louise Brown's birth, which can be summed up in the phase "men should not replace God/nature," have lost their sting, but have been "passed on" nonetheless, almost intact, to other debates, including reproductive and therapeutic somatic nuclear transfer. A (heterosexual) couple's right to ART when such is medically advised is rarely questioned (although some techniques, or how much of the procedure is covered by health plans/medical insurance, may be). In fact, it is an industry in expansion, either because it is more readily available, or because of decreases in "natural" fertility in the western world (and elsewhere), due to environmental or social factors, such as decreases in sperm production in industrial nations, in the former case, or the increase in the average age of first pregnancy and a decrease in number of offspring/couple, in the latter (Aitken et al. 2004; Butler 2004; Sharpe and Franks 2002). There are certainly more important issues on a global scale, for example, those related to reproductive health and reproductive rights (el Feki 2004). But global insignificancies may also be terrible personal anguish, as anyone who has known an ART patient can testify (Franklin 1998).

SCIENCE AS IT HAPPENS

But what does all this have to do with the so-called science/culture wars, with all the tension that opposes "the sciences" and "the humanities," and that has begotten so many notable tomes? It might be useful to paint a rough picture of a slippery discussion with many slopes. One of them (clearly stated, for example, in Santos 1987, 1992) has to do with the way many social scientists question the production of knowledge in the "natural/hard sciences" (focusing mostly on physics and chemistry, more recently on biology). This involves identifying aspects that, according to their analysis, are more dependent on the nature and organization of the scientific community than on an impartial quest for truth, the nature of which is also questioned (hence the label of "scientific relativism"). Another common thread of this type of discourse (again, see Santos 1987, 1992) is the suggestion that other forms of knowledge, and not just scientific knowledge, may be considered, at different levels, in the decision-making process, especially in times of crisis. It is this issue that raises eyebrows, not only among scientists, but also thinkers who tend to view science as the sole (or, at least, privileged) universal and true language. The only one capable of providing impartial, unbiased, answers to fundamental questions, and whose characteristics must therefore be preserved, at all costs (see, for example, Wolpert 1993, or, for a philosopher's slightly different perspective, Haack 1998, 2003).

These two opposite perspectives (easily identified by the positive or negative value they place on terms such as "multiculturalism". . .) have been know to trigger much debate and discussion, and the noise and entropy that are often generated may obscure the simple fact that one might often not be talking about the same things. And this is a truly important issue: not only are language and methodology very diverse, but the object of analysis itself may also differ. While researchers working in the "natural" sciences attempt to decipher vital mechanisms in their respective fields of study, framed within established paradigms (which can be validated, adjusted or shattered in the course of any given investigation), sociologists who observe them as case studies are more concerned with the way the mechanisms are studied, and therefore with the manner in which paradigms are constructed and evaluated within the scientific community. In other words, while some worry about the ends, the others concern themselves with the means. To complicate things even more (or to stimulate discussion, it all depends on the point of view) there are two other issues that warrant mentioning.

Firstly, that it is probably not very pleasant to be the object of research (the proverbial tasting of one's own medicine, see Saulson 2001), and this may be the simple reason why many scientists are uncomfortable with sociology of science. Of course, one could also quote the famous apocryphal statement (attributed to Richard Feynman or Steven Weinberg), according to which "Sociology of Science is about as useful to scientists as Ornithology is to birds." On a day-to-day basis, this may well be true (at least as far as most scientists are concerned). Indeed, even such moderate attempts to frame science logically in a sociophilosophical context as made by a firm admirer of the scientific language (Haack 2003) will unavoidably slam into objections that science succeeded exactly because it was never "logically" framed, actually providing rather counterintuitive accounts of how the world works (Wolpert 1993), and also that science is, at its core, undemocratic. If a theory is right, no amount of votes from concerned citizens will "prove" it wrong. In that light such frameworks are thus, at best, useless. But one should also be wary of such sweeping statements. Although the implications are profound, on a day-to-day basis it is also totally irrelevant to the vast majority of people (and scientists) whether the sun goes around the earth or vice versa. It is (always) a question of scale, a question of subject, and what one takes (or is willing to take) for granted. Although the "blasé" attitude of scientists can, in turn, lead to disasters such as the ceeping in of creationism/intelligent design into mainstream education, in an attempt to "vote out" evolution.

Secondly, and more importantly, one has to consider what exactly is being investigated. For example, in their much debated book *The Golem: What You Should Know about Science* (1993, revised second edition in 1998) sociologists Harry Collins and Trevor Pinch focus on large and (at the time) major paradigm shifts, that is, profound changes in the way we view the world and our place in it. The point is that these shifts (such as the consolidation of Einstein's theory of relativity, or Pasteur's hypothesis of the biotic origin of life) are considered in the book at *the exact moment they took place*, and the behavior of the scientists involved (as well as that of the larger community), in terms of the planning and interpretation of crucial experiments, is analyzed according to *what was known at the time*; that is, without the benefit of hindsight. Again, it is important to clarify the object being studied and, in this case, the object has a crucial time frame. Corroborating information that was obtained with further research conducted after the fact (and which we now take for granted) was simply not available when the experiments discussed were being carried out.

This is exactly what Collins and Pinch are interested in, the pursuit (and validation) of knowledge at the very borders of a paradigm. Accordingly, they analyze several examples of what textbooks usually describe as "definitive experiments," meaning experiments that are described as unequivocally resolving an issue and thus shifting a paradigm, by decisively tipping the scales in favor of one of the hypothesis. Their main thesis, in a nutshell, is that some experiments usually cited as crucial were not as conclusive as would be expected. Therefore, that the reason they were acclaimed as unequivocal successes at the time had to do with other factors, either the many other pieces of data that pointed in the same direction (though no individual experiment was definitive), or the prestige of the scientists involved (Einstein and Pasteur, in the examples mentioned earlier), or the willingness of the scientific community to embrace the transition ("it was an idea whose time had come"). This does not mean, and herein lays the major misunderstanding, that the authors reject the experiments they are analyzing (or the knowledge associated with them) as false or unreliable. What they question is the manner in which those experiments established themselves as unequivocal. There is no incompatibility here, and most scientists would confirm that "definitive experiments" are often misnamed. This is eloquently obvious in the dialogue Collins and Pinch establish with physicist David Mermin, concerning diverging interpretations of experiments that confirmed the theory of relativity (Mermin 2001). One of the vocal critics of the first edition of *The Golem* (1993), Mermin recognizes the validity of Collins and Pinch's approach, while these authors clarify their discussion on the topic in the second edition (1998).

There are several reasons why other scientists may find it difficult to follow in Mermin's dialoguing footsteps. First, scientists are used to evaluating science outside their immediate field of expertise by way of textbooks, assays or review articles. These works usually provide very clear pictures of the advances in a given field of knowledge; but their gaze is mostly historic in nature, and often gives rise to what appears to be a logical, cumulative, progression, which may not represent the whole historical truth. Indeed, much of the tricky issues and heated debates that took place along the way are routinely simplified (or even omitted), as they are not useful for the linear clearness at which these sorts of narratives usually aim. Second, most scientists are not familiar with the language of sociology, and presume (rightly or wrongly) that sociologists are equally unfamiliar with scientific terminology. This is a very real and serious issue, and most of the acrimony surrounding the work of Alan Sokal and Jean Bricmont on how scientific concepts are (mis)interpreted by linguists, sociologists and philosophers can be traced back to it (Sokal and Bricmont 1998). However, it should also be stressed that language (Laurie Anderson's virus) is a convenient hiding place in which both sides are more than willing to take refuge. Third, scientists feel that any sociohistorical analysis that is not openly celebratory of science is either useless, or an assault on their form of knowledge (Wolpert 1993). Fourth, and this is related to the previous point, scientists fear that their position may be compromised in terms of public perception, eventually leading to loss of prestige and possibly even power in the decision making processes that involve scientific issues. In this regard, it is useful to remember that some scientists have also called attention to the dangers scientists themselves have incurred by mythifying science, and thus creating lofty standards that are neither realistic nor true (Goodstein 1996). Fifth, the scientific community often feels that sociologists are only interested in controversial, sensationalist, findings, where they are more likely to find problems in the validation of scientific knowledge. Although this criticism can be countered by mentioning the tradition of analyzing day-to-day laboratory activities (see Latour and Woolgar 1986; Nunes 2001), even if it were true, aren't most researchers (in any field) naturally attracted to controversial issues? In this regard, the only obligation is to identify the object of analysis, and the context in which it is inserted.

Differences in approach and methodology are fairly obvious and can be noted using the "Cold Fusion" case as an example. In 1989, two scientists at the University of Utah, Martin Fleismann and Stanley Pons, announced that they had obtained fusion of hydrogen at room temperature, using a very simple apparatus. If true, this would constitute a powerful (and cheap) way of obtaining energy, "the sun in a test tube" (see Huizenga 1993; Collins and Pinch 1998; Simon 2002). It seemed too good to be true. And it apparently was. Nobody was able to repeat the experiments, and the two researchers were discredited, their careers essentially over. As far as scientists are concerned, this is all that matters. In the end, the mechanisms of validation worked, and bad science was punished. The same reasoning is applied in the more spectacular cases of scientific fraud, which have also been used to question how scientific knowledge is established. Something the scientific community is well aware of, as the recent repercussions of two large frauds in physics attest, notably because physics was deemed immune to these sorts of manipulations (Dalton 2002; Service 2002). But again, although these are troublesome events (one of the scientists involved, Jan Hendrik Schön, was highly regarded as an exceptional rising star), most scientists would take solace in the bottom line, and reaffirm that, if some parts may malfunction, the whole remains functional, and that is all that really matters. End of story.

The thing is, sociologists such as Collins and Pinch may not be interested in the end of these stories, but rather in the flow of events that surrounded them. Despite their ultimate downfall, for a while Pons and Fleismann were riding high, as debates raged in different campuses, specialty journals, and the media. The reason for this was quite simple. There were many other researchers attempting similar things, and the experiments were not as ridiculously naïve as they may seem in hindsight to a casual reader (even one with a scientific background). Truth be told, in the months following the initial announcement there were many different experiments, contradictory findings, diverging interpretations, heated discussions for funding to finance further work, papers in journals and press conferences. Indeed, despite the more prevalent (negative) view on Cold Fusion (Huizenga 1993), some scientists believe that the whole field was given a bad rap (Mallove 1991), and research continues, although much more subdued (Simon 2002). Again, this is not to say Collins and Pinch actually defend the methodology or experimental approaches carried out by Fleismann and Pons as good science that was not given the proper consideration or benefit of the doubt. What they are interested in are the reactions of the scientific community at the time, what data were available for them to act on (and when were they available), and what where the mechanisms that finally led to the dismissal of the data and hypothesis. However, "final" is a dangerous word to use in science, and continuing research on this subject (Simon 2002) may mean that the "final" verdict on Cold Fusion would indeed have to be revised, at some level. Or not, we'll have to see the data first. In other words, the analysis of Collins and Pinch ends at the time Cold Fusion was dismissed by the scientific community. And it is this end, and not the means to arrive at it, that interests most scientists. So the same story may be approached in different ways, yielding different readings that are not mutually exclusive

It is important to note that the examples used to illustrate these points (and many could be mentioned) usually have already been historically resolved, and are evaluated with the full benefit of hindsight. Not that this means that, as mentioned, they cannot be revisited and reevaluated (another common event in science), but the fact that there is a time buffer between the end of the story (even if only a temporary ending) and its interpretation suggests an obvious question. How should one react in the exact moment a controversy is brewing, when all the information needed is not yet available or hasn't been adequately processed? What to do in the absence of some healthy historical hindsight, especially if one must face a pressing need to act, and when those actions may influence policies, the society, and future decades of research? Who can (should) make those decisions? And what kind of decisions should be made based on what criteria? In other words: what to do in the *now*?

Polarizing debates of this nature are rampant, from transgenic crops (are they safe? do they influence ecosystems and traditional crops? in what way?), to the potential uses of stem cells (are embryonic stem cells needed, or can optimal results be obtained with adult cells?). In many such cases, the scientists involved are accused of falling into the same traps as Fleismann and Pons, such as describing results in press conferences, rather than in scientific forums, where closer scrutiny is possible. Or publishing incomplete or inconclusive data as if they were complete and conclusive; and here one also has to factor in the role of scientific journals, interested in publishing the latest breakthroughs, and whose press releases often simplify the actual findings, which can result in those findings sounding more definitive than they actually are. From experiment, to scientific report, to press release in a specialized journal, to the general media, to the reader; the levels at which scientific information is translated are many. On top of all this, there can also be financial or ideological interests involved. For example, at a time when it is almost impossible to find influential scientists in key fields that have no direct or indirect vested interests in commercial enterprises (and may thus be biased when expressing what should be neutral opinions), this issue is particularly important (Krimsky 2004). On the other hand, discussions on the potential of embryonic or adult stem cells to regenerate human tissues ravaged by disease or aging are as much a debate on actual scientific findings with either cell type, as they are ideological debates on the use of human embryos for research (Reichhardt 2004).

Another example of a debate that will only be settled in the long term has to do with the safety of assisted reproductive technologies (ARTs). By definition, these techniques facilitate human fertilization in situations where it wouldn't normally take place. If a physical barrier between gametes is sometimes the issue (a problem in the fallopian tubes, for example), on other occasions ART is basically helping to viabilize what would have been unviable gametes *in vivo*. Although there are concerns related to the manipulation of gametes and embryos outside the female reproductive tract, classical in vitro fertilization (IVF) bypasses several barriers imposed to the incoming sperm by the female genital tract. However, the success of these techniques implies many viable, motile sperm. By directly injecting a sperm inside an oocyte, intracytoplasmic sperm injection (ICSI) goes one step further, bypassing barriers imposed by the oocyte itself, and by the protective layers that surround it (Schatten et al. 1998; Hewitson et al. 1999; Ramalho-Santos et al. 2000b; Schatten 2002). Sperm motility is not an issue here, quite the opposite, as the manipulator artificially selects the individual sperm, by capturing it in the injection pipette. Again, the end result is that gametes that were nonfunctional (for a variety of reasons) can be rendered functional. But does this have any long-term consequences for the children born through ART procedures in general and ICSI in particular? The safest answer at this point might be: "we don't know yet, but we are looking into it" (Schatten 2002; Schultz and Williams 2002; Winston and Hardy 2002).

To this day, the stunning success of ART has led to a highly rewording activity, both in terms of results, research and finances (Katz et al. 2002). The importance of these achievements cannot be underestimated. If the final goal (often years in the making, amid all sorts of personal sacrifices) is to obtain a healthy ("perfect") baby, and if the end result (when obtained) is so obvious that it can't help but meet expectations, what else can one ask as empirical validation for the procedure? In the case of ICSI (but not IVF) empirical validation was the only possible kind. The end result of a slight, but decisive, adjustment in a previous assay (as noted above), ICSI has another curious characteristic: it works very well in primates but requires much more effort, and technical tricks, in other species normally used as preclinical animal models. The truth is that animal models to study ICSI systematically were essentially developed *after* the technique was firmly implanted as a standard clinical procedure, in an obvious inversion of what might be considered the scientific "method" (Schatten et al. 1998; Schatten 2002). This would probably be unthinkable in any other therapeutic context, but the empirical success seemed to render any criticism irrelevant. As a side note, it might be useful to stress that, although the issue of human reproductive cloning is very much present in the media and public consciousness, at this stage, technical constraints currently make it a nonissue, as far as reproductive biologists are concerned (Rhind et al. 2003; Schatten et al. 2003).

If the importance of ART has revolutionized how we think about reproduction, careful analysis of ART children, as well as the development of animal models have raised some questions as to possible collateral effects of the different ART techniques (recent examples can be found in Hansen et al. 2002; Stromberg et al. 2002; discussed in Katz et al. 2002; Lambert 2002; Schatten 2002; Schultz and Williams 2002; Winston and Hardy 2002). Although they are probably minor and not very significant, these effects can vary from genetic and morphophysiological (relatively easy to detect) to behavioral (much harder do quantify, also given the fact that most ART children, notably ICSI children, are very young). If these findings had the merit of triggering research on what had been, up to that point, merely a service provided to couples, they were received with what could be described as aggressive indifference by some members of the community, likely also worried about the possible implications in terms of liability (Lambert 2002).

Even if these factors are removed from the equation, analysis on the long-term effects of ART technologies (and how, if there are effects, they can be reduced) will likely take decades of complex interpretations. It is crucial to know, for example, what characteristics were already present in the parents (and so could "logically" be expected to appear, at some level, percentage-wise, in the offspring), and which ones might have resulted from gamete and embryo manipulations themselves. The use of animal models also raises tricky issues. For example, even if mice born after ICSI can be shown to perform poorly in tasks that require some form of learning, does this have any bearing in the (much more complex) behavior of human children?

There can be little doubt that, in the long run, the odds are good that the mechanisms of science validation will also work in this case. But they can be painfully slow. And we live *today*. In the meantime, there are patients to advise, destinies and expectations that depend on the "best possible decision." Scholarships and research funds to be handed out, policies to implement. Graduate student careers that can vanish if they dedicate some of their most creative years to the "wrong" project (the history of science is also brimming with such nonstories).

This is the crux of the matter, and it all boils down to, not necessarily knowledge, but power. Collins and Pinch, among others, maintain that in times of crisis, an intelligent and informed citizen (informed not only on the issues, but also on the vested interests that may be at stake) is as able to participate in the discussion as a scientist, even if said scientist specializes in the field in question (and it is well known how overspecialization in science has greatly diminished the number of people theoretically able to debate any given issue). This is especially obvious when it is possible to find respected scientists on both sides of any given polarized barricade. As Collins and Pinch argue, if a decision is being made under these circumstances it is likely to be affected, not only by the scientific facts, but also by the prestige of the scientists involved in defending one side or the other, by their access to the media, by their networking, etc. In other words, by the exact same mechanisms that regulate the way any community is organized, at many different levels. Most scientists have a profound admiration for the way science works, at least at an abstract level. But no scientist working in a field where polarizing debates with social implications take place, and in which the scientific evidence is not yet unequivocal, will reject the validity of this analysis without being hopelessly naïve or hypocritical (or both). Even a staunch, but intelligent, critic of any form of "scientific relativism" will concede this point, to a certain extent (Haack 1998, 2003). However, a vast majority of those scientists will rarely, if ever, even acknowledge the problem, let alone concede the analysis.

The goal must be, therefore, to create a balance that allows science to find appropriate and (as much as possible) impartial answers for the decision-making process in a timely fashion, while maintaining enough scrutiny so as to determine that that is exactly what is going on. And also to acknowledge that, although major decisions must be made and lived with (as undecided immobility can be as lethal as bad decision making), they must always be opened to adjustments, as new information is made available. In science yesterday's doubts are often today's certainties, but the opposite is just as often true. This ongoing and fluid process requires effort, dialogue, understanding of different languages, humor, politeness, as well as the identification of useless discussions. Sometimes it is better not to talk.

This is especially evident in many rude exchanges, in which two things become obvious. Firstly, that is quite comfortable to split people and opinions into two camps, and then have them often treated as caricatures of complex realities. Black/White. Left/Right. Genes/Environment ("Nature"/ "Nurture"). Day/Night. Sciences/Humanities. Good/Evil. Secondly, for those identifying themselves with either side of each of these dichotomies, and despite the much-publicized virtues of dialogue and compromise, the extreme positions are always easier to listen to, to build careers upon, and even to understand. Why? Because they are usually simpler, and less contaminated by the engine of all research that doubt is? Because, in psychological terms, the stress of having too many choices may actually outweigh the perceived benefits (Schwartz 2004)? Because, since humans have bilateral symmetry (that is, the body can be divided into two approximately equivalent halves), this separation of two extremes makes sense, intrinsically? It would be interesting to ask a starfish (or another creature with radiated symmetry) if each issue suggests five distinct extreme positions, while humans are stuck with two. . . . In this context, one of the most worrisome trends is the tendency to relate elements of two distinct dichotomies, as if they were related. For example, science in general (and Darwinism in particular) is often associated with right-wing positions, an "alliance" that is neither "natural" nor unavoidable (see Gould 1980,1997; Lewontin 2000; Singer 1999).

Of course there are no guarantees that sociopolitical powers will react "appropriately," even after being presented with unequivocal scientific evidence of any given phenomenon. The (slow) reaction of industrialized nations to global warming is a case in point. There can be little doubt that an effective participation in society requires both science and sociology, to define objective information, and, then, to decide what to do with it. As scientists, we can interrogate our systems, and then let "society decide how to use" the answers obtained. This is the well-established position of the vast majority within the scientific community, strengthened by the well-known fact that the implications of most research projects cannot be entirely evaluated at the time they are being carried out. It is the most obvious position, and it is also the easiest. Personally, as a citizen I find it unacceptable, and, as a scientist, understandable. Nobody is perfect. But taking more responsibility for what is done with the answers one obtains is also an imperative of the active citizenship all scientists should practice. To me, this is the essence of what Santos calls "prudent knowledge for a decent life" (1987, 1992).

DIALOGUES WITHOUT BORDERS

As all epochs before it, ours also fulfills the ancient Chinese curse, "may you live in interesting times." In the case in point, part of the interest resides in the way languages related to different fields of knowledge seem to merge to create novel insights. Even though it should also be noted that this results as much from constructive dialogue, as from knee-jerk reactions of what some perceive as an "invasion" of an exclusive territory, which must therefore be "countered." Sociological interpretations of how "hard" scientific knowledge is constructed (e.g., Collins and Pinch 1993) are just one example. In fact, one could suggest that biological interpretations of the way individuals interact in society (e.g., Buss 1994; Daly and Wilson 1999; Thornhill and Palmer 2000) could constitute the exact mirror situation. In other words, on the one hand, it is argued that scientists interact as social beings and are influenced by their environment, and that these elements (and not only the quest for "pure" truths) can have a deciding influence on the conclusions reached or the paradigms accepted at a given historical juncture. On the other, it is argued that human beings unknowingly pursue the same Darwinist ideals of survival and the passing of genes unto the next generation as other species, and all forms of behavior (however complex, however grand) can be explained within this framework, which allows no escape (or a mere temporary respite) from a deeply engrained biological heritage. These two key concepts, that of the gene as basic element for natural selection (Dawkins 1976), and that of the evolutionary importance of behaviors, on a par with morphophysiological characteristics (Wilson 1975, Alcock 2001), eventually morphed into other approaches, often much more radical than what was originally proposed (see, for example, the discussion of Wilson by Hrdy 1981). The controversy (and media coverage) increases when analogies are drawn to explain (at least partially) the biological basis of behaviors such as jealousy (Buss 1994), violence towards adopted children (Daly and Wilson 1999) or rape (Thornhill and Palmer 2000). However, one should also remember that the importance of genes in predetermining complex realities is questioned by scientists who would clearly identify themselves as Darwinists (e.g., Gould 1980; Lewontin 2000).

For better or worse, analogies and metaphors have long been a staple of scientific discourse (Keller 2002), and they have been used in regard to almost anything. For example, and on the subject of human reproduction, is there any meaning associated with the fact that the oocyte is large and immotile, while the sperm is small and motile? That there are more of the latter than of the former? That the female gamete contributes most in terms of mass and energy to the embryo? That it actually digests many sperm components? That one sex has a clear menopause, but not the other? That it is the sperm that "activates" the oocyte, signaling that it has been fertilized, or that it is the male-borne centrosome that organizes the joining of the two genomes? These are simple (scientific) facts, but do they somehow reflect the way we organize ourselves as individuals that belong to a given sex? Indeed, if many of these examples seem related to clichés normally heard when classifying human sexes in terms of characteristics, reproductive strategies, behavior and contributions to embryo viability, it is because those correlations have already been made (see, for example, Birkhead 2000; Buss 1994, Burnham and Phelan 2000). For some authors they are truly relevant (Buss 1994; Burnham and Phelan 2000; Thornhill and

Palmer 2000), a sort of survival manual in an unavoidable world shaped by our biology and, ultimately, by our genes. Of course, other researchers also note that, similarly to biblical parables or popular sayings, animal reproductive behaviors run a full gamut of possibility, strategies are so diverse that analogies are very easy to find for virtually anything, and it remains to be determined, what, if anything, is relevant for humans (Zuk 2002). Regardless, it will likely be on the field of neo-sociobiology/evolutionary psychology that many future heated debates will be held, not only because they will take place on the very edges of biology, philosophy, psychology, anthropology and sociology, but also because the difficulties in obtaining, analyzing and interpreting data likely will mean that there will be much open ideology, mixed in with Science. For this kind of discussion many intelligent and open-minded partners will be needed.

As all who preceded and will follow we live history unknowingly. Therefore, we must accept that others decipher it on our behalf. To be sure, with twists and turns (and words, and concepts) that may not be part of our immediate references. But that may be useful. Or not. History is also made of conversations. An excellent example is the interesting "The One Culture? A Conversation about Science" (2001), organized by Jay Labinger and Harry Collins. The title is the program: having researchers with distinct backgrounds and convictions exchange views. And the results are also obvious, from mutual enrichment (as in the aforementioned contribution by David Mermin), to monologues that are more or less enclosed within themselves. But the essence is revealed in a quote by Stephen Hawking: "just keep talking."

CONCLUSION

A scientist working in one of the more promising, competitive and scrutinized fields of contemporary biology once confided that she felt the need to change her area of research. Tired of all the media attention, of the (direct or indirect, obvious or not so obvious) pressures from sociopolitical-industrial-biotech lobbies, of her own institution in constant search for patentable knowledge. Tired of reading half-baked reports that had been published way before they were ready (and that would never have been submitted in a field that was less "hot"), in an attempt to establish precedence, justify funding, feed media and public curiosity, or even the egos of scientists hooked on personal fame and screen time (still a cardinal sin, as far as the community is concerned). My friend complained about the continuing saga of announced "major breakthroughs" in her field (some indeed major, some relevant, some barely publishable, some not at all, as usual), and mused ironically at how such constant accumulation of certainty had led to so little progress in terms of the basic questions and problems at stake, and actually had raised even more issues (but isn't that always the case?). On more than one occasion, she confessed the yearning for simpler times, the will to work on a more obscure topic, where her results would be analyzed by less people but, paradoxically, with an Okham-like clarity she felt lacking under the spotlight of the major journals, such as *Nature* and *Science*. Needless to say, my friend and colleague remains hard at work in the exact same competitive field. Searching for simplicity and answers under the noise that envelops all exciting research topics.

Because that is the place we all wish, or should wish, to be. In the midst of the organized chaos of discovery. Finding out things about the world, or about ourselves, it matters little (and could actually be the same thing). Right here, in places like this book. Surrounded by observations, doubts, exaggerations, imprecisions, certainties, insights, errors, hypotheses. Hoping someday someone might logically organize an entropic path we never had the time to characterize. Because we were too busy walking it. Right there. Right here. On the edge.

REFERENCES

Aitken, R. J., P. Koopman, and S. E. M. Lewis. 2004. "Seeds of concern." *Nature* 432:48–52.

Alcock, J. 2001. *Animal Behavior: An Evolutionary Approach*, 7th ed. Sinauer.

Birkhead, T. 2000. *Promiscuity: An Evolutionary History of Sperm Competition and Sexual Conflict*. New York: Farber & Farber.

Burnham, T., and J. Phelan. 2000. *Mean Genes*, New York: Penguin Books.

Buss, D. M. 1994. *The Evolution of Desire*. Basic Books.

Butler, D. 2004. "The fertility riddle." *Nature* 432:38–39.

Charo, R. A. 2002. "Children by Choice: Reproductive Technologies and the Boundaries of Personal Autonomy." *Nature Cell Biology* 4:s23–s28.

Cohen, J., R. Scott, M. Alikani, T. Schimmel, S. Munne, J. Levron, L. Wu, C. Brenner, C. Warner, and S. Willadsen. 1998. "Ooplasmic Transfer in Mature Human Oocytes." *Molecular Human Reproduction* 4:269–80.

Collins, H., and T. Pinch. 1993. *The Golem. What You Should Know about Science* (second edition, 1998). Cambridge: Cambridge University Press.

Dalton, R. 2002. "Misconduct: The Stars Who Fell to Earth." Nature 420:728–29.

Daly, M., and M. Wilson. 1999. *The Truth about Cinderella: A Darwinian View of Parental Love.* Darwinism Today Series. New Haven, CT: Yale University Press.

Dawkins, R. 1976. *The Selfish Gene.* London: Oxford University Press.

El Feki, S. 2004. "The Birth of Reproductive Health: A Difficult Delivery." *PLOS Medicine* 1:e4–e9.

Franklin, S. 1998. Making Miracles: Scientific Progress and the Facts of Life. "In *Reproducing Reproduction*, ed. S. Franklin and H. Ragoné." University of Pennsylvania Press, 102–17.

Goodstein, D. 1996. Conduct and Misconduct in Science. "In *The Flight from Science and Reason*, ed. Gross, Levitt and Lewis." New York: New York Academy of Sciences, 31–38.

Gould, S. J. 1980. *The Panda's Thumb: More Reflections in Natural History*. New York: W. W. Norton & Company.
———. 1997. *Full House: The Spread of Excellence from Plato to Darwin*. New York: Random House.
Haack, S. 1998. *Manifesto of a Passionate Moderate*. Chicago: The University of Chicago Press.
———. 2003. *Defending Science—Within Reason*. Prometheus Books.
Hansen, M., J. J. Kurinczuk, C. Bower, and S. Webb. 2002. "The Risk of Major Birth defects after Intracytoplasmic Sperm Injection and In Vitro Fertilization." *New England Journal of Medicine* 346:725–30.
Hewitson, L., T. Dominko, D. Takahashi, C. Martinovich, J. Ramalho-Santos, P. Sutovsky, J. Fanton, D. Jacob, D. Monteith, M. Neuringer, D. Battaglia, C. Simerly, and G. Schatten. 1999. "Unique Checkpoints during the First Cell Cycle of Fertilization after Intracytoplasmic Sperm Injection in Rhesus Monkeys." *Nature Medicine* 5:431–33.
Hrdy, S. B. 1981. *The Woman That Never Evolved* (2nd edition, 1999). Cambridge, MA: Harvard University Press.
Huizenga, J. R. 1993. *Cold Fusion: The Scientific Fiasco of the Century*. University of Rochester Press.
Johnson J., J. Canning, T. Kaneko, J. K. Pru, and J. L. Tilly. 2004. "Germline Stem Cells and Follicular Renewal in the Postnatal Mammalian Ovary." *Nature* 428:145–50. Erratum in: *Nature* 430:1062.
Katz, P., R. Nachtigall, and J. Showstack. 2002. "The Economic Impact of the Assisted Reproductive Technologies." Nature Cell Biology 4:s29–s32.
Keller, E. F. 2002. *Making Sense of Life. Explaining Biological Development with Models, Metaphors and Machines*. Cambridge, MA: Harvard University Press.
Krimsky, S. 2004. *Science in the Private Interest: Has the Lure of Profits Corrupted Biomedical Research?* Lanham, MD: Rowman & Littlefield Publishers.
Kupker, W., S. Al-Hasani, R. Johannisson, I. Sandmann, M. Ludwig, D. Jocham, K. Diedrich. 2002. "The Use of Cryopreserved Mature and Immature Testicular Spermatozoa for Intracytoplasmic Sperm Injection: Risks and Limitations." *Seminars in Reproductive Medicine* 20:25–35.
Labinger, J. A., and H. Collins, orgs. 2001. *The One Culture? A Conversation about Science*. Chicago: Chicago University Press.
Lambert, R. D. 2002. Safety Issues in Assisted Reproduction Technology: The Children of Assisted Reproduction Confront the Responsible Conduct of Assisted Reproductive Technologies. *Human Reproduction* 17:3011–15.
Latour, B., and S. Woolgar. 1986. *Laboratory Life: The Construction of Scientific Facts* 2nd ed., Princeton, NJ: Princeton University Press.
Lewontin, R. 2000. *The Triple Helix: Gene, Organism and Environment*. Cambridge, MA: Harvard University Press.
Mallove, E. J. 1991. *Fire from Ice: Searching for the Truth behind the Cold Fusion Furor*. Infinite Energy Press; Reissue edition.
Mermin, N. D. 2001. "Conversing Seriously with Sociologists." In *The One Culture? A Conversation about Science*, org. Labinger and Collins. Chicago: Chicago University Press, 83–98.
Nunes, J. A. 2001. "Laboratórios, escalas e mediações na investigação biomédica. A oncobiologia entre o global e o local." In *Enteados de Galileu? A semiperiferia no*

sistema mundial da ciência, org. Nunes and Gonçalves. Edições Afrontamento, 33–75.

Palermo, G. D., J. Cohen, M. Alikani, A. Adler, and Z. Rosenwaks. 1995. "Intracytoplasmic Sperm Injection: A Novel Treatment for All Forms of Male Factor Infertility." *Fertility and Sterility* 63:1231–40.

Palermo, G. D., H. Joris, P. Devroey, A. C. Van Steirteghem. 1992. "Pregnancies after Intracytoplasmic Injection of Single Spermatozoon into an Oocyte." *Lancet* 340:17–18.

Powell, K. 2004. "Age Is No Barrier." *Nature* 432:40–42.

Ramalho-Santos, J., R. D. Moreno, P. Sutovsky, A. W. Chan, L. Hewitson, G. M. Wessel, C. R. Simerly, and G. Schatten. 2000a. "SNAREs in Mammalian Sperm: Possible Implications for Fertilization." *Developmental Biology* 223:54–69.

Ramalho-Santos, J., G. Schatten, and R. D. Moreno. 2002. "Control of Membrane Fusion during Spermiogenesis and the Acrosome Reaction." *Biology of Reproduction* 67:1043–51.

Ramalho-Santos, J., P. Sutovsky, C. Simerly, R. Oko, G. M. Wessel, L. Hewitson, and G. Schatten. 2000b. "ICSI Choreography: Fate of Sperm Structures after Monospermic Rhesus ICSI and First Cell Cycle Implications." *Human Reproduction* 15:2610–20.

Reichhardt, T. 2004. "Studies of Faith." *Nature* 432: 666–69.

Rhind, S. M., J. E. Taylor, P. A. DeSousa, T. J. King, M. McGarry, and I. Wilmut. 2003. "Human Cloning: Can It Be Made Safe?" *Nature Reviews Genetics* 4:855–64.

Santos, B. S. 1987. *Um discurso sobre as Ciências.* Afrontamento.

———. 1992. "A Discourse on the Sciences." *Review Fernand Braudel Center* 15 (1):9–47.

Saulson, P. R. 2001. "My Life Inside a Case Study." In: *The One Culture? A Conversation about Science,* org. Labinger & Collins, pp. 73–82. Chicago: Chicago University Press.

Schatten, G. 1994. "The Centrosome and Its Mode of Inheritance: The Reduction of the Centrosome during Gametogenesis and Its Restoration during Fertilization." *Developmental Biology* 165:299–335.

———. 2002. "Safeguarding ART." *Nature Cell Biology* 4 Suppl:s19–22.

Schatten, G., L. Hewitson, C. Simerly, P. Sutovsky, and G. Huszar. 1998. "Cell and Molecular Biological Challenges of ICSI: ART before Science?" *Journal of Law and Medical Ethics* 26 (1):29–37, 3.

Schatten, G., R. Prather, and I. Wilmut. 2003. "Cloning Claim is Science Fiction, Not Science." *Science* 299:344.

Schultz, R. M., and C. Williams. 2002. "The Science of ART." *Science* 296:2188–90.

Schwartz, B. 2004. "The Tyranny of Choice." *Scientific American,* April 2004, 42–47.

Service, R. F. 2002. "Breakdown of the Year: Physics Fraud." *Science,* 298:2303.

Sharpe, R. M., and S. Franks. 2002. "Environment, Lifestyle and Infertility—an Inter-generational Issue." *Nature Cell Biology* 4:s33–s40.

Simerly, C., C. Navara, S. Hwan Hyun, B. Chun Lee, S. Keun Kang, S. Capuano, G. Gosman, T. Dominko, K. Y. Chong, D. Compton, W. Suk Hwang, and G. Schatten. 2004. "Embryogenesis and Blastocyst Development after Somatic Cell Nuclear Transfer in Nonhuman Primates: Overcoming Defects Caused by Meiotic Spindle Extraction." *Developmental Biology* 276:237–52.

Simon, B. 2002. *Undead Science: Science Studies and the Afterlife of Cold Fusion*. Rutgers University Press.

Singer, P. 1999. *A Darwinian Left: Politics, Evolution and Cooperation*. New Haven, CT: Yale University Press.

Sokal, A., and J. Bricmont. 1998. *Fashionable Nonsence: Postmodern Intellectual's Abuse of Science*. Picador.

Steptoe, P. C., and R. G. Edwards. 1978. "Birth after the Reimplantation of a Human Embryo." *Lancet* 2:366.

Stromberg, B., G. Dahlquist, A. Ericson, O. Finnstrom, M. Koster, and Stjernqvist K. 2002. "Neurological Sequelae in Children Born after In-Vitro Fertilisation: A Population-Based Study." *Lancet* 359:461–65.

Tesarik, J. 2002. "Reproductive Semi-cloning Respecting Biparental Embryo Origin." *Human Reproduction* 17:1933–37

Thornhill, R., and C. T. Palmer. 2000. *A Natural History of Rape: Biological Bases of Sexual Coercion*. Cambridge, MA: MIT Press.

Wassarman, P. M., L. Jovine, and E. S. Litscher. 2001. "A Profile of Fertilization in Mammals." *Nature Cell Biology* 3:E59–64.

Wilmut, I., N. Beaujean, P. A. De Sousa, A. Dinnyes, T. J. King, L. A. Paterson, D. N. Wells, and L. Young. 2002. "Somatic Cell Nuclear Transfer." *Nature* 419:583–87.

Wilmut, I., A. E. Schnieke, J. McWhir, A. J. Kind, and K. H. Campbell. 1997. "Viable Offspring Derived from Fetal and Adult Mammalian Cells." *Nature* 385:810–13.

Wilson, E. O. 1975. *Sociobiology: The New Synthesis*. Cambridge, MA: Harvard University Press.

Winston, R. M. L., and K. Hardy. 2002. "Are We Ignoring Potential Dangers of *In Vitro* Fertilization and Related Treatments?" *Nature Cell Biology* 4:s14–s18.

Wolpert, L. 1993. *The Unnatural Nature of Science*. Faber and Faber Limited.

Yanagimachi, R. 1994. Mammalian fertilization. In *The Physiology of Reproduction*, ed. E. Knobil, and J. D. Neill. New York, Raven Press, 189–317.

Zuk, M. 2002. *Sexual Selections: What We Can and Can't Learn about Sex from Animals*. University of California Press.

III

GLOBAL COGNITIVE JUSTICE: RECONSTRUCTING KNOWLEDGES AND WORLD MAKING

13

Actors, Networks, and New Knowledge Producers

Social Movements and the Paradigmatic Transition in the Sciences

Arturo Escobar

INTRODUCTION: THE PARADIGMATIC TRANSITION REVISITED

Originally published in 1988 in Portuguese, "A Discourse on the Sciences" (Santos 1992) crystallized a series of important trends and presaged trends still to come. It elegantly gathered critiques of the sciences and new proposals, historical and epistemological arguments, and social and institutional analyses concerning what at that point in time was a seemingly mounting evidence pointing at the crisis of the dominant scientific paradigm. To this extent, "A Discourse" was one of the very first works in the social sciences to announce the start of a paradigm shift that was already being broadcast with great excitement by a group of pioneers in the natural and physical sciences, from Jantsch and Bohm to Prigogine and Stengers. Its author's aim was to build on "the loss of epistemological confidence" caused by the crisis in order to articulate a new model for the production of knowledge. If the dominant scientific paradigm had been one that pitted "proper sciences" against both common sense and the humanities, in the new model it would be these latter that would be accorded a place of pride. Once conventional criteria of rationality and objectivity were given up, once the subject was brought back into the picture—after scientists realized that "the object is the continuation of the subject by other means," as he provocatively put it (1992, 41)—then there would be a revaluation of the humanities, a call for methodological plurality, and a vindication of common sense understanding as crucial to all knowledge. Two other things would follow this resubjectification of the sciences: the necessarily situated character of all knowledge; and, concomitantly, the fact that knowledge would be freed from the straightjacket of objectivity and rationality and enabled to broach the crucial question of how we can live ethical lives in an increasingly complex and ambiguous world.

The present moment, argued Santos in the 1980s, was one of interparadigmatic transition, with the social sciences playing a central role. First, by rejecting the binarism between nature and culture, subject and object, mind and body they would trigger the dissolution of the boundary between the natural and the social sciences once and for all. As we shall see, this feature seems to be validate by recent developments, such as actor-network theory and political ecology. Second, the emergent paradigm would highlight the simultaneous local and total character of knowledge. Recent trends such as the feminist epistemology of situated knowledge, which was being articulated in the late 1980s (e.g., Haraway 1988), the entire revaluation of local knowledge in ecology and anthropology (e.g., Hobart ed. 1993), the emphasis on place-based knowledge as a corrective to globalocentric discourses (e.g., Casey 1997; Prazniak and Dirlik 2000; Escobar 2001; Harcourt and Escobar 2002), and the apparent come back of phenomenology in a variety of fields with its emphasis on nondualist conceptions of knowledge (e.g., Ingold 2000), all seem to validate this observation, even if the question of the allegedly total character of knowledge remains controversial or undertheorized. The emphasis on the local and partial aspect of all knowledge is linked to the third dimension of the emergent paradigm, the fact that "all knowledge is self-knowledge" (Santos 1992, 40), that is, that all knowledge produces a historical identity that is far from the universal subject posited by modern logocentric perspectives. In this respect, the poststructuralist prescription that knowledge (discourse) is constitutive of social reality is another reminder of the intersubjective dimensions of all knowledge.

Finally, for Santos the new paradigm also entailed a return to the question of common sense understanding, one that would shake off the entrenched and naturalized common sense of modernity anchored in rationality and alleged objectivity. What we need instead, Santos argued, is a new common sense that is pragmatic, situated, place-based, and grounded in experience, even if of course this common sense would have to engage with scientific knowledge and, in doing so, produce a common sense that avoids both the limitations of traditional common sense and the arrogant and equally provincial claims of modern rationality as the only possible common sense for the world. This last feature is probably the aspect that has been given least attention in the last twenty years among those highlighted in Santos' conceptualization. As we shall see, however, it has been implicit in debates on local knowledge and in phenomenological approaches. But it has been perhaps in the terrain of nonacademic actors where the issue of common sense knowledge has been brought to the fore. This argument can be made by looking at the kinds of knowledge that social movement activists, and a host of intellectual-activists linked to social movements and NGOs, have been producing in the context of struggles that are localized and globalized at one and the same time.

Not everything is going, of course, quite in accordance with Santos' crystal ball. One important qualification to be made, in my mind, concerns the dynamic of the transition, including the main agents driving it. At a social level, opposition to the transition has gained strength in the past twenty years. It is only now that the tremendous depth and pervasiveness of the neoliberal revolution kicked off by the Thatcher-Reagan-Bush years are being realized. The reideologization of the dominant model—resulting in what the liberation theologian Frank Hinkelamert has called "the age of the total market"—is not only enabling the creation of a worldwide order that favors the world's elite at the expense of the world's poor and the world's environments, it is renormalizing academic institutions worldwide and putting a brake on many of the pioneering agendas identified by Santos. At this level, at least in the United States and despite certain encouraging trends (e.g., some trends towards internationalization of curricula, experiential learning, etc.) the academy has shied away from deepening the paradigmatic transition and has retrenched into more conventional agendas. Calls for new paradigms seem to be falling in deaf ears (see also the report of the Gulbenkian Commission in this regard, 1996). Social scientists have been in retreat. If in the 1980s the social sciences were infusing the natural sciences with new idioms and ideas, today it seems to be the other way around. Notions of complexity, webs, networks, self-organization, etc., are now being more actively developed in the natural sciences, although of course there are attempts to bring it all back to the social sciences again. The shift in the humanities towards the production of critical intersubjective knowledge for social transformation—while important in some fields such as cultural and so-called postcolonial studies, and feminist and critical race theories—has floundered in the persistent Achilles' heel of their engagement with extra-academic worlds. In this context, nonacademic knowledge producers seem to have taken the lead, as I shall argue in this chapter.

The following pages introduce several interrelated arguments as a way to discuss some of these predicaments. Part I outlines a provisional interpretation of anti-globalization social movements by appealing to concepts in complexity theory, particularly self-organizing networks. Nonacademic social actors are seen as triggering a social and epistemological transformation and pushing academics to search for new languages of description and explanation, which they may find in the natural sciences, among other sources. This first part thus illustrates the first aspect of the transition in Santos' conceptualization. Part II pushes this argument in the direction of the production of local knowledge and new common sense by discussing the case of a Colombian ethnic and environmental movement. It examines the kinds of knowledge that activists of this movements are producing about their situation and struggle, and the ways in which this knowledge de facto confounds the distinctions between social, natural, and human sciences. The activists' knowledge is characterized from the perspective of place. In this way, the chapter addresses primarily the first and last dimensions of the paradigmatic change envisaged in "A Discourse on the Sciences." The chapter concludes by emphasizing the need to take into account new knowledge producers and epistemic perspectives in revisiting the character of the transition. What this means also is that we need to pay attention to the changes that are happening in the world at large, and particularly to those subaltern perspectives that are articulating challenges to the new neoliberal hegemony, if we are to rearticulate the question of a new paradigm for the sciences from a perspective of the social transformations that could make it viable.

NETWORKS, MESHWORKS, AND ANTI-GLOBALIZATION SOCIAL MOVEMENT

The seeming consolidation of neoliberal capitalism as the main mode of globalization after 1990 was accompanied by a concomitant change in the mode of operation of social movements. While some see in the ascendancy of globalization and the disappearance of "real socialism" the demise of social movements, others perceive in the incipient forms of global protest the beginning of a new round of resistance that may or may not generate its own theoretical and political momentum. Will the movements that started to gain visibility in the mid-1990s result in the sustained construction of imaginaries for alternative modernities and perhaps even non-Eurocentric modes of analysis and social life? The answer to this question will depend on the character these social movements adopt and on the extent to which they might be able to generate their own "sustainable" structures for the production of knowledge. Both requirements demand new ways of thinking and activism on our part as analysts, academics, or intellectual-activists.

It is already a common place to assert that the social movements of today operate under different rules, and that even theories of the most recent past are not necessarily applicable to the today's collective action. In other words, we do not have good models to describe and understand how social movements construct their strategies and articulate their claims. If there is any degree of consensus on how anti-globalization social movements (AGSMs) operate is that (a) they operate at various scales (from the local to the global); (b) they do not have a centralized structure, command center, or shared set of demands even if at some level they can be said to have a "common enemy" (neoliberal globalization); (c) consequently, they are tremendously pluralistic, which is seen by some as a strength, as a weakness by others; and (d) that the most apt metaphor to describe them is that of networks. "Networks" is in fact a concept that enjoys great favor in a variety of fields and in connection to a variety of problems; it has been invoked as the central organizing principle of the information society; as the basis on which society and technoscience are structured and function (the actor-network theory); and as an explanation for transnational social movements organizations. Networks are also central to some strands of complexity theory arising from the natural sciences.[1]

I would like to focus in this section on a certain understanding of networks in complexity theory and its applicability to anti-globalization social movements.[2] Arising chiefly in the biological sciences, theories of complexity and self-organization have not yet generated a social theory per se, although some scholars have extended complexity into the realm of the social. One of the basic insights is the need to account for "structure" as the emergent property of some dynamical system, rather than assuming an already constituted system. In Manuel de Landa's work (1997), the same basic nonlinear dynamic can be seen at play in geological, biological, and social systems producing structures (e.g., economic, linguistic, geological, or biological structures) that are the result of the self-organization of matter and energy (and information and culture). Self-organization entails structure-generating processes that result in self-consistent aggregates or systems. There are two main structures of this type: hierarchies and meshworks. Hierarchies have conscious goals and overt control mechanisms; they are formed by sorting out elements into classes and ranks that become consolidated into permanent states (strata) with emergent properties of their own (examples: sedimentary rocks, biological species, social classes); hierarchies have tended to dominate social life in the last three hundred years. Meshworks, on the other hand, are articulations of heterogeneous elements in terms of their functional complementarities. There are three elements that characterize a meshwork as a structure-generating process: a) the bringing together of a set of heterogenous elements in terms of functional complementarities (the interconnection of diverse but overlapping elements); b) the presence of intercalary elements (catalysts) that facilitate the interconnection; and c) a stable pattern of behavior, endogenously generated, that results from the interlocked heterogeneities (examples: granite, rainforest ecosystems; small town self-organizing markets).

Meshworks are self-organizing and grow in unplanned directions (akin to Deleuze/Guattari's rhizomes, 1987); they are made up of diverse elements, including of course human and nonhumans, organisms and machines; they usually exist hybridized with other meshworks and hierarchies; they accomplish the articulation of heterogenous elements without imposing uniformity; and they are determined by the degree of connectivity that enables them to become self-sustaining. There are two moments in the construction of a meshwork: a localization strategy, which increases the heterogeneity between localities while that within localities decreases (increased local homogeneity, at the interior of a node or site); and a strategy of interweaving, by which the heterogeneity of each locality increases, while that among them decreases. Meshworks may form loops where the product of one node serves as a catalyst for another; a meshwork might start with two nodes and incorporate new nodes as long as internal consistency ("autopoiesis") is maintained. As the meshwork complexifies, new nodes are added following a self-organizing dynamic. At some point, there obtains a self-sustained intensification maintained by new nodes inserting themselves into a growing autocatalytic loop. Meshworks are often fostered by mutually reinforcing and autocatalytic innovations (these are well known in the history of technology; e.g., the system coal-steam-money-mill at the eve of the industrial revolution). Meshworks may be formed by motions of destratification or deterritorialization of intermediate intensity, thus connecting flows and nodes that might otherwise remain unconnected (e.g., symbiotic nets of small producers involved in volatile trade and in import substitution may have created the possibility of nonhierarchical local and regional economies in history, and may do so again in the future). Although meshworks result from the action of many individual and collective decision makers, they take on a life of their own. They form wholes that "add themselves to an existing population of individual structures operating at different scales" (de Landa 1997, 271). However, it is a whole that does not totalize, nor unifies, the parts but that is rather added to the parts "as a new part fabricated separately"[3].

When applied to social domains, this view of meshworks has the following political implications:

a) The political character of the meshwork will depend on the nature of the heterogeneous elements brought together and on the types of articulations established among them (à la Laclau and Mouffe 1985), and not only on who does the meshworking or meshweaving (i.e., not because women or subaltern groups do the meshworking will the meshworks be necessarily progressive; this will depend on the de/restratifications the meshwork might be able to effect). Conversely, some destratifying meshworks, such as computer networks, might also lead to reactionary restratifications.

b) Meshworks may perform a de-stratifying function in relation to dominant forms of power-knowledge; they enact a de-stratified power operating through a multiplicity of informal constraints. These constraints may operate as catalysts or triggers in the formation of meshworks.

c) Strategies of domination may also form meshworks, such as the ensemble of political technologies that resulted in the formation of the disciplinary society discussed by Foucault (1979). These technologies might form a blueprint or general method (Foucault's panopticon), or the congealment of nonlinear flows into a set of practices, institutions and discourses (their "mineralization," in de Landa's terms).

Many of today's social movements can be seen as either confronting globalization directly or as implicated in struggles that have to do in one way or another with the social antagonisms created by globalization. We can thus properly speak of a set of AGSMs that would include those explicitly self-defined as such (e.g., the People's Global Action Against Free Trade, PGA; anti-GATT and anti-WTO protests, from India and Chiapas to Seattle and Prague) as well as those seemingly localized struggles that in one way or another confront aspects of neoliberal capitalist globalization. There is no longer any strictly localized subalternity; this also means that no movements are in a position of pure resistance or exteriority, if they ever were, but in a position of pursuing politically enriched forms of alterity. In some cases, they might foster alternative modernities, and in the best of cases non-Eurocentric modes of knowledge and forms of social and ecological life in general. For some, AGSMs may be seen as constituting counter-hegemonic globalizations.

Anti-globalization movements can be seen as horizontal, self-organizing meshworks of heterogeneous struggles brought together by diverse interfaces and catalysts, particularly NGOs and pioneering social movements (e.g., EZLN, PGA, Direct Action Network, MST, Porto Alegre, others). New nodes are brought into the meshworks through strategies of interweaving that enable the construction of collective agendas and fronts of struggle, which subsequently become part of the localizing strategies of the particular social movements making up the meshwork. This dialectic of localization and interweaving is at times fueled by nonlinear dynamics that produce alternations in the scale and visibility of mobilization (e.g., Geneva and Montréal in May 1998, Seattle in November 1999, and the many irruptions that followed in various parts of the world; these act as attractors in this regard). Meshworks engage with hierarchical, dominant actor-networks (DANs) as part of their struggle (e.g., networks established by organizations such as the WTO and the World Bank, GMOs research and development, TNCs such as Monsanto or Novartis, G-8, etc.); these confrontations may also have a catalytic function, "galvanizing opposition. Meshworks might also hybridize with transnational social movement organizations or advocacy networks, but are rarely reducible to them. Meshworks are actor-networks (actors in a net/mesh doing some work) involving human and nonhuman elements of all kinds. They can be seen as apparatuses for the production of discourses and practices, that is, as creating discursive fields of action that harbor a number of sites extending over vast aspects of the social field. Meshworks create flows that link sites which, operating more like fractal structures than fixed architectures, enable diverse couplings and forms of connectivity (structural, strategic, conjunctural) with other sites and networks.

Anti-globalization meshworks are not based on the purificatory operation of dividing nature/culture and us/them (Latour 1993). Unlike Latour's proposal for a nonmodern constitution (one that would finally separate nature and culture, that is, a strictly antiecological constitution that presupposes a postnatural world), many AGSM meshworks struggle for ecological practices of local and regional world making, that is, for an integration of human and biological processes that does not wreck natural systems. Some meshworks, then, can be seen as pursuing a strategy of ecological design; this is so because for many actors making up networks of this sort it is patently clear that the environment is a central contradiction of the kind of logocentric capitalist modernity that is still at the basis of neoliberal globalization. Many movements struggle against various forms of power simultaneously (e.g., ethnic domination, ecological degradation, patriarchy and economic exploitation). For the same reason, meshwork strategies are pluralistic; they take difference-in-equality as a point of departure and render obsolete the us/them distinction, although without erasing it for obvious political reasons. For this task, they bring to the fore existing local models of nature, social relations and economy and project them in transformative directions for sustainable and pluralistic strategies of world making. At their best, these movements enact a cultural politics of difference. Finally, meshworks of this kind are always weaving and cutting the network at the same time in light of political strategy, the dangers of repression, or political expediency (Strathern 1996).

It is not easy to fully characterize the dynamic that characterizes the formation of these meshworks—the linking-up of new nodes and sites, the types of flows that are of importance, the scalar politics in which they engage. From the principles of nonlinear dynamics, some additional aspects could be highlighted in this regard. Nodes are brought together in terms of certain types of functional complementary; this bringing together entails a dialectic of sameness and difference that activists have to take into account. Since nodes with/from different place-based cultural worlds are brought into contact, there is an inevitable effect of meshworking on the background practices of the subworlds involved (see Spinosa, Flores and Dreyfus 1997 for a discussion of this aspect in a different context). Conversely, meshworking enable social movements to effect important destratifications and reterritorializations of places, territories, regions, identities. It is likely that meshworks evolve both "by drift" and according to carefully thought out political strategies; up to a point, it appears that they tend to establish autocatalytic loops that drive the dynamic of self-organization. But again more research would be needed to ascertain these aspects of meshwork dynamics.

From the perspective of complexity, in sum, one could make the following *provisional* argument about AGSM:

a) AGSN meshwork can generally be described as a nonhierarchical, self-organizing, malleable structures which are only partly visible. What happens between events/mobilizations, and among sites, remains (and should remain?) largely invisible. In a general way, it can be said that underlying all meshworks are the "submerged networks" (Melucci, 1989) of meanings, frameworks, and practices of difference out of which the social movements themselves emerge. Some see in network ethnographies an important way to understand these phenomena and for researchers to position themselves politically in this regard (King 2000).

b) The goal of many of the anti-globalization struggles can be seen as the defense of particular, place-based historical conceptions of the world and practices of world making—more precisely, as a defense of particularly constructions of place, including the reorganizations of place that might be deemed necessary according to the power struggles within place. These struggles are place based, yet transnationalized (Harcourt and Escobar 2002; Escobar 2001). It is possible to talk about them as constituting *subaltern strategies of localization* (opposed to the *strategies of localization by capital,* which operate through quite different means) with three different components or dimensions: 1) place-based strategies of localization for the defense of local models of social life (e.g., local models of nature and economy, ethnic and gender relations, etc.); 2) global strategies of localization through an active engagement with both dominant networks and resistance meshworks; and 3) shifting, inter-scalar political strategies that link identity, place, nature, and culture at local, regional, national, and global levels. Social movements often engage in a novel *politics of scale*; they jump from the local to the global producing "scaling-up" effects in terms of length and size of the networks, as in the Latourian conception.

c) AGSM meshworks face a double, interrelated struggle: for the defense of place-based cultural models of social and natural life against the delocalizing effect of global culture/economy; and against the ongoing restructuring of local worlds by global capital and culture. For this double task, they engage in two types of activism place-based mobilizations in their node/site; and translocal mobilizations (virtual and face to face) to resist, negotiate, or disrupt the conditions of the restructuring (hence the centrality of WTO, the World Bank and the IMF as a target of struggle at present). The tension between this double set of demands is only resolved in the practice, by articulating activism in place with activism at a distance. This articulation between place politics and cyber-politics is a characteristic feature of many contemporary social movements (Harcourt 1999; Ribeiro 1998; Escobar 1999). Said differently, AGSMs must struggle for (a) the defense of place and constructions of place; (b) the transformation of places away from place-based entrenched forms of power and domination (particularly gender and ethnic domination; what Mignolo (2000), following others, has called double critique); (c) the construction of coalitions through the engagement with translocal actor-networks, including cyberspace. As Dianne Rochelau put it, it is a matter of "the combination of people-in-place and people-in-networks, and the portability (or not) of people's ways of being-in-place and being-in-relation with humans and other beings" (personal communication).

(d) The political character of the meshwork is not given by the identity of the meshworker or meshweaver (after Alvarez in press). The politics of meshworking, thus, has to be found *at the intersection of the emergent identities* (women, ethnic and indigenous groups, peasants, new worker organizations, gay and lesbians, environmentalists, anarchists, socialists, etc., all of which are by their very engagement with the meshwork), the *scaling effects* the meshworks are able to produce, and of course the concrete power relations and practices they effect.

(e) Finally, the result of meshwork practices and actions can be seen as the creation of alternative configurations of culture, economy, nature, and identity, that is, of socionatural orders (no matter how fleeting and precarious) that do not conform to the dominant organizing principles of neoliberal capitalism and dominant modernity. The utopia of AGSMs is that of reconceiving and reconstructing the world from the perspective of manifold place-based cultural, ecological, and economic practices of difference. While it would be, of course, naive and ahistorical to think that subaltern-produced meshworks might be able of fully "crowding out" dominant networks, this utopia is being reconstituted.[4]

As part of their struggle, social movements produce flows of information and knowledge that often times amounts to a veritable theoretico-political framework. A well-known example is the Zapatista theory of autonomy. There are also theories of nonviolent resistance, social theories of anarchism, alternative theories of development, sustainability and conservation, critical theories of GMOs and transgenic agriculture, and even alternative frameworks of the economy and markets are been produced. Some of these forms of knowledge concern matters of strategy, others the nature of domination, others the defense and reconstruction of local and regional worlds. In some cases, these knowledge is produced at the intersection of critical conversations in the disciplines (geography, anthropology, ecology, political economy, feminist theory) and political-intellectual conversations within/among social movements. This knowledge production, as we shall see in the next section, is governed by considerations that are of course more pragmatic than scholarly. The knowledge produced by the meshworks should be an important part of academics' own theorizing and research agendas. It is no longer the case that some produce knowledge (academics, intellectuals) that others apply (social movements); these boundaries are completely disrupted at present, as movements become knowledge producers and intellectuals are called upon to engage more and more in activism. This important aspect of the paradigmatic transition of the sciences is further explored in the next section.

MESHWORKS IN PRACTICE. A BRIEF COLOMBIAN EXAMPLE

It is early to claim a firm understanding of how these meshworks operate (or even if this is the most apt structural/functioning model to describe them), and it is clear that we need detailed ethnographies before we can do so. A Colombian example might help to bring some of these features into sharper focus. Based on an ongoing ethnography and collaboration with a group of activists over a number of years (1993–present), the following description examines the movement as a knowledge producer, its discursive field of action, and its networking activities as it ventured out of the region where it emerged into the world at large.

The social movement of black communities emerged in the early 1990s in the Pacific rainforest region of Colombia in the midst of a historical conjuncture marked by three main factors: the neoliberal opening of the country to the global economy after 1990; the reform of the constitution in 1991 which, among other things, granted collective territorial rights to the ethnic minorities of the region; and the discovery of the Pacific as one of the most important "hot-spots" of world biodiversity. Inhabited by close to a million people (90 percent Afro-Colombians, 5 percent indigenous peoples, 5 percent whites and mestizos; 60 percent of them still living along the region's innumerable rivers and with cultural, ecological, and economic practices that contrast sharply with those of the country's Andean majority), the region has seen the growth of political activism for the implementation of the law of territorial rights, and for the defense of the territory, culture, nature and identity. One of they most visible groups in this movement is the Process of Black Communities (PCN, Proceso de Comunidades Negras), a network of about 120 local organizations active especially in the southern Pacific. The region has been the object of conventional development plans since the mid-1980s, and has seen an increased pace of capitalist activities in some areas, including the expansion of African palm plantations, industrial shrimp cultivation, gold mining, and increased timber extraction. Local groups have mobilized in response to the developmentalist and capitalist onslaught unleashed especially after the mid-1980s.[5]

I will highlight in this brief exposition two aspects of the movement: its role as knowledge producer; and its networking activities. Activists of the movement have developed what might be described as an alternative political ecology framework that includes alternative conceptualizations of development, conservation, and sustainability. The framework starts by seeing the territory as the space of *effective appropriation of the ecosystems* by the local communities (involving, say, an entire river basin and interriver territorial usage). Considering the Pacific as a whole, activists have introduced the notion of the Pacific as *a region-territory of ethnic groups*, that is, an ecological and cultural unit that has been laboriously constructed throughout the centuries by black and indigenous groups. If the territory embodies *the historical life project* of the communities, the region territory is a *political construction* for the defense of that project and the specific territories. In this way, the region-territory is a strategy of sustainability, and vice versa, sustainability is a strategy for the defense and construction of the region-territory. The struggle for territory is thus a cultural struggle for autonomy and self-determination and an ecological struggle for the defense of local ecosystems and models of nature. In this way, "biodiversity conservation" is seen as obtained only when territory and culture are brought together in the practices of the communities. Activists have also developed notions of alternative development, local "traditional production systems," and forms of governability that can support the effective defense of the territory.[6]

PCN activists have not produced it in isolation but in the encounter with a host of actors at all levels, from the local to the global, including the following: the State, NGOs, universities and experts, the global biodiversity network, and other social movements. The most prominent of these actors is the State with its development and conservation strategies. Experts of all kinds started to "parachute" on the region in the mid-1980s, and they have produced particular representations and knowledge about the region and its "problems." A small, but symbolically crucial, project was the Biodiversity Conservation Project for the Pacific (Proyecto Biopacifico, PBP) in effect between 1993 and 1998. PCN and indigenous activists became the main interlocutors of the project and some of them became part of the project's staff. It was in the context of the dialogue between the PCN and the PBP staff that notions such as "traditional production systems," "territory," "and region-territory" started to be conceptualized by biologists and activists and subsequently refined in productive, even if oftentimes difficult and tension-ridden, dialogues. Second, local and international NGOs have been important to the knowledge production practices of the movement—through funding, participation in joint events, discussion of specific concepts, and partial political support. This has been particularly true of environmental organizations (including local chapters of, say, WWF, but especially a number of Colombian NGOs). Universities and research centers provided a third important space of engagement for activists, particularly in anthropology, geography, landscape ecology, conservation biology, and cultural and gender studies; these took the form of exchanges and interaction with a growing number of black, nonblack Colombian and nonblack foreign professionals. Fourth, and very importantly, the global biodiversity dominant actor-network was crucial in the engagement of PCN actors with dominant representations of forests, biodiversity, genetic resources and sustainability and in catalyzing the development of alternative conceptions.[7]

What have been the main practices for the production of this knowledge? These range from work with the river communities to international traveling including: (a) exercises of several days with river communities for the demarcation of territories and application for collective land titles. These exercises included reconstructing the history of settlement of the river, mapping the spatial distribution, collection of "traditional knowledge practices," and so forth; (b) more specialized *talleres* or workshops with local leaders on aspects ranging from the political aspects of identity to the more technical issues of river basis design, territorial appropriation, management plans, etc.; (c) participation in an infinity of negotiation exercises with the government *(concertación)*, especially around development and conservation projects; (d) political/conceptual presentations at these and many other meetings; (e) writing of documents for other movements, the government, or NGOs; (f) preparation of elaborate papers for presentation in academic spaces, chiefly nationally but also internationally.

In this context, the knowledge production by the PCN can be said to have the following features:

(a) It is conjunctural without being punctual—to a certain degree it is cumulative and progressively refined.

(b) It is developed "on the run," so to speak; there is not much time to pause and think, although the internal discussion and debate never stop. There is a rich and at times conflictual internal conversation among activists at the national and regional levels on conceptual, political, and pragmatic issues. Since this internal conversation tends to obey to the pressures of the day, it means that there is no time to create the knowledge-production infrastructure that could made knowledge production more systematic and effective.

(c) It is pragmatic without being just utilitarian or functional to the struggle; knowledge is seen as crucial to the political strategy. It is geared towards the articulation of demands, but always with a sense of the long run goal, namely, the *defense of the historical life project of the communities*. In this way, it has a powerful vision of the future.

(d) It is recursive to the extent that the same themes (territory, identity, biodiversity, cultural practices, etc.) are worked on, and worked out, at many levels, in different ways, from the local to the global. That is, they have a fractal recursivity that continuously brings about new ways of looking at the same processes.

(e) It is "epistemologically dirty"—it grabs what it can from whatever sources are at hand. It cares little or nothing about disciplines, and proceeds more through bricolage than through systematic theory building. There is, however, a knowledge build-up that is not negligible.

(f) It is profoundly "interdisciplinary" by necessity, although disciplinary forms of knowledge have been important at particular moments (e.g., anthropology at the moment of the formulation of the cultural and territorial rights law; geography and landscape ecology in negotiations of conservation projects, river mapping, etc.; gender studies in negotiations about strategies for black women).

Even if so sketchily characterized, it can be said that this case exemplifies some of the features highlighted by Santos as marks of the new paradigm. This knowledge is "gathered around themes adopted by concrete social groups at a given time as projects of local life" (1992, 38); similarly, activists reconstitute "the local cognitive project(s), stressing their exemplarity and thus turning them into enlightened local thinking" (38). The knowledge so produced serves as material that circulates through networks and might be adopted and adapted in other interested sites. It is also clearly the case that "such knowledge is relatively unmethodical, since it springs from methodological plurality" (39), even if, as we pointed out, there is a recursivity that enables some building up of knowledge. Above all, it can be said that activist knowledge is self-knowledge in that it broaches the issue of "not [only] how to survive, but how to live" (42). While PCN's framework is by no means exclusively place-based knowledge, it has a crucial place-based dimension in its orientation towards the defense of local socionatural and cultural worlds. It is then guided by a sense of prudence for a decent life, that is, by a return to wisdom for ethical living in a complex, interconnected world. What I mean by wisdom here is type of knowledge that finds its raison d'être in its deep connection to place—a knowledge that is embedded in locality and that is responsive and accountable to place-based constituencies—as opposed to the detached expert knowledge of modernity. It is, in sum, a kind of "ethical know-how" as opposed to an expert's "know-what" (Varela 1999; see also Spinosa, Flores and Dreyfus 1997 on this point).

In sum, it can be said that activist knowledge today, at least in some cases, approaches Santos' condition of common sense as a knowledge that is "practical and pragmatic. It produces knowledge drawn from the life trajectories and experiences of a given social group, and asserts that this link to group experience renders it reliable and reassuring" (44). In accepting "what exists as is," activist common sense knowledge is at its best when facing the world in all of its complexity, without logocentric reductions demanded by theory, method, or epistemology; it does so not out of any academic injunction for holism or even integrative perspective, but precisely because in seeing this complexity lies the only possibility of survival and the chance for living a decent life. This common sense, finally, articulates with expert knowledge in multiple ways, and in doing so fosters the creation of new rationalties and knowledge practices (see, e.g., Leff 2000 for the creation of new environmental and production rationalities).

The political need to face the complexity of the present world—in particular the complexity of power and domination—leads activists to a politics of meshworking and articulation. Let me finally give a sense of the transnational dimension of the movement to illustrate this point. First, it should be emphasized that the PCN can be seen as creating both a powerful discursive field of action (Alvarez, in press), as well as a network of sites; both entail connections with other networks and hierarchies and, conversely, constitute a space for other actors to construct their own linkages. For instance, the PCN network and discourse (e.g., the concept of region-territory of ethnic groups) becomes an obligatory passage point for some other actors, for example, for some government projects. In other words, the PCN network does not exist by itself, but as inevitably imbricated with other networks, from the global biodiversity network to other social movement networks and even the networks of repression.

The internationalization of PCN started in the mid-1990s and has oscillated between periods of intense and sparse activity according to domestic and international conjunctures. This internationalization process has taken place in four main areas:[8]

(a) Transnational anti-globalization mobilizations, including participation in well-known events in Europe (e.g., large demonstrations in Geneva and Prague) and lesser known events (e.g., anti-globalization mobilizations in India).

(b) Latin America-wide meetings of black movements, particularly the Afro-Latin American Network and the Afro-Latin American Women's Network (e.g., events in Colombia, Uruguay, Peru, Bolivia, Ecuador, and Chiapas). This includes the binational *encuentros* of black peoples of Colombia and Ecuador, and a number of preparatory meetings toward the World Conference on Racism, Xenophobia and Other Forms of Discrimination held in Johannesburg in 2001.

(c) Environmental/development-related events and mobilizations, particularly in the context of the Convention of Biological Diversity (e.g., COP3 in Buenos Aires, November 1997; meetings of Andean Pact countries on genetic resources); forest problems and policy (including green certifications of forest products, opposition to the impact of free trade measures on deforestation, etc.; meetings in Chile and Costa Rica); genetic resources and intellectual property rights regimes (particularly TRIPs); and around the experience of collective territories (e.g., meeting at the Inter-American Development Bank in Washington, DC to discuss report on collective territories in Latin America, 1999).

d) Academic events at which PCN activists present formal papers, even if not necessarily adhering to conventional academic canons and standards (e.g., LASA meetings in 1995, 2001; occasional conferences and talks at universities in the United States and Latin America).

e) Events around particular conjunctural situations seen as key to either enforcing globalization or articulating resistance to it (e.g., Chiapas, and events around Plan Colombia, that has been taken up as a banner by a number of anti-globalization groups, the Red Afro-Americana, and other organizations).

It can be said, then, that the movement creates a "network of its own" with geographical basis on the rivers and towns of the Pacific and with a number of regional, national, and transnational nodes. What circulates in the meshwork are activists, cultural models, information, frameworks, e-mail messages, communiqués, declarations, agreements, and concrete actions and mobilizations. This network becomes increasingly tied to other dominant and subaltern actor-networks. The presence of some catalysts have been important for the construction of the PCN network, including some progressive church organizations, environmental, and popular communications NGOs (and, by default or *ab absurdum*, some dominant NGOs and government programs); ICTs themselves, particularly the Internet; some scholarly spaces, theories and specific works; and particular groups of activists beyond those belonging to the PCN themselves. The aim of the PCN in joining the AGSM meshwork can be described as (a) to contribute to resist global transformations and policies that might hinder (or, in some cases, be favorable to) the defense of the territory and the cultural project of the communities in the Pacific itself; (b) to contribute to strengthening particular struggles or redress particular situations (e.g., struggle against Narmada; Chiapas; Plan Colombia) that may affect the course of neoliberal globalization, which in turn might have broader effects. This can be said to be the double political principle guiding most movements' politics of meshworking.

CONCLUSION

A "Discourse on the Sciences" announced a paradigm shift in the production of scientific knowledge; it outlined the main features of the new paradigm; and it posited a stage of inter-paradigmatic transition without specifying in detail the dynamics of the transition. At the level of the diagnosis of the shift, it built on pioneering arguments in the natural and physical sciences. However, the new view of knowledge production it suggested was, in many ways, unique. As we have seen in this paper, some of the trends envisaged by Santos in the late 1980s have become more acute (e.g., the dissolution of boundaries between the natural and the social sciences), even if the direction of the changes have changed somewhat (e.g., with the natural sciences leading the change at present). The study of anti-globalization social movements through perspectives derived in part from the natural sciences is a case in point. Other trends have moved in unexpected directions. As I have argued in this chapter, the production of place-based knowledge has been led by nonacademic knowledge producers, pretty much in accordance with the conditions hypothesized in "A Discourse" but which remain relatively invisible even today in most conventional social science research.

As much as the social movements, the sciences should be more concerned with the state of the world at present. Today, it is even more clear that neoliberal globalization is intended to defend the privileges of the world's rich at the expense of the world's poor and the environment. The structure and practices of global capitalism have certainly changed in ways that we no longer adequately understand. We are thus facing a double challenge: to bring our knowledge up to date with the changing character of global hegemonies, in part by building on the most recent epistemological trends (thus deepening the transition identified by Santos in a way that is responsive to the historical conditions of the day); and to orient this knowledge—and, at least partially, our own selves as knowledge producers—towards projects of emancipation and social transformation alongside, and in critical support of, social movements. Only in this way can a plural political ecology of knowledge, and plural regimes of world making, be fostered and maintained.

ACKNOWLEDGMENTS

I would like to thank the group of PCN activist with whom I have had the privilege to work over the years, particularly Libia Grueso and Carlos Rosero, from the PCN's National Coordinating Committee, for innumerable and enlightening conversations about the PCN; Eduardo Restrepo for sharing his knowledge and analysis of the Pacific; Wendy Harcourt and the SID "Women and the Politics of Place" research group for enlightening discussions on place; Manuel de Landa for his pioneering work on complexity in social domains; and Boaventura de Sousa Santos for the opportunity to revisit some of the issues he raised about the sciences in the context of my own work. Most of this chapter had been written as a conference paper (Escobar 2000). The research on which this article is based was funded partly by grants from the Guggenheim Foundation and the MacArthur Foundation's Global Security and Sustainability Program, whose support I greatly appreciate.

NOTES

1. I cannot discuss the various theories of networks here. Suffice it to say that one can differentiate between two kinds of theories of networks. In the first type, the concept of network fits into an existing social theory (examples are new theories of society, but not of the social, such as Castells' (1996); theories of activist networks, such as Smith, Chatfield and Pagnucco, eds. (1997), Keck/Sikkink (1998); conceptualizations of "netwar" (Arquilla and Ronfeldt 2001); and theories of social movements that incorporate aspects of networks or webs, such as Alvarez 1998, in press). In the second group, social theory itself is re/constructed out of, or on the basis of, the concept of network (Latour and Callon's actor-network theory; theories of complexity). As Latour put it, actor-network theory (ANT) "aims at accounting for the very essence of societies and natures. It does not wish to add social networks to social theory but to rebuild social theory out of networks" ("The Trouble with Actor-Network Theory," from Latour's Web page, While both sets of theories are necessary for thinking about AGSMs, the latter provides a set of insights for thinking about what might be most intractable and radical in the AGSMs of today. For a more detailed explanation, see Escobar (2000; also at the UNC Anthropology department Web page). The study of social movements from network perspectives are coming of age in fields such as anthropology, sociology, and political science. Several anthropology dissertations in progress or recently completed approach aspects of anti-globalization social movements from network perspectives (See dissertations in progress by Mary King on networks in the field of biodiversity and by Chaia Heller on anti-GMOs mobilizations, both at University of Massachusetts, Amherst; by Michal Osterweil and Vinci Daro on Italian movements, at Chapel Hill. See also the dissertations on Zapatista struggles by Leyva-Solano (2002) and Thomas Olesen (2002); and Mendiola's (c. 2001) dissertation that develops a theory of social movements from the perspective of theories of actor-networks and rhizomes).

2. It would be impossible to provide even a simple primer on complexity here. A basic bibliography is included in an earlier piece (Escobar 1994; see Kauffman 1995, Solé and Goodwin 2000 for excellent works on complexity in biology with some application to social systems). A central feature of this body of work is its focus on complex adaptive systems that display emergent behavior; emergence refers to behavior arising out of a multiplicity of localized interactions that often involve nonlinear dynamics; in cases where this behavior involves some learning, one can speak of complex adaptive systems. Many of these cases can be seen in bottom-up, rather than top-down, systems.

3. There are other bodies of work to which de Landa's conceptualization can be related. For instance, with Maturana and Varela (1981, 1987) it can be said that meshworks of heterogenous elements evolve by drift; the environment only acts as a trigger for changes within the meshwork; there is structural coupling between meshwork and environment with preservation of internal organization, that is, of the meshwork's autopoiesis.

4. The "crowding-out' effect of meshworks can be visualized with the help of Stuart Kaufmann's representation of chemical autocatalysis in complex chemical reaction networks; see Kauffman (2000, 36).

5. For a description of the movement and its strategies, see Grueso, Rosero and Escobar 1998; for general background on the movement and the region, Escobar and Pedrosa 1996; Camacho and Restrepo 1999; Pardo 2001. I discuss the political ecology of the movement in Escobar (1998). There are already a few ethnographic investigations of this process for the 1990s, including the making of ethnic identities (Restrepo 2002), the role of the State in such a process (Alvarez 2002), and of articulation of gender, ethnicity, and biodiversity (Asher 1998). The recent massive displacement from this region as a result of armed conflict is analyzed in Escobar (2003).

6. This framework is presented here in its barest way. It is given a lengthy treatment in a book I am currently completing.

7. The global biodiversity network constitutes a good example of the entanglement of meshworks and hierarchies. Its history goes back to the establishment in the early 1980s of a network by a group of European NGOs intended to enable the use of information and computer technologies for alternative social action. In contrast to the military and corporate computer networks of the 1980s, this network (called Interdoc) was deliberately designed as decentralized and nonhierarchical (small scattered units with creative linkages). One of the NGOs making up this network was the International Coalition for Development Action (ICDA), which eventually became GRAIN, Genetic Resources Action International, based in Barcelona. Along with a handful of other progressive NGOs, this organization was to play a prominent role in online debates leading up to the Earth Summit in Rio de Janeiro in 1992, where the Convention on Biological Diversity (CBD) was signed. This history is traced by Mary King (2001), who argues that some of the key issues incorporated into the CBD (e.g., the question of indigenous knowledge, the role of women in conservation, provisions for local conservation methods, etc.) were achieved because of the work of this network. The meshwork character of the internet as a whole in its early stages is also described by Murphy (2001). According to Murphy and King, Interdoc played a formative role in the configuration of a global activist culture. Interdoc could be seen as an early experience of meshworking in the computer age.

8. This map of PCN transnational activities is chiefly based on interviews with Libia Grueso in October, 2000 in Chapel Hill; it is also based on my collaboration with the movement over the years in particular projects, and my participation at some of the events mentioned. The first international experience was a tour through several European countries in the fall of 1995, sponsored by progressive church-affiliated and environmental European NGOs, which yielded two initial contacts that were to prove of great salience in the ensuing years. The first was with the group of activists in Geneva that were to create the PGA in 1997; the second with the International Working Group for Indigenous Initiatives (IWGIA), in Copenhagen. This latter contact became the entry point for the participation of PCN activists in meetings of the United Nations Working Group on Indigenous Affairs on the status of groups such as the black river communities of the Pacific. The contact with IGWIA was important in securing funding for several PCN projects in the Pacific from 1998 to the present.

9. This included PGA-organized events and other anti-globalization mobilizations (e.g., anti-WTO, World Bank, IMF, TRIPs, GMOs, and G-8 mobilizations). This started with a February 1998 organizing meeting and the May 1998 European mobilization, PGA's first great mobilization that convened over 50,000 in Geneva alone. It also included, in 1998–2000, a series of events in Europe, Asia and Latin America, including a Caravan of 500 activists, mostly farmers from India but also a couple of PCN people, who toured Europe in 1998; a large meeting of AGSM activist in Bangalore (1999); regional PGA meetings in Latin America; a PGA "gender" meeting, held in Mexico in 2000 ("Neither men not women, but the other way around"; *Ni hombres ni mujeres sino todo lo contrario*); the Prague mobilization plus follow-up meetings in Berlin (for a brief history of the PGA, see PGA 2000). Anti-globalization events also included the International Group for Grassroots Initiatives (IGGRI)-sponsored meeting of social movement network representatives in Helsinki (September 98), in which a PCN representative participated as part of a PGA delegation, and the second global EZLN-initiated meeting against neoliberalism held in Bahia, Brazil in 1998.

REFERENCES

Alvarez, Manuela. 2002. *Altered States: Culture and Politics in the Colombian Pacific.* Master's thesis, University of Massachusetts, Amherst.

Alvarez, Sonia E. 1998. "Latin American Feminisms 'Go Global': Trends in the 1990s and Challenges for the New Millennium." In *Cultures of Politics/Politics of Cultures. Revisioning Latin American Social Movements*, ed. Sonia E. Alvarez, Evelina Dagnino, and Arturo Escobar. Boulder, CO: Westview Press, 293–3245.

———. In Press. *Contentious Feminisms: Critical Readings of Social Movements, NGOs, and Transnational Organizing in Latin America.* Duke University Press.

Arquilla, John, and David Ronfeldt. 2001. *Networks and Netwars.* Santa Monica, CA: Rand Corporation.

Asher, Kiran. 1998. *Constructing Afro-Colombia: Ethnicity and Territory in the Pacific Lowlands.* PhD Diss., University of Florida, Gainesville.

Camacho, Juana, and Eduardo Restrepo, eds. 1999. *De montes, rios y ciudades.* Bogota: Natura/ICAN/Ecofondo.

Casey, William. 1997. *The Fate of Place.* Berkeley: University of California Press.

Castells, Manuel. 1996. *The Rise of the Network Society.* Oxford: Blackwell.

de Landa, Manuel. 1997. *A Thousand Years of Nonlinear History.* New York: Zone Books.

Deleuze, Gilles, and Félix Guattari. 1987. *A Thousand Plateaus.* Minneapolis: University of Minnesota Press.

Dirlik, Arif. 2001. "Place-Based Imagination: Globalism and the Politics of Place." In *Places and Politics in the Age of Globalization*, ed. Roxanne Prazniak and Arif Dirlik. New York: Rowman and Littlefield, 15–52.

Escobar, Arturo. 1994. "Welcome to Cyberia. Notes on the Anthropology of Cyberculture." *Current Anthropology* 35 (3):211–32.

———. 1998. "Whose Knowledge, Whose Nature" Biodiversity Conservation and the Political Ecology of Social Movements." *Journal of Political Ecology* 5:53–82.

———. 1999. "Gender, Place, and Networks. A Political Ecology of Cyberculture." In *Women@Internet. Creating New Cultures in Cyberspace*, ed. Wendy Harcourt. London: Zed Books, 31–54.

———. 2000. *Notes on Networks and Anti-Globalization Social Movements*. Presented at the Session on Actors, Networks, Meanings: Environmental Social Movements and the Anthropology of Activism. AAA Annual Meeting, San Francisco, November 15–19.

———. 2001. "Culture Sits in Places: Reflections on Globalism and Subaltern Strategies of Localization." *Political Geography* 20 (2001):139–74.

———. 2003. "Displacement, Development and Modernity in the Colombian Pacific." *International Social Science Journal* 175:157–67.

Escobar, Arturo, and Alvaro Pedrosa, eds. 1996. *Pacífico, desarrollo o diversidad?* Bogota: CEREC/Ecofondo.

Grueso, Libia, Carlos Rosero, and Arturo Escobar. 1998. "The Process of Black Community Organizing in the Southern Pacific Coast of Colombia." In *Cultures of Politics/Politics of Cultures. Revisioning Latin American Social Movements*, ed. Sonia E. Alvarez, Evelina Dagnino, and Arturo Escobar. Boulder: Westview Press, 196–219.

Gulbenkian Commission. 1996. *Open the Social Sciences*. Stanford, CA: Stanford University Press.

Haraway, Donna. 1988. "Situated Knowledge: The Science Question in Feminism and the Privilege of Partial Perspective." *Feminist Studies* 14 (3):575–99.

Harcourt, Wendy, ed. 1999. *Women@Internet. Creating New Cultures in Cyberspace*. London: Zed Books.

Harcourt, Wendy, and Arturo Escobar. 2002. "Lead Article: Women and the Politics of Place." *Development* 45 (1):7–14.

Hobart, Mark, ed. 1993. *Anthropological Critique of Development*. London: Routledge.

Ingold, Tim. 2000. *The perception of the environment*. London: Routledge.

Kauffman, Stuart. 1995. *At Home in the Universe*. New York: Oxford University Press.

———. 2000. *Investigations*. Oxford: Oxford University Press.

Keck, Margaret, and Kathryn Sikkink. 1998. *Activists beyond Borders. Advocacy Networks in International Politics*. Ithaca, NY: Cornell University Press.

King, Mary. 2000. *Of Unknown Quantity: NGO Network Organizing and Global Environmental Politics*. Presented at the Session on Actors, Networks, Meanings: Environmental Social Movements and the Anthropology of Activism. AAA Annual Meeting, San Francisco, November 15–19.

———. 2001. "Translating Risk and Safety through the Transnational Biodiversity Network. Presented at the session, *Producing Knowledge, Producing Value: New Political Ecology and the Global Optics of Nature*. AAA Annual Meeting, Washington, DC, November 28–December 2.

Laclau, Ernesto, and Chantal Mouffe. 1985. *Hegemony and Socialist Strategy*. London: Verso.

Latour, Bruno. 1983. *We Have Never Been Modern*. Cambridge, MA: Harvard University Press.

Law, John. 2000[1992]. "Notes on the Theory of the Actor Network," downloaded from http://www.lancs.ac.uk/fss/sociology/papers/law-notes-on-ant.pdf.

Leff, Enrique. 2000. *Saber ambiental*. Mexico: Siglo XXI.

Leyva-Solano, Xochitl. 2002. *Neo-Zapatismo: Networks of Power and War*. PhD dissertation, Department of Anthropology, University of Manchester.

Maturana, Humberto, and Francisco Varela. 1981. *Autopoiesis and Cognition*. Boston: D. Riedel Publishing Company.
——. 1987. *The Tree of Knowledge*. Berkeley, CA: Shambhala.
Mendiola, Ignacio C. 2001. Movimientos sociales y trayectos soniológicos: hacia una teoría práxica y multidimensional de lo social." PhD Diss., Universidad del País Vasco.
Melucci, Alberto. 1989. *Nomads of the Present*. Philadelphia: Temple University Press.
Mignolo, Walter. 2000. *Local Histories and Global Designs*. Princeton, NJ: Princeton University Press.
Murphy, Brian. 2001. *Towards a Critical History of the Internet*. Unpublished manuscript, University of Massachusetts, Amherst.
Olesen, Thomas. 2002. *Long Distance Zapatismo. Globalization and the Construction of Solidarity*. PhD Diss., University of Aarhus.
Pardo, Mauricio, ed. 2001. *Acción colectiva, estado y etnicidad en el Pacífico colombiano*. Bogota: ICANH/Colciencias.
PGA (People's Global Action). 2000. *Bulletin No. 5*, UK Edition. London: Reclaim the Street.
Restrepo, Eduardo. 2002. *Memories, Identities, and Ethnicity: Making the Black Community in Colombia*. Master's thesis, University of North Carolina, Chapel Hill.
Ribeiro, Gustavo Lins. 1998. "Cybercultual Politics: Political Activism at a Distance in a Transnational World." *Cultures of Politics/Politics of Cultures. Revisioning Latin American Social Movements*, ed. Sonia E. Alvarez, Evelina Dagnino, and Arturo Escobar. Boulder, CO: Westview Press, 325–52.
——. 1992. "A Discourse on the Sciences." *Review* 15 (1): 9–47.
Santos, Boaventura de Sousa. 2002. *Towards a New Legal Common Sense*. London: Butterworth.
Smith, Jackie, Charles Chatfield, and Ron Pagnucco, eds. 1997. *Transnational Social Movements and Global Politics*. Syracuse, NY: Syracuse University Press.
Solé, Ricard, and Biran Goodwin. 2000. *Signs of Life. How Complexity Pervades Biology*. New York: Basic Goods.
Spinosa, Charles, Fernando Flores, and Hybert Dreyfus. 1997. *Disclosing New Worlds*. Cambridge, MA: MIT Press.
Strathern, Marilyn. 1996. "Cutting the Network." *Journal of the Royal Anthropological Institute* (n.s.) 2: 517–35.
Varela, Francisco. 1999. *Ethical Know-How*. Stanford, CA: Stanford University Press.

14

Reconstructing Unruly Ecological Complexity:

Science, Interpretation, and Critical, Reflective Practice

Peter Taylor

Environmental activism in Australia during the 1970s led me to study ecological science. Having a mathematical disposition, I chose to focus less on field studies and more on quantitative analysis and theoretical modeling. I soon developed an interest, which continues to this day, in ecological complexity as a challenge to conventional scientific ways of knowing. As I explored this challenge, my work in ecology and socioenvironmental studies opened out to interpretive studies of science and then to facilitation of critical, reflective practice. The common thread has been to *problematize boundaries* used by researchers to partition of complex situations into *well-bounded systems* and backgrounded or hidden processes.

When researchers assume that there are systems with clearly defined boundaries, coherent internal dynamics, and simply mediated relations with their external context, they can locate themselves outside the systems and seek generalizations and principles affording a natural or economical reduction of complexity. In contrast, we can view well-bounded systems—when they are encountered—as special cases of *unruly complexity* arising from *ongoing change in the structure of situations* that have built up over time from *heterogeneous components* and are *embedded* or *situated* within wider dynamics. Boundaries and categories are problematic, levels and scales are not clearly separable, structures are subject to restructuring, and components undergo ongoing differentiation in relation to each other. Control and generalization are difficult, and no privileged standpoint exists. Researchers who want to discipline unruly complexity—but not suppress it—must, I believe, pay more attention to their own *agency* within the *participatory restructuring of knowledge-making and social change*. Such a position resonates with de Sousa Santos's "discourse on the sciences."

This chapter reconstructs my intellectual journey towards this position (a journey elaborated in Taylor 2005[1]). The episodes are less about participatory restructuring, however, than they are about *exploring concepts* used to address—or discount—heterogeneity, embeddedness, and ongoing restructuring. Three strands are interwoven: disciplining unruly complexity; linking knowledge-making to changing diverse social relations; and wrestling with the potential and limitations of conceptual exploration.

Although the detail I present will be specific to my own inquiries, the themes developed are meant to stimulate readers to problematize analogous boundaries and address analogous complexities in their own fields of inquiry and practice. I believe that the concepts and issues I raise should be taken up more broadly, but let me admit the *heuristic* intent of the chapter. I recognize that analogies can be applied in circumstances for which they do not serve well or can misguide the theorist. As will become evident, I am especially interested in conceptual moves that open up issues about addressing complexity, but do so in ways that point to further work that needs to be undertaken to deal with particular cases.

PROBLEMS OF BOUNDEDNESS IN MODELING ECOLOGICAL SYSTEMS

The Construction of Complexity

My undergraduate mathematical ecology text conveyed succinctly the quality of ecological complexity (Pielou 1969, 1): Organisms come from a range of species; within any species they differ in age, sex, genetics, experience, and so on; and any particular individual changes over its lifetime. Any situation an ecologist might study is continually altered by births and deaths, by migratory exchanges with other places, and by seasons and climatic change. Even so, ecological regularities persist long enough for some order to be recognized, such as, an oak-maple forest or the sequence of plants encountered as one moves inland from the seashore.

How could ecologists account for such ordered complexity? Ecology is not like thermodynamics; it does not yield the statistical complexity of large numbers of identically behaving components. Moreover, progress in the physical sciences depends greatly on controlled experiments in which systems are isolated from their context; this strategy is not so clearly appropriate for understanding organisms in their ecological context, where they respond to a multiplicity of hazards and resources distributed in various ways across place and time. Yet, analysis of observations from nonexperimental situations is beset by circularity—much needs to be known in advance about the causal factors involved in order to design methods of data analysis capable of exposing those very factors (Austin 1987).

Given the character of ecological complexity and its analysis by ecologists, could any general theories of its structure apply? During the 1960s and 1970s many academic ecologists, especially in the United States, thought such theories could be developed (Kingsland 1995). In this tradition, the age-old idea of a balance of nature (which many environmentalists still invoke) led to a search for a significant relationship between the complexity of ecological communities and their stability. However, mathematical ecologists, using computers to analyze large samples of possible model communities, discovered that the proportion of the sample that is stable decreases with larger communities and stronger interactions among populations. If complexity in itself does not promote stability, May (1973, 174) asked, what are the "devious strategies, which make for stability in enduring natural systems"? A considerable body of work then explored this line of thinking or disputed the idea that complex natural communities were stable, however that was defined (DeAngelis and Waterhouse 1987).

Yet complex ecological communities are generally not the outcome of some sampling process; they arise through development over time involving the addition, growth, decline, and elimination of populations. My approach to the complexity-stability question, which built on work by Tregonning and Roberts, showed that, whereas stable communities may be extremely rare as a fraction of the model communities being sampled, they can be readily constructed over time by the addition of populations from a pool of populations and the elimination of populations from systems not at a steady state (Taylor 2005). (This pool of potential entrants can be visualized as neighboring communities or patches, refuges in which species persist undetected at low abundances, the seed bank, and so on.) Moreover, the result of this ongoing construction and turnover depends on the order in which populations were added (Drake 1991).

This result indicated that ecologists who are interested in explaining the persistence of complex communities need to examine not only the stability and structure of the current configuration observed but also its construction over time from diverse components—its contingent history of becoming structured and its ongoing restructuring in a wider spatial context or "landscape."

The Hidden Complexity of Simple Models

Field ecologists are perennially skeptical about textbook mathematical models, but Vandermeer's (1969) laboratory data on four competing protozoan populations fit well to relatively simple equations that generated correct predictions about which populations could coexist. I noticed, however, an anomaly in Vandermeer's results that he and subsequent discussants had not addressed. The model he fitted to his observations indicated that three of the six pairs of interactions between the pairs of protozoan populations were positive for one population and negative for the other. One would expect this of predator-prey relations, not of competitive interactions. Were these interactions actually predator-prey? Indeed, were those pairs with negative-negative interactions actually competitors?

My questions arose in a context in the 1980s in which ecologists had begun to pay attention to the effects mediated through the dynamics of populations not immediately in focus (Strauss 1991). Note that Vandermeer's equations had not specified all the components of the community. Each day he had removed a sample from his experimental tubes and added an equal volume of culture medium with bacteria. The bacterial populations were alive and able to grow until consumed by the protozoa—they had dynamics of their own not referred to in the equations. Because there was no explicit reference to the relationships with the hidden part of the community, the interaction values in the model had to incorporate these and any direct interactions. The combination could be called "apparent" interactions. Although ecologists would think that the protozoan populations should be competitors because they share a food resource, Vandermeer's study showed that some counterintuitive apparent predator-prey interaction values were needed to fit the observations.

I developed a mathematical way to investigate apparent interactions more generally and concluded that they often deviate significantly from direct observations of interactions and from ecologists' intuition about plausible interactions among populations (Taylor 2005, chap. 1B). It seemed that when simple models fit observations well this may be because hidden variables happen to remain within narrow bounds, not because the model approximates the actual interactions. Principles proposed for simple subcommunities, such as the non-coexistence or "competitive exclusion" of species with similar requirements, can be confounded by the dynamics of populations with which those subcommunities interact in naturally variable and complex ecological situations. In short, boundaries that discount the dynamics of nonmodeled variables become problematic.

The notion of *inseparable dynamics*, which link the (sub-)system that is the focus of research with the backgrounded processes, has potentially profound implications for thinking about knowledge-making. It makes clear that controlled experiments in which systems are isolated from their context are not appropriate for understanding organisms embedded in a dynamic ecological context and responding to consumers and resources that are unevenly distributed across place and time. Inseparable dynamics, together with the picture above of ongoing construction and turnover, suggest that theorists should not assume that ecological complexity can be partitioned into communities or systems that have clearly defined boundaries, coherent internal dynamics, and simply mediated relations with their external context (Taylor 2005, chap. 1B).

Problematizing boundaries in these ways should be relevant beyond ecology. Indeed, in the early 1980s the anthropologist Eric Wolf proposed a conceptual inversion: Whenever theory has built on the dynamic unity and coherency of structures or units—in Wolf's case, societies or cultures—consider what would follow if those units were to be explained as contingent outcomes of "intersections" among processes that implicate or span a range of spatial and temporal scales (Wolf 1982, 387). As will emerge in sects. 4 and 6, socioenvironmental studies proved to be a more fertile field than ecology proper for me to elaborate on Wolf's conceptual inversion and his image of *intersecting processes*.

STRATEGIES OF MODELING

The perspectives developed in sect. 1 are especially challenging for mathematical modelers—building and analyzing a mathematical model almost requires the assumption of a fixed, delimited set of components. Moreover, these theoretical explorations seemed to work against aspirations for general principles about ecological patterns supported by mathematical models. Those aspirations were also being questioned, but from a different angle, by ecologists in the early 1980s who found that many of the models of community ecology fit the data no better than alternatives that assumed random interactions among populations. Simberloff (1982), deeply skeptical about the possibilities of general ecological theory, argued that many factors operate in nature and in any particular case at least some of them will be significant. A model cannot capture the relevant factors and still have general application. Instead, Simberloff contended, ecologists should intensively investigate the natural history of particular situations and test specific hypotheses about these situations experimentally. They may be guided by knowledge about similar cases, and they may end up adding to that knowledge, but they should not expect their results to be extrapolated readily to many other situations.

The modeling experience described in sect. 1 led me, however, to resist the emphasis on testing specific hypotheses about particular situations and to be less skeptical about theory built through modeling. The use of mathematical models for conceptual exploration had allowed me to reformulate issues and generate new questions, which seemed worthwhile. In response to the philosophical positions ecologists like Simberloff were enunciating, I noted that, although models necessarily simplify reality, they are not all designed and applied according to the same standard for correspondence with observations. I classified models according to the level of correspondence with observations established for a model's distinguishing feature and independently for its assumptions (Lloyd 1987; Taylor 2005, chap. 2). My taxonomy secured a place for the generation of theory through "exploratory" modeling (see also Levins 1966), at the same time noting that the insights so derived are about a mathematical system or conceptual schema. Their relevance to biology is yet to be established; there remains an uneasy tension between mathematical and conceptual tractability and the demand that the exploratory model should eventually yield something that can be more strictly evaluated against observations. In this light, the results of exploratory modeling are best thought of as heuristics—propositions that can stimulate and orient research, but which may turn out to have been applied in circumstances for which they do not serve well or otherwise misguided the theorist.

Equally importantly, this account of modeling allowed me to identify several *open sites* where considerations other than explicit analysis of a model's correspondence with evidence must come into play. Modeling and theory-building necessarily operate within webs of social and technical decisions that ecologists make about which questions to put to nature, categories to use, observations to construct, analyses to perform, degree of confirmation to require, ways to revise models, and so on. The recognition that knowledge-making must always extend beyond the dialogue between models and evidence—that this dialogue is embedded in a larger dynamic context of influences shaping the ecologists' decisions—opened my inquiry to include investigation of the "sociality" of ecology and environmental studies.

SOCIAL-PERSONAL-SCIENTIFIC CORRELATIONS

Conventional accounts of scientists and philosophers of science assume that the scientist's dialogue between models and evidence can be separated from the dynamics of their dialogue with other scientists to establish what counts as knowledge. They would contend, for example, that decisions made at the open sites (see sect. 2) can lead some scientists to tackle anomalies that others had dismissed as negligible, and thus science progresses. In any community of scientists disputes may be resolved when one scientist's biases are countered by others'. Social influences, such as research funding, may primarily inhibit or accelerate improvement in scientific knowledge. In these senses science's sociality does not prevent its "referentiality"—the correspondence of models or theories with evidence—from determining what counts as knowledge—if not immediately, at least in the long run.

There is a deeper sense of sociality, however, that is harder to reconcile with conventional views. All scientists engage in various arenas of social activity—they build careers and institutions, use and transform language, facilitate policy formulation, and so on. Given this context, scientists select problems, define categories, collect data, and present findings not only to develop models of their subject matter, but also to secure the support of colleagues, collaborators and institutions, and to enable others to act upon their conclusions. This might happen in idiosyncratic ways, but I was interested in the possibility that the simultaneous pursuit of referentiality and sociality could sometimes lead to systematic effects on the content of scientific knowledge.

I began to explore such effects by examining the history of systems ecology, a field that emphasizes nutrient and energy flows in entire systems. H. T. Odum (1924–2002), a pioneer in systems ecology, conceptualized ecological complexity in terms of energy circuits (like electrical circuits) subject to feedback that ensured homeostasis of the overall system. The principles that ecologists might discover through collecting data on the energy flows in mature, productive ecological systems, such as tropical rainforests, could guide the design of "systems of man and nature." In this vein, Odum began to analyze energy flows in social systems; at the same time his theoretical propositions pointed to an important role in systems for high-quality, low-energy circuits (Odum 1971).

Odum's work during the 1950s and 1960s reduced the complexity of social and ecological relations to a single currency— energy—whose flows could be adjusted or redesigned. As I interpreted Odum, the high-quality, low-energy circuits were allegorical; he had found "in nature" a special role for systems engineers, such as himself, working in the service of society (Taylor 2005, chap. 3). This overall interpretation built from noting that for Odum the social, personal, and scientific realms reinforced each other during the postwar years of "technocratic optimism." This reinforcement occurred from many angles. Government funding and organization of science under military imperatives during World War II had produced significant results, giving currency to the belief that intervention on a large scale could be practically realized. Moreover, scientific control of complex systems seemed necessary to prevent further social upheavals or holocaust. During the postwar decades optimism about the benefits of technocratic systems management overshadowed possible doubts about its implications for democratic political life. Science in the service of social progress was also a theme stressed to Odum by his father, the sociologist H. W. Odum, who promoted the cooperation of intellectuals and other social elements in the reintegration of the American South into the nation. In the vision of H. W. Odum and many others during the Great Depression, when the social "organism" is unhealthy the natural division of labor needs to be restored. In the realm of science, the young H. T. Odum explored electrical circuitry and the harnessing of energy (which in terms of wider social attention were the mid-century equivalent to computers at the century's end). After wartime service in tropical meteorology, Odum was recruited by the Yale ecologist G. Evelyn Hutchinson, whose wide-ranging work included synthesis of data on the stocks and flows of specific chemical elements through the biosphere, exploration of mathematical approaches to ecological theory, and participation in the influential Macy conferences on cybernetics and feedback systems. These strands are all evident in the work Odum undertook as a young professor—in the diagrams he drew of energy stocks and flows, the analogous electrical circuits he built and manipulated, and the whole-system research projects he had funded generously by the major institutions of postwar science.

Historians of science, seeking to illuminate why certain categories are plausible and certain lines of inquiry are pursued, have often identified underlying, perhaps implicit, patterns of thought and metaphors shared among different sciences and social thought more generally (Stepan 1986). The scientist in me was interested in the mechanism producing such correlations—how could scientists do their actual research in ways that historians could interpret as resonating with the scientists' concerns about social order? By considering Odum's practice, which included not only his concepts and production of theory but also his methods and organization of research, I saw Odum as someone working to make the overlapping realms he inhabited—the social, personal, and scientific—reinforce each other, so that efforts made and directions pursued in one realm did not undermine those in the others. This view of Odum's practice expanded the historian's idea of a shared metaphor—the various *social-personal-scientific correlations* enabled Odum not simply to *think* that ecological complexity is like electrical circuits, but to *act as if* it were.

HETEROGENEOUS CONSTRUCTION

It was clear to me that the socialty of science may have systematic effects on the content of scientific knowledge, but the scientist in me wanted ways to bring such interpretations to bear productively on subsequent research. In this regard, the case of Odum was a limited model. Personal, scientific, and social considerations reinforced each other so consistently in Odum's life that it was difficult to see how he could have done anything differently. At best, I could have used my interpretation of Odum to suggest a very broad lesson: Scientists opposed to technocratic rationality should not treat ecological complexity as if it were made up of well-bounded systems that could be analyzed in terms of a single currency. Yet, any scientists who wanted to heed such a lesson would still need specific ways to arrange or alter the personal, scientific and social facilitations of their work. To provide insights about how that might be achieved, a finer-grained analysis than the broad historical interpretation of Odum seemed to be called for. With this in mind I chose to consider shorter-term projects of socioenvironmental assessment likely to be governed by more complex and contested pragmatics.

The first case was modeling work I had undertaken in Australia to analyze the possible futures of a salt-affected agricultural region; the second involved U.S. researchers in the mid-1970s building computer models of nomadic pastoralists in drought-stricken sub-Saharan Africa. In both cases I traced diverse interconnections between the various so-called technical decisions of the scientists and the social considerations that influence how scientists perform the resulting tasks. To address the question of how something comes to be established as knowledge, I assessed what would be entailed *in practice* to modify that knowledge (Latour 1987; Taylor 2005, chap. 4). The modeler in the U.S. project on sub-Saharan Africa, as I interpreted his work, had to deal with diverse considerations such as the available computer compiler, published data, the short length of time both in the field and for the project as a whole, the work relations within the research team, the relationship of the United States efforts to other international involvement in the region, the terms of reference set by the funding agency and its contradictory expectations of the project, and so on. The practical considerations that were "resources" for the modeler's knowledge-making were also commitments to certain actions; these actions implicated many other social agents and spanned local and wider social realms (Taylor 2005, chap. 4). In the diverse, particular ways, I contended, the modeler had been imagining and engaging in social change at the same time as making knowledge.

My image of scientists working in a social context had evolved from social-personal-scientific correlations into one where scientists harnessed heterogeneous resources as they simultaneously represented and engaged in socioenvironmental situations. One virtue of interpreting scientific knowledge-making as such *heterogeneous construction* is the emphasis on multiple points at which scientists could engage differently in scientific practice and try to modify its outcomes. Whether any specific modifications—in the case above: working with a different compiler; spending more time in the field; and so on—are doable depends on the position and resources of the specific scientists as they enter into negotiations with other relevant social agents. Scientists' ability in practice to make knowledge is *distributed* beyond their persons, not concentrated mentally inside them; it depends on intersecting processes (Taylor 2005, chap. 4).

ECOLOGISTS MAPPING THEIR OWN SOCIALITY

If ideas about multiple specific modifications were to feed productively back into science, scientists would have to become interested in analyzing their diverse resources and paying attention to their distributed agency—to become *practically reflexive*. It would be inconsistent for interpreters of science to take responsibility for analyzing the full complexity of any scientist's resources, let alone for delivering additional resources needed for the scientists to modify this complexity. One project I undertook to foster practical reflexivity was to convene some pilot *mapping workshops*, in which the scientist-participants (mostly advanced graduate students) were encouraged to be more explicit and strategic about modifying their social context and their research together. Interactions among the researchers centered on "maps"—pictorial representations they drew of the things—theoretical themes, methodological tactics, organisms, events, institutional facilities, disputes, and so on—that motivated, facilitated, or constrained their inquiry and action (Taylor 2005, chap. 5B).

Although I will not claim the pilot workshops were followed by dramatic changes in the work of the participants, the experience provided me material for further theoretical reflection. I noted that participants articulated connections that had previously been unexamined, unspoken, or discounted. The researchers showed that, when encouraged or prodded by interaction with others, they knew more about their social situation than they usually acknowledged. In their usual practice and discourse, however, researchers discount their awareness of their complex, distributed "situatedness" in favor of a *concentrated* view of their own agency and a partitioning of the realms of science and interpretation. This implies, however—according to the perspective of heterogeneous construction—that researchers' simple formulations about their social situatedness were serving as one resource among many for them in their knowledge-making.

This tension between the simple things that researchers say (or tell themselves) about what shapes their work and the complex situatedness that they could at times acknowledge suggested a reframing of practical reflexivity. The ideal need not be that scientists at all times keep in mind a systematic account of the resources that enable them to do their research, but that they seek out situations and conditions that enable them to keep the backgrounded processes in view and *periodically* bring them back into play when mobilizing different resources or organizing them in new directions. The tension between concentrated and distributed agency could be kept active and productive; the boundaries problematic, not taken for granted.

A FRAMEWORK TO KEEP TENSIONS ACTIVE AND PRODUCTIVE

If agency is distributed, reflecting on one's situatedness is no guarantee that a researcher will be able to mobilize different resources to significant effect. Stanley Fish (1989), an influential interpreter of the situatedness of legal and literary texts, took this insight a step further and asserted that reflection on one's situatedness is irrelevant to changing it. In contrast, I have treated it as an empirical matter—one to be established through experiment and experience—which kinds of reflection and workshop processes, which modes of interaction and support contribute most to scientists modifying the situations in which they make knowledge. In this spirit, I sought out opportunities to develop my experience and skills in workshop facilitation (Taylor 2002).

More generally, I began to conceptualize a realm of critical, reflective practice in which researchers—not only scientists, but also interpreters of science—would address self-consciously the complexities of the situations they study and their own social situatedness as they affected knowledge-making and social change. (Such "social change" might range from the level of global environmental politics or to the more modest level of teaching students and influencing colleagues.) In this realm, an important variant or analog of the simple-complex tension (see sect. 5) is evident. Simple themes, such as "Biodiversity is important to the balance of nature" or "Population growth leads to environmental degradation" are easier to communicate to a general audience than the complex intersecting processes in particular environmental situations. (More difficult still is to convey the heterogeneous construction of the practice of particular researchers or a combination of intersecting processes and heterogeneous construction.) The simple themes seem to be more potent resources for mobilizing others to think in your terms and to act accordingly. However, simpler, more memorable, and adaptable accounts are only apparently simple (analogous to the apparent interactions in ecology described in sect. 1). Their impact and importance must depend on the ways they are linked to other resources by scientists and others who are negotiating how to contribute to changing knowledge, society, and ecology.

To address this variant of the simple-complex tension I developed the following framework (Taylor 2005, chap. 6), which synthesizes the passages I have presented in the preceding sections. Consider three angles—like facets of a crystal—from which to view the practice of researchers as knowledge-makers:

1. their study of complex situations (sects. 1 and 2);
2. their interactions with other social agents to establish what counts as knowledge (sects. 3 and 4); or
3. their pursuit of social change through attention to the complexities of the situations studied and the researchers' own social situatedness (sects. 5 and 6).

For each angle, I distinguish three broad ways to address complexity. The first relies on simple formulations of well-bounded systems, which have coherent internal dynamics and simply mediated relations with their external context. In contrast, we can look at work based on dynamics that develop over time among particular, unequal units or agents whose actions implicate or span a range of social realms. Although simple formulations are easier to communicate than reconstructions of particular complex situations, their simplicity can be confounded by the dynamics they hide. I find it is useful in between these two formulations to introduce scenarios and heuristics that are readily communicated, which open up issues and greater complexity and, at the same time, point to further work needed to be undertaken to deal with particular cases.

Before discussing the full framework of three formulations for each of the three angles, let me contrast the three kinds of formulation for angle A—researchers studying complex situations—by rehearsing a lesson I often take students through (Taylor 2001).

Many environmentalists point with concern to the rising global population, for example, "hold(ing) it to be self-evident (and) undeniable that prospects for the future would be more favorable if there were fewer people on earth" (Okoye and Smith 1994, 11). The greater the population, the greater the erosion of arable lands, consumption of nonrenewable resources, and production of greenhouse gases and pollutants in general. This is an example of the first kind of formulation—a simple, well-bounded system.

Now consider a simple scenario: There are two countries. Each has the same amount and quality of arable land, the same population size, the same level of technical capacity, and the same population growth rate of 3 percent per year. Country X, however, has a relatively equal land distribution, while country Y has a typical 1970s Central American land distribution: 2 percent of the people own 60 percent of the land; 70 percent live on only 2 percent. Both countries double their populations very rapidly, but five generations—120 years—before anyone would be malnourished in country X, all of the poorest 70 percent in country Y would be—unless they acted to change their situation.

This is not just an issue of the timing of the crisis in the two countries. Y's poor would probably first experience what others call population pressure in the form of food shortages. They might link these shortages to inequity in land distribution (Durham 1979; Vandermeer 1977) and attempt to take over the underutilized land of the wealthy. The wealthy, anticipating this possibility, might fund paramilitary operations that target leaders of campaigns for land reform. Or build factories that employ some of the land-starved poor. And so on. The nature of foreign trade and availability of military aid would influence the actual choices in specific instances.

No real country is like country X. The crises to which actual people have to respond come well before and in different forms from the crises predicted on the basis of aggregate population growth rates and ultimate biological and physical limits to growth. Although many environmentalists stress that affluent countries and people have disproportionate effect on the environment, the scenario says something deeper. The point is about dynamics—groups with different wealth and power exist, change, and become involved in crises because of their dynamic interrelations.

The lesson so far can be summarized in a heuristic: the analysis of causes and the implications of the analysis change qualitatively if uniform units are replaced by unequal units subject to further differentiation as a result of their linked economic, social and political dynamics. This heuristic points to the need to examine a range of issues that complicate the demographic ones at the heart of the first formulation, but further work would be needed to analyze the complexity of any particular case. In this vein, let me sketch the picture of severe soil erosion in a mountainous agricultural region in central Mexico produced by Raúl and Luis García-Barrios (1990) after extensive field and archival work.

The erosion in the region studied turned out not to be the first occurrence of such a problem. After the Spanish conquest, when the indigenous population collapsed from disease, the communities abandoned their terraced lands, which then eroded. The remaining populations moved to the valleys and adopted laborsaving practices from the Spanish, such as cultivating wheat and using plows. As the population recovered during the eighteenth and nineteenth centuries, collective institutions evolved that reestablished terraces. Erosion was reduced, soil dynamics were stabilized, and perhaps some soil accumulation was stimulated. But this type of landscape transformation needed continuous and proper maintenance. If a terrace were allowed to erode the soil would wash down and damage lower terraces; there was the potential for severe slope instability. The necessary maintenance was made possible by the collective institutions mentioned, which first revolved around the Church and then, after independence from Spain, around rich Indians called *caciques*. These institutions mobilized peasant labor for key activities—not only maintaining terraces, but also sowing corn in work teams and maintaining a diversity of maize varieties and cultivation techniques. The caciques benefited from what was produced, but were expected to look after the peasants in hard times. Given that the peasants felt security in proportion to the wealth and prestige of their cacique, and given that prestige attached directly to each person's role in the collective labor, the labor tended to be very efficient. In addition, peasants were kept indebted to caciques, and could not readily break their unequal relationship. The caciques insulated this relationship from change by resisting potential laborsaving technologies and ties to outside markets in maize.

The Mexican revolution ruptured the closed system of reciprocal obligations and benefits by taking away the power of the caciques and opening the communities to the changing outside world. Many peasants migrated to industrial areas, sending cash back or bringing it with them when they returned to the community for periods of time. Rural population declined; transactions became monetarized; and prestige no longer derived from one's place in the collective labor. With the monetarization and loss of labor, the collective institutions collapsed and terraces began to erode. National food-pricing policies favored urban consumers, which meant that corn was grown only for subsistence needs in this area. Little incentive remained for intensive agricultural production. New laborsaving activities, such as goat herding, which contributes in its own way to erosion, were taken up without new local institutions to regulate them.

The work of the García-Barrioses illustrates well a formulation of socioenvironmental change in terms of intersecting processes (or inseparable dynamics) (see also Little 1987; Taylor and García-Barrios 1995; Peet and Watts 1996). Population, agro-ecology, and socioeconomic institutions are interlinked and local changes and continuities interlink with wider developments within Mexico. No single strand on its own, such as population growth (or, during the twentieth century, population decline), could be sufficient to explain the currently eroded hillsides.

The population-environment lesson spells out for angle A the use of an in-between formulation (the two islands scenario) that disturbs understandings based on simple, well-bounded system (population growth) and points to the need for work to be done in particular cases on dynamics among unequal agents (or units) whose actions implicate or span a range of social realms (the soil erosion case). I can now move to the full framework I developed (table 14.1). Notice that the same three formulations are implicit in the three angles on the practice of researchers. The conventional formulation of science as a dialogue with the situations studied (angle A) is disturbed by interpretations of the researchers' interactions with other social agents to establish what counts as knowledge (angle B). Interpretations, such as those in sections 3 and 4, which emphasize the diverse practical considerations researchers address, suggest heuristically that researchers are always already imagining and engaging in social change at the same time as making knowledge. This opens up the possibility of people pushing the knowledge-making/social-changing connection self-consciously, that is, of critical, reflective practice (angle C).

Within the realm of critical, reflective practice the three kinds of formulation also apply. Reflexive researchers who want to impart their knowledge to others could decide, for example, that it is more effective to use terms familiar to other researchers in their audience. Conventionally, these terms center on marshalling concepts and evidence about the situation studied (angle A) and leave most aspects of the researcher's social situatedness (angle B) taken for granted. This formulation is disturbed by mapping workshops, which show that researchers are able to speak more about their so-

Table 14.1. Three angles from which to view researchers' practice (A, B, C) and three kinds of formulation of each angle (1, 2, 3), with examples of each combination drawn from the chapter

	A. Researchers' study of complex situations	B. Researchers' interactions with other social agents to establish what counts as knowledge	C. Researchers' pursuit of social change through attention to the complexities of both the situations studied and the researchers' own social situatedness
1. Simple formulations of well-bounded systems (coherent internal dynamics and simply mediated relations with their external context)	e.g., Stability (or instability) of ecological communities derives from their complexity	e.g., Science's sociality has at most a transient effect on what counts as knowledge	e.g., Researchers preserve terms familiar to their audience and thus choose not to discuss their social situatedness
2. Scenarios and heuristics readily communicated, which open up issues and point to further work to deal with particular cases	e.g., Exploratory modeling of construction and turnover of ecological communities (sect. 1a)	e.g., Social personal scientific correlations in the work of systems ecologist, H. T. Odum (sect. 3)	e.g., Mapping workshops (scientist-participants expose and explore their social context of research and their research at the same time; sect. 5)
3. Work based on dynamics that develop over time among particular, unequal units or agents whose actions implicate or span a range of social realms	e.g., Case of soil erosion in central Mexico (sect. 6)	e.g., Heterogeneous construction of socioenvironmental assessment (knowledge-making requires diverse considerations in practice; sect. 4)	participatory restructuring of knowledge-making and social change

cial situatedness and relate it to possible modifications of their research. But the participants in the mapping workshops were self-selected and mostly at a similar early stage of their careers. More work would be needed to bring together a diverse set of researchers and other participants and, given that anyone's ability to make knowledge is *distributed* beyond their persons, to sustain the participants' interaction until new complexity-addressing collaborations emerged. These interactions would need to acknowledge and mobilize the tension between using simple, digestible models and addressing particular complexities of both the situations about which they want to make knowledge and their social situatedness.

I continue to struggle with awareness that many readers would like me to present an exemplary case of such a process. To do so, however, would be to concede to the angle-A idea that the power of this essay should rest on my ability to produce a faithful account of the complexity of the situation I study. (This "situation" is now one that includes interpretation of science and reflective practice as well as the ecological complexities with which I began.) Such an expository move would leave it to readers to mobilize the collaborators, sources of funding, and diverse, particular resources needed to contribute to work that matches the exemplary case. Too much weight would be placed on the concentrated, not distributed, agency of author and readers alike. Instead, the final cell in the framework—the intersecting processes/inseparable dynamics formulation of critical, reflective practice—should remain unfilled. This move leaves *opened and active* questions about the role of individuals and their knowledge, heuristics, and other awareness of complex situations and situatedness. Clearly, more work is needed on what I and other agents can do—but not alone nor through our accounts of the world alone—to contribute self-consciously to the ongoing restructuring of the dynamics among particular, unequal knowledge-making agents whose actions implicate or span a range of social realms.

EPILOGUE: FLEXIBLE ENGAGEMENT

The openness of such an ending cannot be expected to make any reader comfortable—or myself for that matter—so let me add a story. I am wary of ways that narrative tends to reinforce our experience of ourselves as concentrated agents, but I am learning that certain stories—told to myself and to others—can help keep distributed agency in view. These are stories that crystallize the challenges of mobilizing resources and organizing them in new directions, that orient the knowledge-makers less towards the product than towards knowledge-making/social-changing collaborations, and that make a virtue of keeping tensions active and opening up question (Taylor 2005, epilogue).

The story, briefly, concerns the Institute of Cultural Affairs (ICA), an international organization whose approach has developed through several decades of experience "facilitating a culture of participation" in community and institutional development. Their work anticipated and now exemplifies the post–Cold War emphasis on a vigorous civil society, that is, of institutions between the individual and, on one hand, the state and, on the other hand, the large corporation (Burbidge 1997). ICA planning workshops involve a neutral facilitator leading participants through four phases of envisioning and revisioning the challenges they face (Stanfield 2002). The goal of ICA workshops is to elicit participation in a way that brings insights to the surface and ensures the full range of participants are invested in collaborating to bring the resulting plans or actions to fruition.

Such investment was evident, for example, after a community-wide planning process in the West Nipissing region of Ontario, 300 kilometers north of Toronto. In 1992, when the regional Economic Development Corporation (EDC) enlisted ICA to facilitate the process, industry closings had increased the traditionally high unemployment to crisis levels. As well as desiring specific plans, the EDC sought significant involvement of community residents. Twenty meetings with over 400 participants moved through the first three phases—vision, obstacles, and strategic directions. The results were synthesized by a steering committee into common statements of the vision, challenges, and directions. A daylong workshop attended by 150 community residents was then held to identify specific projects and action plans, and to engage various groups in carrying out projects relevant to them.

A follow-up evaluation five years later found that they could not simply check off plans that had been realized. The initial projects had spawned many others; indeed, the EDC had been able to shift from the role of initiating projects to that of supporting them. When the accomplishments were assembled, over 150 specific developments could be cited: together they demonstrated a stronger and more diversified economic base and a diminished dependence on provincial and national government social welfare programs. Equally importantly, the community now saw itself as responsible for these initiatives and developments, eclipsing the initial catalytic role of the EDC-ICA planning process. Still, the EDC appreciated the importance of that process and initiated a new round of facilitated community planning in 1999 (West Nipissing Economic Development Corporation 1993, 1999).

The West Nipissing plan stands in contrast to the dominant model of relying on specialists to analyze a situation and formulate policy options. The plan built from straightforward knowledge that the varied community members had been able to express through the facilitated participatory process. The process was repeated, which allowed participants to factor in changes and contingencies, such as those that must have flowed from the start of the North American Free Trade Association and the decline in the exchange rate with the United States. And, most importantly, the process has led community members to become invested in carrying out their plans and to participate beyond the ICA-facilitated planning process in shaping their own future.

This contrast opened up some difficult questions for me, given that my own ecological and socioenvironmental research had drawn primarily on my skills in quantitative methods. What role remained for researchers to insert into participatory planning the "translocal," that is, to insert analyses of changes that arise beyond the local region or at a larger scale than the local? (Indeed, the "local" for professional knowledge-makers cannot be as place-based or fixed as it would be for most community members.) What would it mean to take seriously the creativity and capacity-building that seems to follow from well-facilitated participation but not to conclude that researchers should "go local" and focus all their efforts on one place? (Taylor 2002).

Using "freewriting" (Elbow 1981) to reflect on this question during a workshop was leading I coined the term "flexible engagement." This seemed to capture the challenge for researchers in any knowledge-making situation of connecting quickly with others who are almost ready to foster—formally or otherwise—participatory processes like those of ICA and to enhance, through the experience that such processes provide their participants, the capacity of others to do likewise. The term plays off the "flexible specialization" that arose during the 1980s, wherein transnational corporations directed production and investment quickly to the most profitable areas, discounting previous commitments to full-time employees, their livelihoods, and their localities. Will flexible engagement constitute resistance or accommodation to flexible specialization? This remains an open question.

NOTE

1. A list of the original publications worked into Taylor (2005) that are relevant to this chapter is available at http://www.faculty.umb.edu/peter_taylor/92crefs.html.

REFERENCES

Austin, Michael. 1987. "Models for the Analysis of Species' Response to Environmental Gradients." *Vegetatio* 69:35–45.

Burbidge, John, ed. 1997. *Beyond Prince and Merchant: Citizen Participation and the Rise of Civil Society*. New York: Pact Publications.

DeAngelis, Donald, and Joyce Waterhouse 1987. "Equilibrium and Non-equilibrium Concepts in Ecological Models." *Ecological Monographs* 57:1–21.

Drake, James. 1991. "Community-Assembly Mechanics and the Structure of an Experimental Species Ensemble." *American Naturalist* 137:1–26.

Durham, William. 1979. *Scarcity and Survival in Central America: Ecological Origins of the Soccer War*. Stanford, CA: Stanford University Press.

Elbow, Peter. 1981. *Writing with Power*. New York: Oxford University Press.

Fish, Stanley. 1989. "Anti-foundationalism, Theory Hope, and the teaching of Composition." In *Doing What Comes Naturally: Change, Rhetoric, and the Practice of Theory in Literary and Legal Studies*. Durham, NC: Duke University Press, 343–55.

García-Barrios, Raúl, and Luis García-Barrios 1990. "Environmental and Technological Degradation in Peasant Agriculture: A Consequence of Development in Mexico." *World Development* 18:1569–85.

Kingsland, Sharon. 1995. *Modeling Nature: Episodes in the History of Population Ecology*. Chicago: University of Chicago Press.

Latour, Bruno. 1987. *Science in Action: How to Follow Scientists and Engineers through Society*. Milton Keynes: Open University Press.

Levins, Richard. 1966. "The strategy of model building in population biology." *American Scientist* 54:421–31.

Little, Peter. 1987. "Land Use Conflicts in the Agricultural/Pastoral Borderlands: The Case of Kenya." In *Lands at Risk in the Third World: Local Level Perspectives*, ed. Peter Little, Michael Horowitz and A. Nyerges. Boulder, CO: Westview, 195–212.

Lloyd, Elisabeth. 1987. "Confirmation of Ecological and Evolutionary Models." *Biology and Philosophy* 2:277–93.

May, Robert. 1973. *Stability and Complexity in Model Ecosystems*. Princeton, NJ: Princeton University Press.

Odum, Howard. 1971. *Environment, Power, and Society*. New York: Wiley-Interscience.

Okoye, Samuel, and Philip Smith 1994. "Introduction." In *The World at the Crossroads: Towards a Sustainable, Equitable and Livable World*, ed. Philip Smith, Samuel Okoye, Jaap de Wilde, and Priya Deshingkar. London: Earthscan.

Peet, Richard, and Michael Watts, eds. 1996. *Liberation Ecologies: Environment, Development, Social Movements*. London: Routledge.

Pielou, E. C. 1969. *An Introduction to Mathematical Ecology*. New York: Wiley-Interscience.

Simberloff, Daniel. 1982. "The Status of Competition Theory in Ecology." *Annales Zoologici Fennici* 19:241–53.

Stanfield, R. Brian. 2002. *The Workshop Book: From Individual Creativity to Group Action*. Toronto: Canadian Institute of Cultural Affairs.

Stepan, Nancy. 1986. "Race and Gender: The Role of Analogy in Science." *Isis* 77:261–77.

Strauss, Sharon. 1991. "Indirect Effects in Community Ecology: Their Definition, Study and Importance." *Trends in Ecology and Evolution* 6:206–10.

Taylor, Peter. 2001. "How Do We Know There Is a Population-Environment Problem?" http://www.faculty.umb.edu/peter_taylor/popdialogue.html.

———. 2002. "We Know More Than We Are, at First, Prepared to Acknowledge: Journeying to Develop Critical Thinking." http://www.faculty.umb.edu/peter_taylor/journey.html.

———. 2005. *Unruly Complexity: Ecology, Interpretation, Engagement.* Chicago: Chicago University Press.

Taylor, Peter, and Raúl García-Barrios. 1995. "The Social Analysis of Ecological Change: From Systems to Intersecting Processes." *Social Science Information* 34:5–30.

Vandermeer, John. 1969. "The Competitive Structure of Communities: An Experimental Approach with Protozoa." *Ecology* 50:362–71.

———. 1977. "Ecological Determinism." In *Biology as a Social Weapon*, ed. Science for the People. Minneapolis, MN: Burgess, 108–22.

West Nipissing Economic Development Corporation. 1993. "Vision 20/20: Shaping Our Futures Together, Executive Summary."

———. 1999. "Vision 2000 Plus, Executive Summary."

Wolf, Eric. 1982. *Europe and the People without History.* Berkeley: University of California Press.

15

The Demise of Critical Theory in Economics

Francisco Louçã

Boaventura de Sousa Santos's lecture, "A Discourse about Sciences" has long traveled since 1985, when it was originally presented at the opening of the academic year in Coimbra, until its first edition in 1987 as a book and from then until the current critical edition the reader has in hands. The "Discourse" has been read, discussed and criticized—and may now be studied at the light of the decade and half passed since its first presentation. This edition is indeed the best demonstration of how a critical theory survives and develops, applying the tools of critique itself.

As a sequence of this fascinating debate, the current chapter proposes to consider the evolution of the economic science, inquiring about the conditions for the local emergence of a new scientific paradigm, for which Sousa Santos has been fighting and for which the "Discourse" became a fundamental manifesto.

First, I reconsider some of the fundamental theses of the "Discourse" in order to question if they are sufficient to orient choices through the essential epistemic crossroads in contemporary economics. And I discuss that reading in a precise context, that of the emergence of the paradigms of modernity in economics, and in particular in econometrics, that is, in the theoretical space in which Newtonian quantification and legislation appropriated elements of quantum physics. In all accounts, it was an exceptional period of great creativeness and ingenuity. Yet a question arises about its evolution: why were critical theories in economics, unlike what happened in the province of other social sciences, marginalized, or, more precisely, why did not these critical theories substitute or transform the dominant paradigm while they contested it, as it happened in physics, and were on the contrary absorbed or destroyed by it, in a recessive movement prolonging until today.

On the other hand, I emphasize in this chapter that this destiny is most paradoxical, since some of the main protagonists of this modernization, the mathematical economists embodying the probabilistic revolution in economics, were precisely moved by societal transformation, but that this motivation was what led them to believe and develop methods preconditioned against change. This is how one of the central *corpus* in of the critical theory in economics waned away and returned to its margins. In other words, the modernization movement born in the last quarter of the nineteenth century dissipated the potentiality of epistemological renovation transported by the great modernisms in economic science, and contemporary difficulties, which we feel today derive from this lost confrontation. Furthermore, I will argue that it was a tragedy that this powerful movement for modernization, neoclassical economics imposed itself as a dogma. Consequently, it alienated from what modernity itself has been requiring: in particular, it avoided and rejected the deepest consequences of the analogy with the quantum revolution and the introduction of the concepts of probability for the description of social processes.

Finally, I ask if there is and what may be the application of the prescriptive methodology of Boaventura Sousa Santos, the argument of his "Discourse." I will answer positively to this first question, emphasizing, in relation to the latter, that the *fin-de-siècle* economics concentrates in its contradictions all the interrogations to which a realist and prudent epistemology must answer, as defined by the postmodern proposal of Sousa Santos.[1]

A RECAPITULATION OF THE LONG NINETEENTH CENTURY

The "Discourse" opens affirming that we still live in the framework of the large traits of nineteenth-century science, in spite of the emergence of the new characteristics of the century emerged—and perhaps this is the reason why the contribution of the twentieth century was less impressive in paradigmatic terms. Once entering the new century, we may now ask ourselves about the validity of this long-term view (p. 5). From that point of view, the imprints of modernization in the long nineteenth century extended until the first decades of the twentieth century: Adam Smith, Ricardo, Lavoisier, Marx, Durkheim, Max Weber, Humboldt, Planck, Darwin, Poincaré, Einstein, as well as Freud. From this point of view, this was the epoch of the founders.

Boaventura de Sousa Santos presents a history of the crisis of the paradigm, the history of our twentieth century, the period during which those foundations were severely questioned: from relativity to statistical mechanics and the principle of uncertainty, from the rediscovery of the Mendelian laws to the formulation of evolutionist orthodoxy, from the preliminary versions of the new paradigm (Capra, Prigogine, Jantsch, Habermas, . . .) to its first texts and manifestos, this short time of liberation out of positivism and Newtonian scientism was simultaneously a long period of approximation and redefinition among sciences.

Moreover, Boaventura discusses the irreversible crisis of the dominant paradigm, under Einstein and quantum mechanics (Santos 1987, 23ff.) and, as a consequence, indicates the desirable characteristics of the emergent paradigm, that of "postmodern science" (Santos 1987, 37ff.). I will return later on to this proposal. For now, the following sections concentrate on a brief description and interpretation of the process of translation of Newtonian mechanics in economics and of the contradictions this process imposed, in particular due to the conditions of the dominant paradigm supposing the existence of specific temporalities of social processes, since "a knowledge based on the formulation of laws requires the implicit theoretical hypothesis of order and stability of the world, the idea that the past repeats itself in the future" (Santos 1987, 17).

THE NEOCLASSICAL REVOLUTION

The vision of social order as the analogue for celestial mechanics and, as a consequence, the adoption of astronomy, first, and in general of physics, as the reference science for the delimitation of protocols, instruments and legitimate results in economics, was the first step for one of the more powerful intellectual movements that ever reformatted this science. At the beginning of the last quarter of the nineteenth century, a number of economists of several countries, as far as it is known without any direct contact among them, protagonized a decisive change, the introduction in economics of methods, concepts and strategies directly metaphorized from the previous generation of energetics and physics. Unlike the classical economists who, after Smith and Ricardo, and at its apogee Marx, had defined economics as a historicized science concerned with the social relation of production and distribution, this new generation, which was to be called neoclassical, emptied the social contents of economics in the name of a "social physics," simultaneously rigorous since mathematized and scientific since reproducing laboratory practices. Yet and evidently, when this new science compared economic agents and social processes to mere incognito particles, it chose to ignore options, strategies, relations, institutions, that is, real economies. And it went on in this mood.

Jevons and Edgeworth, in England, Menger, Walras and Pareto, in the continent, and later on Irving Fisher, in the United States, were the protagonists of this paradigmatic revolution, which had a devastating effect and redefined the frontiers of economics. These authors assumed the first Law of Thermodynamics to be the unifying mathematical pattern allowing for the analysis, among other processes, of the creation of value ("utility"): in the space of commodities, subject to the restriction of conservation of total energy, Lagrangean maximization could be applied considering the objective function. In this framework, the concept of equilibrium is densified as that point in which dynamics collapses as the coordinates of a fixed point, whereas the agents—the atoms—are supposed not to modify their preferences or attitudes. Moreover, these economists added a normative concept that was naturally ignored by energetic physics, that such equilibrium would correspond to the best possible social situation, to the perfection of economic relation established by a multitude of agents who, egoistically, endeavor to maximize their own utility and obtain a social optimum. Consequently, we are blessed simultaneously by a method of computation and an ideology in the full sense of the word. Pareto optimization constitutes an undisguised justification of the presumed stability of the emergent social relations in the capitalist market.

The founders of this new economy were confident men: Jevons praised Laplace, arguing that the perfect knowledge of the universe was possible given the use of the tools of mechanics and the understanding of the general laws ruling all movement. In this framework, chance is merely the translation of our ignorance, eventually transitory and surely to be dissipated through the progress of one science able to know everything. The general laws embodying this capacity of science were supposed to correspond to the description of a reproductive mechanism under rigid determinism, as causality was understood: in the imported mechanic tradition, the engine was the paradigm for this science. This concept reinforced both the notion of conservation of energy and the certainty that causality revealed a preestablished order. Science could ignore nothing and would know everything. Newton was the hero of this triumphant scientism: under his spell, positivism conquered neoclassical economics.

Neoclassical economics was consequently the first inaugural moment of naturalization of social sciences, and that was the program of its modernization. But if we consider this movement of modernization as a whole, its slow and difficult maturation finally led to a devastating victory: the alternative programs of research were not only cursed (Marxism as well as the whole classical economics), but a restless court of illegitimating and exclusion from the domain of science was put into action, and it worked without truce through the twentieth century—against the German Historical school and then against Institutionalism.

Yet this drive was challenged by resistances and skirmishes, and undermined by alternatives and difficult interrogations. In fact, the neoclassical revolution was based upon a deficient metaphor, since the translation of the First Law into "social physics" required the acceptance of Draconian conditions, first of all that of an ontological equivalence between money and utility, what no one seemed to be able to argue for. Nevertheless, such a contradiction was largely ignored, since it was supplanted by the scale production effect (the creation of a "normal" science in the Kuhnian sense) and of community self-identification, as it was guaranteed by the prestigious belonging to the scientific tribe. The drama, because it was a drama, was that this feeling was based upon an illusion, and its collapse provoked the crisis of this project of naturalization of social sciences—and it was indeed an illusion since it immediately revealed its maladjustment in relation to physics itself, its origin.

While this *blitzkrieg* was pursued, the Second Law was already being discussed in physics, abandoning the assumption of a closed space for the process of conservation of energy, making possible the maximization of an objective function, since both transmutability and conservation were simultaneously required. The reason for the change is particularly important: the research on the efficiency of engines had revealed the dimension of the problem of dissipation and emphasized the implausibility of the hypothesis of conservation of energy in real movements. Consequently, the Second Law suggested a new reading of the universe and the existence of entropy came to be considered a powerful universal law. In one word, natural processes generate disorder—an understanding that, combined with the affirmation of evolution theory in biology, made clear that the concept of equilibrium as used in economics is meaningless and mystical.

The introduction of the concept of entropy had another deep consequence on the project of naturalization of social sciences: it paved the war for the adoption of more realistic concepts applied to social processes that could, consequently, reunify the two main branches of sciences, if the willing existed—and it did not. What previously approximated both branches, the colonization by natural sciences and the ignorance of the specific object and purpose of social sciences (under the rule of the 1st Law), then prevented the capacity of enunciation by the subjects of social sciences (under the 2nd Law). The previously acquired method of computation (Lagrangean maximization and the application of calculus) was challenged and economics was left without an operational equivalent to that of the core sciences. The crisis deepened as a consequence, since economists did not know how to deal with Heisenberg's principle of uncertainty, with quantum mechanics, with the definition of intrinsic probability and nonlinearities. The condition for emancipation, allowing for a dialogue between the two branches of sciences precisely when physics began to socialize, was disturbed by the lack of epistemic autonomy of economics, which did not cease to naturalize itself.

INCOMPLETE MODERNIZATION

Since the early days of the twentieth century, quantum theories and relativity reformulated physics. In spite of not abandoning its strong legislative ambition, which was based upon the concept of reality as the reproduction of a general causal mechanism, physics began defining itself as an interpretation of irregularities, in particular as part of the creation of order in entropic processes. Consequently, physics gave a step forward in the direction of evolutionary biology and used statistical mechanics in order to interpret nature.

As this turmoil went on, neoclassical economics was not able to vanquish its main opponents, the Historic School (Germany), Institutionalism (United States) and even in Britain it was challenged by a strong national specific tradition, the Cambridge school, inspired by Alfred Marshall. As an abstract theory, neoclassical economics was languishing as a metaphysical imitation to physics, until it was saved by a new generation of economists, who championed empirical analysis, quantification and inductive verification of general laws. Indeed, this landscape suffered a sweeping revolution only when a group of young mathematical economists gives birth to a second neoclassical generation, adopting a new program: the introduction of operational concepts in economics in the name of a rigorous formalism allied to experimental capacity simulating a laboratory, able to use the statistical techniques inspired by quantum physics, rigidly standardized procedures and a new logic language. These economists were the protagonists, in spite of their important differences, of a deep transformation of their science, as they reinforced its explanatory capacity, allowing for the introduction of objective methods of selection of hypotheses and theories and adopting a modern lexicon and a new grammar.

Through the following sections, I will summarize previous research about the nature and motivation of this generation, and then I will confront it with the process of modernization it gave birth to, in order to conclude about the adequacy of the new modernist program (or postmodernist, accepting the terms of Sousa Santos) for economics.

I will emphasize that they aimed at having an immediate impact of their research on political decisions, given in particular their intention to shape economics as that science able to tame unemployment after the Great Depression and considering the menace of war in the thirties. That took these young economists to adopt the neoclassical paradigm, whose hypotheses precisely rejected their program. Since they intended to adopt authoritative forms of computation, they tried to reconcile the previous vision of a conservative mechanism with the artificial and heterogeneous superimposition of stochastic impulses, eclectically including the universe of the first Law into that of the second, and lightly ignoring the deep epistemic contradiction they were generating.

Moreover, according to this synthesis—in spite of the challenge it suffered ever since—these economists moved away from their contemporary process of modernization of sciences, the creation of what has been recently dubbed "cyborg sciences," for which John von Neumann was the main character (Mirowski 2002). Impressed by the results by Kurt Godel on undecidability of mathematical logic and the research on computation by Alain Turing, von Neumann and his collaborators argued for the abandonment of the paradigm of the engine and adopted the model of the computer as the reference for all sciences: the metaphor of thought as computation, and the analogy between the brain and the computer allowed for the definition of the lemma of this program—Command, Control, Communication-Information, or C^3I. Given it was generously financed as few others, in particular for the expected benefits in the framework of the World War and then of the Cold War, as well as for the important results it obtained with the creation of operational research, game theory, experimental economics, computational economics, with the extension of the macroeconomic models and the transformation of microeconomics, this cyborg modernization deeply troubled the province of economics. Yet the neoclassical paradigm resisted almost woundless: this resistance may be understood in the framework of the turn in the thirties, thanks to the foundation and development of the econometric movement. This is the topic for next sections.

POLITICS AND SCIENCE FOR THE GENERATION OF FOUNDERS OF ECONOMETRICS

After the First World War, the European political crisis attained new dimensions with the victory of governments under fascist inspiration, such as Mussolini and Salazar or Franco, and the rise of Nazis in Germany in the context of hyperinflation. The main world economies sunk in an unprecedented crisis, the first generalized recession of the century, imposing unemployment rates of two digits both in the United States and in Europe.

The generation of economists that began to take position in the twenties had a mathematical preparation largely superior to that of their predecessors—some of them graduated in mathematics and physics, migrating afterwards to social sciences—and a strong motivation for a fight against the causes and effects of recession. Econometrics, in particular, originated as a program for a rigorous science and therefore a quantified one, claiming to be able to choose among activist policies against depression: from that point of view, it was initially a non-neoclassical program, as it required the active regulation and intervention of public authorities, frankly at odds with the notion of *laissez-faire*. The common characteristic of this generation, in coherence with this vision of economics and eventually with the exceptions of Amoroso and Schumpeter, was its leftist leanings, precisely what drove them to filiate their scientific priorities to the ambition of creating full employment and fighting the evil of liberalism.

The econometric movement was naturally populated by distinct currents and opinions about the purpose of economics and the contribution of social sciences to the normative determination of policies. But an important part of these young men had chosen economics precisely for its disputes on influential choices ahead: they intended, and quite openly stated so, to help avoiding the continuation of crises and the specter of unemployment. This was the case of Keynes, Tinbergen, Marschak, Lange, Koopmans, Neyman, Meade and others, but before all that of the founder of the Econometric Society, Frisch. This group of scientists marked the twentieth century in economics: Keynes was probably the most influential economists of that period, and Frisch shared with Tinbergen the first Nobel Prize in economics, in 1969, precisely for their work during the foundational period of econometrics.

Some of them, as Tinbergen, Marschak and Koopmans had got their graduation in other branches of sciences: Tinbergen and Koopmans as physicists, Marschak as a mechanical engineer. They opted for economics for ideological reasons. Tinbergen was part of the young socialists" organization and was then an active member of the Labor Party. A political attitude involuntarily helped him to come into economics: as he refused to serve in the army in the World War I, Tinbergen was forced to work in the civil service as compensation. After working in the administration of a prison, his father—without his knowledge—managed to obtain his transfer to the Dutch Central Bureau of Statistics (Boumans 1992, 18): a career in economics had begun. Through all his life, Tinbergen insisted on development and planning as the main themes for economics after the war:

> What kind of useful work can be done for the socialist movement by someone who has received an economic-statistical training? . . . I have passed my doctoral examination in mathematics in physics, but I believe that I can not serve the socialist movement with these subjects. Because my professor is very much in favor that I continue my study, I am working on economics and statistics. I arrived at this choice because of the economic character of many socialistic reforms and by the conviction that socialization will need economically trained men to carry it through (letter to the socialist leader F. Wibaut 1926, quoted in Boumans 1992, 11).

This option was not at all incompatible and was even in solidarity with his wish to naturalize economics. Indeed that was why his migration was so successful, since he was instrumental for the transposition of the techniques of Hamiltonian dynamics into economics, and that technical capacity was most welcome. Tinbergen insisted and persisted: econometrics should be the "future science of engineers" (Tinbergen 1936, quoted in Boumans 1992, 4): social engineers would come to be, he thought, the best promoters of progress since they have the indispensable scientific knowledge adequate for those transformations the world urgently requires. By that time, Tinbergen read Marx's Capital and thought the extension of rigorous methods from physics to economics was possible and desirable. His own career testifies that possibility: Tinbergen was one of the builders of the econometric movement in Europe in the thirties and authored the first applied econometric research, the investigation ordered by the League of Nations on the statistical test of the different models of business cycle.

That engagement in political economy for social motives was also the case for younger members of the econometric community, although they lately followed different options. Engineer by education, Marschak had been a minister of the socialist government in Ukraine in his youth and then left for Germany. After the rise of Nazism, he went to Britain and later on to the United States. There, his socialist views were quite well known—some of his students still remember his late participation in demonstrations against the Vietnam War at Stanford.

Marschak's political attitudes were quite well known among his colleagues since his early days. There is an early illustrative exchange of letters between Frisch and Schumpeter on these views: Frisch criticized Schumpeter's opposition against the acceptance of Marschak as a member of the Econometric Society for apparent anti-Semitic or antisocialistic reasons. Schumpeter answered that he did not object to Marschak for being a Jew and a socialist, but for being both at the same time, since he could eventually bring other socialist Jews to the Society (Frisch to Schumpeter 13 November 1932 and 13 March 1933, Schumpeter to Frisch 3 December 1932 and 22 February 1933). Four years later, Schumpeter used again the same argument against the selection of Marschak as the research director of the Cowles Commission: Marschak is a "highly competent and highly trained economist and statistician. (. . .) Of course, he is a Jew and a socialist and it is more than likely that he will try to draw other Jews and socialists after him. But the weight to be attached to considerations of this kind it is beyond my competence to pronounce upon. In the normal course of things I do not apprehend any difficulty on that score" (6 May 1937, Schumpeter to Cowles). It was nevertheless obvious that the crucial point was his organized political activity rather than any other motive: Schumpeter recommended instead Lange, who were also a socialist and whose wife was a Jew (which would mean trouble if returning to Europe, Schumpeter rightly argued). Later on, Schumpeter wrote letters of recommendation of Marschak: to Columbia (22 March 1939) and to Berkeley (6 April 1939).[2]

Marschak was later on a leader of the transformation of this econometric movement in order to conforming it to the neoclassical canon. The discussion on the processes of interpretation and the strategy which led these authors to favor or at least to conform to computation as it was possible under the neoclassical framework of general equilibrium is outside the scope of this chapter and was debated elsewhere (Louçã 1999, 2000, 2001).[3] What is relevant for the current purposes is to emphasize this obvious contradiction between the option to act against the structural disequilibria in capitalist societies, which provoked the crises of the late twenties and early thirties, and the use of a mathematical toolbox based on the presupposition of equilibrium in order to suggest the necessary policies to avoid or to circumvent the crisis. It was the central character of this young econometric movement in the thirties, Ragnar Frisch, one of the founders of the Econometric Society with Fisher and Roos, that exceeded all the others in political determination and best represented this paradox. During the thirties, he was the major driving force behind econometrics, and the reason for his activity was decisively the consideration of the political function of economics.

Frisch was politically active most of his life. As a *compagnon de route* of the Labor Party in Norway in the thirties, he helped to draft an electoral program, participated in parliamentary committees and advised ministers and high officials. But in the forties his relations with the party began to deteriorate, since he was convinced that it followed a wrong path and accepted the rules of the "monetary plutocracy." That divergence came to a rupture, and Frisch did not participate any more in high level cooperation with the Norwegian government. He engaged in large planning work in India and in the United Arab Republic in the fifties and sixties and became very involved in development studies. Later on, he published some articles in the paper of the left socialist party, a split of the LP, and was very active against the participation of Norway in the EEC, which represented for him a maximum expression of "unenlightened financialism."

In the late sixties, he participated in several activities and publicly voiced his opinions against the Vietnam War. The influence of Frisch's arguments against the participation of Norway in the EEC was still felt when the country voted in a new referendum rejecting the adhesion to the EU (1994); by that time, Haavelmo maintained the same opposition (Tinbergen did not follow Frisch on that, and argued for the participation of Netherlands in the EEC, in the seventies).

For this chapter, the essential question is not to value the scope of the political arguments, but to register the importance of this motivation for these young economists, both for the choice of economics itself as the terrain of their activity, and for the selection of problems and priorities as well.

MOBILIZED AGAINST WAR AND CAPITAL

The common ground for most of the United States and European founders of econometrics was the definition of their endeavor as the attempt to prevent unemployment and other socially damaging consequences of business cycles. This was much more accentuated than for the previous generation, in particular for the neoclassical economists (Walras's vague argument for social reform and in particular for distribution of land is normally referred to as an example of an extreme position of the previous generation). For the following generation, there is also vast evidence to indicate that many of them were activists and, furthermore, that many of them considered a program of social transformation to be feasible. A pamphlet—"Econometrics—Towards Making Economics a More Exact Science"—published by the Cowles Commission in 1932, which was the first public presentation of the project, bluntly argued that it was indeed more important than the creation of the telephone and radio, and as viable as these inventions:

> The objectives of the mathematical researchers now being undertaken by members of the Econometric Society are vastly more important than the radio or the telephone. Certainly, an appreciation of the practical value in understanding the causes of the business cycle has of late been amply driven home. And it is altogether possible that, through econometrics, a solution of this problem, which has baffled the world, thrown millions out of world, and caused untold distress, may ultimately be reached. It is the purpose of those who are interested in developing econometrics to turn to the powerful weapons of mathematical analysis on the business cycle, indeed on every aspect of our economic life, in an effort to establish or demolish existing theories, and to discover new economic truths. (Cowles Commission 1932, 7)

At least for some of the econometricians and certainly for those more influential, the search for practical solutions to the problem of unemployment extended to the opposition against war. Tinbergen, for one, was very impressed with the call made by 350 psychologists on October 1935 in favor of peace. On 20 March 1936, he wrote to Frisch in order to suggest a new call, this one by the *Econometric Society* as such. According to Tinbergen, it should declare that the "econometricians feel it as a first duty to raise their voices against the tendencies leading to the largest wholesale destruction of human welfare, the war." In that sense, the econometricians should offer their services to all governments and international institutions, providing neutral scientific procedures for the division of world supplies and consequently preventing the war. It goes without saying that Frisch wisely considered the initiative to be doomed from the beginning, and convinced his colleague to leave the Society out of that mess (Frisch to Tinbergen, 26

March 1936). Yet he fully agreed with Tinbergen's views and both shared the same hostility against war.

Their final choice in relation to the implication of the Society in this debate did not imply that any of them considered not to be concerned by the danger of these days. On the contrary, they thought that formal rigor, leading to mathematically treatable models allowing for concrete predictions and sound policy advice, were the requisite for preventing social tensions, unfairness, despair, conflict and eventually war. Tinbergen's work for the League of Nations on the comparison of the models of business cycles reviewed by Haberler, which was at the origin of the famous debate with Keynes, is one of these contributions. Indeed, Frisch and Tinbergen were close scientific and political allies for all their lives since they shared the same preference function.

Frisch had already suggested, in 1934, that the root of the main social problems was the very organization of the market economy, since it stood over numerous independent decisions taken by too many interdependent agents. In that case, the lack of coordination could lead to the collapse of the system—and he thought that indeed that was the case since 1929.[4] The 1934 paper, "Circulation Planning," argued for a new system of multilateral barter trade, as an experimental method for preventing sectoral or regional mismatches, under the rule of a multilateral clearing agency. By that time, Frisch was already deeply convinced that the indirect steering mechanisms in a Keynesian mood was insufficient and inappropriate.[5] Furthermore, believing that the financial and banking system was highly responsible for the economic crises, he naively favored nonmonetary exchange among agents. This general stance was far more radical than that of Cambridge: Frisch and others defended econometrics as the means for computation as required by planning (Andvig, 1985, 285).

Yet, in spite of the vigor of the argument, the paper did not impress the political decision-makers or even his academic colleagues. The article was just noted for its length—it is still the largest paper ever published in *Econometrica*—and was either criticized or ignored by the econometricians. Roos, the secretary of the association, wrote to Frisch, who was himself the editor of the journal, protesting the inclusion of a so long paper in *Econometrica* and stating that it is not "a very vital contribution to economic knowledge" (letter from Roos to Frisch, 13 February 1935). And as the evidence from the 1935 debate among Frisch, Tinbergen and Koopmans proves, there was a widespread skepticism against this line of research (Louçã 2000).

But Frisch wanted to pursue it. In a letter to Cowles (11 July 1937), he suggested a research project on "economic control" and then presented a memorandum to the Cowles Commission on this very topic. The outline of the research project suggested the recourse to a simultaneous equations system, given that all variables influence each other, and the use of experiments and inquiries in order to define and compare preferences. As Andvig noted, there was a mounting contradiction between these two forms of assessment: by the late 1930s, Frisch was more and more skeptical about the role and use of highly aggregated macrodynamic models, as his own 1933 model, and increasingly favored detailed direct steering mechanisms and planning. But, concentrated as he was in the definition of the social preferences from microdata, he never developed the macro side of his policy alternative (Andvig 1985, 285). Furthermore, his vision of the crisis as the emergent property of noncoordination of agents' decisions promoted this mismatch between the micro and macro levels of analysis and intervention.

In spite of the doubts of his colleagues, this was not felt as a contradiction by this founder of econometrics. This is why the "circulation planning" project reemerged much later when, in July 1962, Joan Robinson joined forces with Ragnar Frisch to present a "Draft of a Multilateral Trade Clearing Agency" to the Commonwealth Conference. As the next section proves, there is widespread evidence to show that he considered the contradictions of capitalism and the market form of social organization to be responsible for war and deprivation. For that reason, a new international organization based on cooperation and fair trade was proposed as a viable alternative—as viable as control of the business cycle was thought to be for the originators of the Cowles Commission thirty years before.

Consequently, when the Econometric Society reestablished its transoceanic functioning after the end of the war, some pressed for its sharp intervention in relation to the pressing agenda of the day. Frisch published an editorial in *Econometrica* explaining why the mathematical economists had the "social responsibility" of solving the problems of misery and unemployment, using their knowledge to prevent the causes for wars and disgrace of the population. But he was then more isolated than previously, and some changes in the agenda of the meetings suggest that the topic did not matter that much for his fellow econometricians.

When the ES meeting was called for September 10–17, 1947, in Washington, the first draft of program (7 January 1947) included a session on "econometrics as an aid to policy decisions," and Frisch, Tinbergen and Richard Stone's papers were scheduled for the discussion. In other sessions, a paper by Fréchet on "the possibility and limitations of the application of probability theory in the field of social and economic phenomena (based on a questionnaire)" was to be discussed by Marschak (who would also present an argument on "statistical inference from nonexperimental observations"). But the second draft of program (20 March 1947) abandoned the idea of having a session on "aid to policy decisions" (March 21 letter by the Program Committee: Koopmans, Leavens, and chaired by Marschak). Topics in the same sense eventually appeared in some of the next meetings of the Society, as that of September 10–12, 1959 in Amsterdam (a session on the "use of econometric methods for policy purposes," with Tinbergen (chair), Frisch, de Wolff, Greniewski, Lesourne), but the focus of attention was already elsewhere.

The Keynesians, on the other hand, shared a similar preoccupation, although arguing for different solutions: in spite of his position as president of the Econometric Society in the forties, Keynes emphatically disagreed with much of the work of its members.[6] Keynes's position and political activities are, of course, quite well known and the recently published monumental biographies by Moggridge and Skidelsky document his trajectory in great detail. Keynes was a center-left supporter of the Liberal Party in the United Kingdom, opponent to the Tories and vaguely sympathetic to Labour. In 1923, in his first editorial of *The Nation*, Keynes presented his politics: "Our own sympathies are for a Liberal Party with its center to the left" (quoted in Skidelsky 1992, 136–37); by that time, he suggested a Labour-Liberal government. He supported the miners during the General Strike of 1926, a social movement that caused a split in the leadership of the Liberal Party; this attitude caused the alienation of his own traditional political relations inside that Party (Skidelsky 1992, 223). In 1931, he was more inclined to support Labour than Liberal candidates; only with the outburst of the World War did he return to the Liberal party tradition (Moggridge 1992, 465). But it must be emphasized that his public position was largely influential and somewhat consensual: in 1937, all the three parties offered Keynes a nomination as independent candidate for MP for Cambridge, which he refused after long hesitation (Moggridge 1992, 628). Among Keynesians, opinions divided between for instance Joan Robinson, a committed socialist, and James Meade, a "committed Keynesian and a moderate socialist" (Harrod 1951, 501).

The younger generation generally shared these points of view. Neyman, for instance, strongly protested against the editorial decision of including in *Econometrica* (Econometrica 6, 1938), a paper by Amoroso on Pareto, which he read as a justification for the regime of Mussolini (Pareto had accepted to be nominated as senator by Mussolini,). Later on, Neyman argued in vain for the Peace Nobel Prize to be awarded to Martin Luther King.

LEARNING FROM THE WAR

War, with its immense horror, proved the danger of economic maladjustments and the importance of economic regulation as part of the solution for social problems—this is how some of the protagonists of this story reacted to the dangerous period they were living through. In particular, Frisch argued for a strongly prescriptive econometrics in this framework. This was quite obvious in his lecture on the "Emploi des Modèles pour l'Elaboration d'une Politique Economique Rationnelle," presented on October 17, 1950, to the Ecole Nationale des Ponts et Chaussées, in Paris. A "rational political economy" was presented as that which could prevent the repetition of war, conflict and social misery: that new sort of economics should lead the way to a reform of the profession and the redefinition of its purpose:

> Les études esquisées ici sont inspirées par une idée directrice: la démocratie, pour survivre à sa crise actuelle—et survivre comme democracie—a besoin de s'appuyer sur un genre assez nouveau d'analyse economique. (Frisch 1950, 474–75)

Now, this "new type of economic analysis" could only benefit from the comparison with the previous great depression and the situation having led to the dramatic devastation of the war:

> Pour bien preciser le problème je considère, comme point de départ, la situation economique telle qu'elle existait dans les années 1930. Le chômage massif, qui régnait alors dans la plupart des pays, conduisit à une situation monstrueuse. Les niveaux de vie baissèrent en plein milieu de l'abondance. Des denrées alimentaires et d'autres objets de consommation furent délibérément détruits, tandis que les gens priaient et espéraient qu'il se passerait quelque chose qui leur permettrait enfin de bénéficier de leur propre travail pour la satisfaction de leurs propres désirs. Cette expérience conduisit à une conclusion simple, mais fondamentale: la nécéssité de prévenir de telles monstruosités. Aucune solution d"un problème économique quel qu'il soit, ne sera admissible si elle ne satisfait pas à cette condition. (Frisch 1950, 475–76)

For Frisch, what had been at stake—and what remained on the agenda—was a life and death question, and that was the measure of the responsibility of the economists and policy-makers:

> Ainsi assistons-nous a une véritable *course entre la recherche economique et les faits économiques*. Il n'est pas exageré de dire que l'enjeu de cette course est une question de vie ou de mort. Ce sera le désastre pour des millions d'hommes si la recherche économique et sociale et ses applications ne triomphent pas. Si elles perdent, nous serons, de temps en temps, les témoins de monstruosités analogues à celles que nous venons de dénoncer. (Frisch 1950, 476–77)

This allowed for the conclusion: to fight unemployment the previous, ten- or twenty-year-old theoretical tools were not adequate any more—just as the use of ten- or twenty-year-old statistics was inadequate for the understanding of the present situation of unemployment (Frisch 1950, 477). This was a typical Keynesian argument in form and content, but this time it was to be used against Keynes himself: although Frisch recognized that economic theory had benefited from Malthus, Wicksell, then Keynes and "his followers," such as Hansen, he also argued that new conditions condemned that theory to oblivion. The "new facts" emerging after the war—the existence of direct forms of economic control in some countries, new patterns of social justice and the modification of the role of international commerce—asked for new theories. In that framework, the Keynesian view of effective demand was inadequate, Frisch argued (Frisch 1950, 477). Planning was the solution.[7] Yet, when Frisch was liberated from the concentration camp and the German occupation of Norway ended and the international communications were reestablished, his position was considered to be eccentric for most of his contemporaries and he was alienated from mainstream econometrics, a process that aggravated his fierce criticism for the latter decades (Frisch 1963, 1970).

This argument favoring the definition of economics as that social science of reforming the market system was, until the war, generally accepted by the European econometricians, with few exceptions, and even by some of the U.S. econometricians. By the end of the forties, and furthermore in the following decades during the Cold War, this position vanished. But the essential factor for its extinction was not the isolation of its champions or event the emergence of a new victorious paradigm. On the contrary, the dominant fact of the history of economics in the twentieth century is that that evolution was determined by the very econometricians who had defined the original program as that of the extermination of the economic cycle through direct action for the regulation of the economies: they considered that, if pure laissez-faire generated economies in danger of coordination collapse producing mass unemployment, then it would be necessary to replace liberalism. In coherence, they conceived econometrics as subsidiary of the mechanics of the first Law of Thermodynamics. This was a conscious choice and it was made before the war.

The debate on the "socialist calculation," so intense between 1920 and 1935, was part of this turn. Originally, the debate emerged out of a confrontation between Von Mises and the philosopher Otto Neurath, who argued, as Frisch would maintain later on, that the events around the First War had proved the possibility and efficiency of concentrating the economic decision in political agents and consequently the viability of socialism as a social model. Von Mises orchestred the campaign to undermine this argument and was assisted by Hayek, who added that information cannot be centralized given its very nature and that its complexity is at odds with the limited computation capacity of the planning system. Marschak, Hurwicz, Lange, Arrow and Koopmans, among others, were part of this debate rejecting the impossibility claimed by Von Mises. To the young econometricians, on the other hand, the aim was not to create an alternative system to that of the market, but on the contrary to affirm that the socialist computer could replace the Walrasian auctioneer in the market itself, solving a system of simultaneous equations: the whole program of structural estimation of the Cowles Commission, during the founding phase of econometrics, is based on this essential option. Modeling business cycles rigorously followed this percept, considering a dissipation mechanism producing a static equilibrium, whereas a simultaneous stochastic process producing small perturbations was hypothesized, although its explanation was declared outside the scope of the theory.

Structural estimation, as a form of empirical statistical analysis of these simultaneous equations, became the leitmotiv of the first years of the Cowles Commission, as a way to give birth to this "market socialism" (Mirowski 2002). Nevertheless, this research was placed in the framework of dominant orthodoxy, because this strategy allegedly permitted the maximization of the capacity to influence the political decision through dialogue and theoretical reasoning.

The same process would involve some of Keynes's collaborators, in particular Harrods, or contemporaries, as Hicks (Louçã 1999): precisely for the importance and urgency of a new economic theory that could be instrumental for fighting unemployment, a compromise with neoclassical formalism was justifiable, since it provided the sole communication platform ensuring simultaneously formal rigor and the capacity of computation of the desired effects of political options. But this implied supposing equilibrium as the theoretical condition to solve the empirical problem of disequilibrium—a contradiction that some Keynesians could not help noticing. This was the motivation for those Keynesians who anticipated the Samuelsonian "synthesis" and in particular the creation of the IS-LM program, for those who were part of the "socialist calculation" debate there was also a supplementary reason: the conviction that the maximization of social well-being could be obtained out of the market mechanisms, if conveniently directed. The *vexatio questio* was precisely the determination of the direction for maximization: that was the answer required from mathematical computation, through the estimation of the equations describing the society and its possible choices. This was the strategy of the Cowles Commission essentially when, under the leadership of Koopmans, it moved to Chicago and adopted an explicitly probabilistic approach.

Transferring its center from Europe to the United States, econometrics got closer to an idea of "market socialism" but also paradoxically to neoclassical economics (Freeman and Louçã 2001). The mystery of the vanishing of critical theory in economics lies in this Pyrrhic victory: after the adaptation of "bastard Keynesianism" and the slow transformation of "market socialists" of econometrics, the power of neoclassical economics was reinstituted.

MODERNIZATION AND MODERNISM

These were two processes of modernization. The first, as it was summarized in the previous pages, was originally propelled by the creation of neoclassical economics and then consolidated with the development of econometrics, reinforcing the empirical claims of the program. This modernization process is marked by the project of naturalization of economics as a form of "social physics," first under the First Law of Thermodynamics, then as a difficult synthesis between the First and the Second Law and the introduction of the concept of entropy, generating new tools that were formed under Quantum statistical mechanics, conceiving society as intrinsically stochastic.

This intellectual movement benefited from the intense participation of several economists or natural scientists immigrated into economics, for whom the priority was the definition of efficient policies against recession and unemployment. In short, the great modernist movements in economics during the twentieth century suicided in the altar of their priorities. In the name of urgency, they opted for the operational concepts they could understand and use, although these were originally translated from the equilibrium framework of conservation of energy and then from probabilistic statistics. Furthermore, the introduction of probability in economics was pregnant of an ambivalence provoking epistemological instability, since it reduced methodological individualism, the condition for these theoreticians for the metaphorization of the agents as atoms, to a senseless emptiness: this methodological individualism, indeed, could not recognize the peculiarities of any individual and, on the contrary, equated the abstract individual to an army of exactly identical clones. Computability demanded the loss of sense and reference in economics: one can compute, but compute nothing at all. In that sense, the concept of paradigmatic crisis, the theme of the "Discourse" by Sousa Santos, fully applies to economics.

The Sisyphean effort of these young mathematical economists becomes obvious and particularly dramatic, since they created a creature that soon escaped: "social physics" demanded an exhaustive determination and consequently generated a monotheism, aspired to an absolute efficacy and consequently eliminated methodological pluralism, required certainty and consequently ignored uncertainty which is intrinsic to social processes. Social physics was physics abdicating from its social being, in spite of the fact that at the same time physics itself was becoming more social. An exemplary effect of this process of epistemic emptying is the acceptance of the *fallacy of exogeneity*, a term coined by Samuel Bowles and developed by Boaventura: the markets are analyzed abstracting from their reality, and the scientist becomes incapable of discriminating the social nets, the relations, the endogeneity of preferences, and consequently radically decontextualizes the analysis of the social process (Santos 2000, 21).

This is the very reason for considering the second dominant process of modernization in economics during the second half of last century, which was a response to this epistemic instability. Its major protagonists were John von Neumann, Alain Turing, Kurt Godel, Claude Shannon, Merrill Flood, John Nash and Herbert Simon, among others.

This cyborg modernization was based on a new conceptual centrality, that of information. It was a contradictory movement: for Shannon to be able to propose a measure of information, it was previously necessary to impose the separation between quantification and qualification, between counting and semantics, since the ontological definition of the concept required the total indifferentiation of the contents of information. Yet the analysis of information potentially resocialized economics, emphasizing the irreversibility. This was a progress, but a limited one, since even if economics adopted this methodology it would still be far away from the analysis of economic agents with their diversity. The history of economics in the second half of the twentieth century is the history of debates about such a transformation.

This process of modernization rapidly achieved important successes. The creation of Operational Research (OR) in England from 1935, as a derivative of pieces of research for the air force, allowed for the quick expansion of this cyborg modernization. Game theory was the next step and the main application in economics, in spite of its severe limitations. One essential difficulty was still not addressed, that which derived from the assumption of the probabilistic nature of social processes. In fact, games are essentially probabilistic if they metaphorize social action, and this implies the adoption of a concept of intrinsic stochasticity—preventing the use of censorship against variation, which was based upon the conception of stochastic processes as noise (error-in-equation), or extrinsic stochasticity.

But games suppress the treatment of information and reduced the action of agents to a strict number of strategies, generally predefined by the program of the game. The concept of Nash equilibrium was consequently suggested as a redefinition of traditional neoclassical equilibrium, in a context in which the action of agents was formally representable. This was indeed the reason for its acceptance by neoclassical economists.

But the concept was strongly challenged for the very same reason. The mathematician Merril Flood, in particular, proposed in 1950 the to-be famous Prisoner's Dilemma: aware that the Nash equilibrium required both players to reject cooperation given the imposed structure of pay-offs, he experienced the game at Rand Corporation and concluded that would be the most implausible result. In fact, Flood asked two Rand colleagues, Almen Alchian, a neoclassical economist, and John Williams, a mathematician, to play sequential games: Alchian cooperated in 68 out of 100 and Williams in 78 out of 100, although the former assumed the rational attitude to be no-cooperation—as a good neoclassical, although his own attitude suggests he was sometimes a nonpractitioner—, whereas the latter assumed the rational behavior to be, on the contrary, to cooperate. Later on, large-scale experimentation and the concourse among alternative programs proved that adaptative behavior, *tit for tat*, would be more powerful than its rivals.[8]

John von Neumann was another voice against the Nash strategy for game theory. Von Neumann, one of the creators of OR and game theory, concentrated essentially in automata theory since he wanted to highlight the role of information, replacing strategies for rules: computation would be based on the development of programs.[9] Computational economics inherits from this movement, but we still are at the early days of this paradigmatic transformation.

The text by Boaventura Sousa Santos suggests a conclusion. Proposing the foundations of the emergent paradigm (p. 36ff.), the "paradigm of prudent knowledge for a decent life," the "Discourse" indicates four main principles orienting such a transformation: "1. All natural-scientific knowledge is social-scientific, 2. All knowledge is local and total, 3. All knowledge is self-knowledge, and 4. All scientific knowledge aims at transforming into common sense." This program contradicts the resistance of a social science such as economics, firmly bounded to the neoclassical dogma, even under the adaptive disguise of the concept of equilibrium following Nash. Therefore, economics of the first phase of modernization remained faithful to a paradigm of science formally metaphorized from nineteenth-century physics and restricting its operations to the available mathematical logic, and resisted the second wave of modernization only assuming formal concessions. What the "Discourse" suggests is, on the contrary, the need to reconstruct a social science based on realistic assumptions: prudent knowledge requires the knowledge of reality and not of fantasistic abstractions such as those of agents and markets in the neoclassical world. But agents and markets can only be understood if information is conceptualized as part of the process of interaction and coordination—and, as information sciences suggested in their cyborg logic, this requires economics to understand the emergence of order in a precise institutional and historic context, describing the complexity of societies.

The verification of this difficulty is also an invitation and an opportunity, since there is today a window of opportunity for change, the possibility of recovering modernisms whose contribution has been repressed in the history of economics and of making new alliances with cyborg sciences, bridging between the natural-scientific and the social-scientific knowledge. This is the avenue for reconstructing economics as a social science, and therefore as part of common sense.

NOTES

1. An old debate I sustain with Sousa Santos concerns the use of the categories of modernity and postmodernity. In short, I prefer to consider postmodernism in a

strict sense, that used for instance by Italo Calvino as he referred to that part of the literary imaginary, as the parodic process of "recycling the used images, inserted in a new context that modifies their sense. One may consider postmodernism as a tendency to use ironically the mass-media imaginary, or then to introduce the taste for the wonderful, inherited from the literary tradition, in the narrative mechanisms reinforcing its power of alienation" (Calvino 1989, 154). In the same sense, I share the main criticism by Fredric Jameson in relation to "really existing" postmodernism. Of course, Sousa Santos intends to create another postmodernist current, that of resistance, embodying an alternative program to that of ephemerization and superficialization of narratives. I adhere to the contents of this proposal, without considering to be forced to assimilate the name of his label.

2. Schumpeter anti-Semitic record is quite well known and was discussed by his biographers. Yet, it should be noted that Schumpeter strongly argued for the inclusion of Samuelson in the economic department of Harvard, against the anti-Jew prejudice of the director.

3. The computation of social processes modeled by neoclassical economics was simple: typically, the computation of the coordinates of the equilibrium solution of the reduced form of a system of simultaneous equations or the determination of the general solution of a difference or differential equation, if linear or linearized. But computation, by itself, could not be restricted to the condition of linearity and simplicity of interaction in social processes. In fact, since one assumes the non-independence of agents, patterns of relation emerge that are computable but not solvable. Some of the young econometricians, in particular Frisch but also Richard Goodwin, opened new avenues in that direction, although following different paths.

4. Frisch formally addressed the lack of coordination not only in this paper (1934), but also in a private debate with Tinbergen and Koopmans next year, during the Namur conference of the Econometric Society. On the details of this debate, Louçã 2000. This example is telling since it represented an attempt to represent in a nonlinear framework the structure of interaction of the different agents. But that was too much for the prevailing mathematical techniques.

5. Although having contributed to the definition of a sort of Keynesian policy, with his "oil brake principle" of anti-cyclical policy, Frisch suspected the viability of the steering control mechanisms, since he believed that no parametric stability existed and therefore that there was no basis for prediction.

6. Perhaps because he distrusted the methods imported from physics, Keynes suspected the econometricians' work and expressed his dissent in the review and debate about Tinbergen's books for the League of Nations. On the other hand, Keynes was not motivated by the need to impose a new methodology for economics and to give neoclassical theories the empirical content econometrics could provide, since he had already discussed with Pigou the problem of unemployment and the liberal proposals, and his own "General Theory" was a critique of that vision.

7. Frisch's proposals for this new line of economic policy were not any more published in *Econometrica*, the journal he had edited for two decades. The "Oslo Refi Interflow Table" was published in the *Bulletin de l'Institut International de Statistique*, after it was presented to a Paris meeting of that Institute in 1961. The same happened to other contributions of the type.

8. As game theory simulates conflict and it was largely financed by the military institutions, some influential economists, in particular from Michigan University, rejected this approach and proposed an alternative vision, which became experimental economics: those were the cases of Anatol Rapoport and Robert Axelrod.

9. This cyborg science is today represented for instance by Wolfram (2002) and suggests the reorganization of the hierarchy and frontiers of sciences.

REFERENCES

Andvig, J. C. 1985. *The Intellectual Background to Early Post World War, II. Economic Policy Regimes in Norway*. Oslo: NUPI, mimeo.

Boumans, Marcel. 1992. "A Case of Limited Physics Transfer—Jan Tinbergen's Resources for Re-shaping Economics." Amsterdam: Thesis Publishers.

Calvino, Italo.1989. "Leçons Américaines." Paris: Gallimard.

Cowles Commission. 1932. *Econometrics—Towards Making Economics a More Exact Science*. Colorado Springs, CO: Cowles Commission.

Freeman, Christopher, and Louçã, Francisco. 2001. *As Time Goes By*. Oxford: Oxford University Press.

Frisch, Ragnar. 1934. "Circulation Planning." *Econometrica* 2:258–336.

———. 1963. "Selection and Implementation—The Econometrics of the Future." In *Pontificiae Academia Scientiarum Scripta Varia* 28:1197–204 and discussion 1205–32; reprinted as "Economic Planning and the Role of Econometrics." *Statsokonomisk Tidsskrift* 1(1964):1–7.

———. 1970. "Econometrics and the World of Today." In *Induction, Growth and Trade: Essays in Honour of Sir Roy Harrod*, ed. W. Eltis, M. Scott, and N. Wolfe, Oxford: Clarendon, 152–66.

Harrod, Roy. 1951. *The Life of John Maynard Keynes*. London: MacMillan.

Louçã, Francisco. 1999. "The Econometric Challenge to Keynes: Arguments and Contradictions in the Early Debates about a Late Issue." *The European Journal of the History of Economic Thought* 6 (3):404–38.

———. 2000. "Is Capitalism Doomed? A Nobel Discussion." In *Why Is Not Economics an Evolutionary Science?* ed. Francisco Louçã and Mark Perlman. Cheltenham: Edward Elgar, 138–52.

———. 2001. "Intriguing Pendula—Founding Metaphors in the Analysis of Economic Fluctuations." *Cambridge Journal of Economics* 25 (1):25–55.

Mirowski, Philip. 2002. *Machine Dreams*. Cambridge: Cambridge University Press.

Moggridge, Daniel. 1992. *John Maynard Keynes—An Economist's Biography*. London: Routledge.

Sousa Santos, Boaventura. 2000. "Towards an Epistemology of Blindness: Why the New Forms of 'Ceremonial Adequacy' Neither Regulate Nor Emancipate?" Coimbra: CES.

Skidelsky, Robert. 1992. *John Maynard Keynes—The Economist as a Saviour, 1920–1937*. London: MacMillan.

Wolfram, Stephen. 2002. *A New Kind of Science*. Champaign: Wolfram Media.

16

An Invitation to a Science War

Shiv Visvanathan

I

The Norwegian sociologist Johan Galtung once talked of a painting that hung in the anteroom of the late Ghanaian leader Kwame Nkrumah. It was a giant picture inevitably of Nkrumah himself struggling loose from his chains. There is thunder and lighting in the air and in one corner of the picture are three men, three white men. The first is the capitalist and he carries a briefcase. The second figure is a missionary and he clutches a Bible. There is a third figure at the remotest corner of the picture. He is holding on to a book whose title can be barely discerned. It is a book entitled *African Political Systems* and third figure is the scientist as the colonial anthropologist.

The iconography of the picture used to fascinate me. It represented the power and fury of nationalism battling against the depredations of capitalism, missionizing religion and colonial science. I would begin my introductory classes in political sociology with a discussion of this picture. Yet looking back, it appears mildewed, embarrassing and outdated. Nationalism, which was a liberating movement, has graduated into the national security state, which is often genocidal. Development, which was meant to create justice and equality, has virtually become a continuation by war by other means against marginals, tribals and peasants in the Third World. Science, which was seen as a magic wand against poverty, was behaving counter intuitively. Modern democracy appeared to be a social contract between science and the nation state to guarantee security and development. It is this new social contract between science, development and the nation state that creates the other science wars that few philosophers talk about.

One needs the passionate scholarship of Ivan Illich to grasp it.[1] Unfortunately the story of the science wars is read today as trumpet sound from *Social Text,* which has brought down the Jericho of social science. If positivism was a quarrel between two university dons, Alan Sokal's article could be considered a superb joke, the work of an almost Jungian prankster. One can add one's laughter and drink to it, even dine out on the story. If it is a spoof on science as a *system* it is a welcome. But the same joke sounds poor when seen from the lifeworld of a tribal, a marginal or a slum. Jokes from the system are not equivalent to laughter in a lifeworld. If the trickster in Sokol is ready to perform a pilgrimage to the lifeworld, a conversation of a different sort could begin. These are notes from one such journey.

I wrote the original jottings for economists and activists talking about sustainable development from a World Bank perspective. I remember at that time we had made a set of pleas. I recollect there were four in all. We had requested (i) that human rights teams be attached to every development project or at least an ombudsman for each technology (ii) that an audit of each development effort be provided in the language and categories of the people involved or affected. Quality of Life (QOL) indicators are adequate but one needs to experience them in the vernacular of a people. The idea of the good life needs its own idiom of rumor and gossip (iii) we pleaded that the scientific epistemology of the project be understood in terms of local theories of knowledge. It was a hint that empowerment in terms of people's voice alone was inadequate. A people's epistemology is as central as people's participation in any discourse on democracy (iv). Finally we had pleaded that one needs some form of insurance, security against development projects. Of course we realized that our pleas were part futile, part utopian, a playful wager against the odds. So we finally suggested as a Sokalian joke that a wailing wall should be instituted in the World Bank office so that we could mourn or grieve together in the aftermath of some projects. No doubt pathologists reports are scientific but one also needs the catharsis of mourning rituals. One mourns over the body of a friend, one performs an autopsy on a corpse.[2]

How do the science wars look from the history of marginality in the third world? This is not to posit the ethnocentrism of the Anglo-Saxon world against the parochialism of the third world. But one has to realize that epistemology is not a remote exotic term. It determines life chances. Science as development, plan, experiment, pedagogy determines the life chances of a variety of people. Here epistemology is politics. The positivist-antipositivist wars need a larger theatre. It is from such a site that one must issue an invitation to a different science war to Sokal and others. Sokal's epigoni have fought a few mild encounters in India to produce pointed but misguided readings of critics of science as sources of communalism but unlike Jericho or *Social Text,* no walls fell. Such readings tend to see fundamentalism and science as incommensurable words without realizing that the specter haunting India is a variant of techno fundamentalism, which hybridizes science with the most blood thirsty of communalism. These notes are one set of reflections on the other science wars. But to understand this we must create a certain relation of map to territory and particularly between science and democracy in discussion issues like development.

In understanding science, development and technology, one feels the same sense of horror and surprise that Hannah Arendt did when she was reporting of the Eichman Trial in Jerusalem[3]. Arendt found it difficult to reconcile the sheer ordinariness of the man with the enormity of his crimes. To answer her nagging question she formulated the idea of the banality of evil, the idea that violence does not require evil, all it requires is normalcy, bureaucratic expertise and a clerical project. The scientific equivalent of a project is a research program and we would like to suggest that the scientist in pursuit a normal science is also capable of radical evil.

The works of Ivan Illich, Vandana Shiva, Ashis Nandy or Boaventura Santos are challenges to the normal science of today[4]. What they introduce is both resistance and the carnivalesque suggesting that a science that diversifies its paradigms attempts to minimize the hegemony of one of them. For instance, Boaventura Santos seminal essay is not only an exercise in intellectual hospitality but also a suggestion that it is not good fences that make good neighbors but good conversation and concepts, which are, disciplined trespassers[5]. Boaventura Santos' suggestion that the flow of concepts across the social and natural sciences has rendered the demon between them a creative gatekeeper. In fact, it has created a powerful hybridity of social science concepts like complexity, uncertainty, risk, and populations seeding the sciences to create a domain that challenges the standard hierarchy that Thomas Kuhn and every modern university impose on the natural sciences, social sciences and humanities. Boaventura Santos work is still the work of a scholar talking to other scholars in a research community. But there is a wider audience and in fact another world of participants that gives it a different meaning. To enter this world, we have to understand the civics of the modern world whose political core is the idea of the transfer of technology.

II

The idea of technology transfer is one of the most quotidian ideas of governance regarding technology. It nestles peacefully in World Bank Reports, Population Council monographs and NGO project reports. In an ordinary sense it captures the hegemony and violence of everyday life in a policy world. John Maynard Keynes claimed years ago that behind every dictator and his policies is a book or a philosopher, the Tyrant had read years ago.

Arendt adds that behind every technocrat and a tyrant is the banality of a project. What we wish to capture is how grassroots movements as philosophers of science have attempted to counter them at the levels of knowledge and justice. One must emphasize that particularly in India the critique of science arose not from university but from human rights activists, ecologists, and feminists fighting a battle against the fundamentalism of official science represented by technology transfer.[6]

The innovation chain as a structured sequence is an attempt to link science, technology and society through the sequences of invention, innovation and development. To quote textbook officialese, invention represents the creation of a scientific idea and its initial visualization as a technological product; innovation, the upscaling and commercialization of an idea, and diffusion its eventual absorption or distribution into the wider society. One might emphasize that the idea of the innovation is not only a research management discourse but one on development and democracy. In a geographical sense, science originates at the center and development occurs at the periphery. It is also a technological vision of democracy where the science is legitimized by experts but consumed by citizens. Participation, local materials and visions appear at the later phases of the innovation chain. The civics of TOT represents a policy map, a model of hegemony, a vision of knowledge and a metaphor for democracy. Science and democracy play out their repertoire of possibilities within an innovation chain. There is a sense of preempted futures here because in official visions the alternative to development is not alternative development but museumization and marginalization.

In terms of a grassroots critique, the model of technology transfer can be read at two angles. The first is dominated by the history and philosophy of science and is basically an internalist historiography, where we wrestle with Kuhn, Popper or Lakatos. Supplementing them within the same frame is the sociology or political economy of science, which emphasizes issues of professionalism, defense pressures or the role of multinationals. But we rarely find a book, which moves from the creativity of the science to the studies of innovation and development dominated by science policy experts, economists of development, or historians of technology. There is a tacit division of labor here that we must ignore. One can't see the first as creativity and the second as productivity. As Steve Fuller demonstrates behind every Thomas Kuhn is a James Conant that Kuhn only transformed and banalized into science what were rules of thumb for cold war politics and strategy[7]. Rules of hygiene prevent an understanding of the connection between the two. Fuller showed that the works of these science-politicians made science safe *for* and *from* democracy. This becomes more apparent in a pedagogic text or transfer of technology model. In these models science is still a "club" driven exercise. What is democratized is not the critique of science and technology but the consumption or distribution of science.

The first thing grass roots groups in India did was to merge the two empires of knowledge. They also helped provide a democratic framework for science by insisting and challenging the fundamental oppositions between (i) lay and expert or science and the mob and (ii) between science and ethnoscience read either as local knowledge or dismissed as folklore or superstition. The science movements performed a brilliant maneuver. They took the science wars across the innovation chain tweaking playfully not only the hierarchies of knowledge within science but the boundary management between science and other knowledges. They insisted that not only was science a new social science but they challenged the dualism of observer and observed in profound ways. Figure 16.1 charts the relation between map and territory for our discussion. In the following section we shall explore the critique of science presented by the science movements.

III

One must mention that science in India began as a positivist celebration where Indian scientists like Meghnad Saha literally dreamt of a society based on the scientific method.[8] India was a society as proud of its sample surveys and its science policy as it was of its flag. A society that dreamt of its laboratories and dams as the new temples of modern India witnessed the

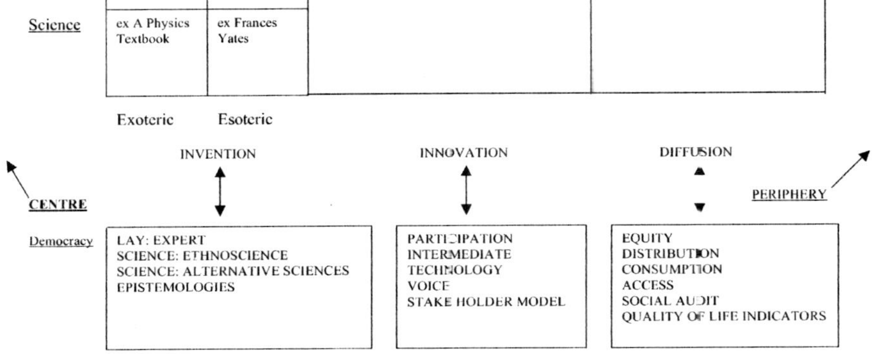

Figure 16.1. Innovation chain

fact that the sacrilege began with science. We faced a set of science projects which were difficult to understand within a positivist science or the old dualism between good science/ bad science policy. We had to face the following facts:

1. Our dams had produced not only energy but an ethnicity of over 40 million refugees. One needs a common base to audit development and displacement within the same discourse.[9]
2. Our scientific forest acts threatened both the animals and people creating a debate between scientific and social forestry threatening one-seventh of our population.
3. The Bhopal gas tragedy created thousands of victims who were subject to the scientific gaze but received neither compensation nor healing.
4. The Green Revolution produced the paradox of an official India self-sufficient in food while increasing the salinity of our soils and decreasing the diversity of our agriculture.[10]
5. The only dictatorship India had was between 1975 and 1977 and it was justified in terms of scientific metaphors which legitimitized compulsory sterilization in family planning and forced demolitions in the name of scientific urban planning.
6. India today is a nuclear state where science has driven the decisions to nuclearization. What is troubling is not just nuclearization but the terms of discourse within which scientific debates are carried out.

The questions the grassroots movements in science had to face as a philosophical and political conundrum was whether the above crises were because of bad science, bad politics and bad technology or was the problem also inherent in the logic of science and technology. This political drama in what I called the politics of knowledge took place at three levels.

1. What are the rules for a scientific controversy in a democracy which includes tribals, marginal fishing groups, shifting cultivators, slums, industrial refugees and a middle class demographically the size of Europe? How does one make decisions about S & T in a society, which is undergoing the first, second, and third industrial revolutions simultaneously?
2. How does one frame the interaction between science and democracy, which looks systematically at the "scienticization" of democracy and "democratization" of science? How does one phenomenologically bracket them so that one can reexamine these taken for granted worlds?
3. How does one create a framework of controversy which neither economizes science by instrumentalizing it or reducing it to a case of a bat-

tle between scientific fundamentalism (positivism and reductionism) and religious fundamentalism. In the pursuit of this thought experiment what concepts do we need which goes beyond rights, cost-benefit, objectivity and efficiency? What one hopes to present is the framework and the repertoire of tactics and concepts generated. One realizes that a wide variety of movements are clubbed under the same rubric but one hopes to capture the unity of this great parliament of science where civil society particularly grassroots groups and dissenting academics built a more democratic framework for science. What emerges is not only a great exercise in democratic theory but a contribution to the philosophy and history of science.

IV

The initial critical moves emphasized the latter half of the innovation chain. The first major critiques came both from the Bernalians within the State and from left movements like the Kerala Sastra Sahitya Parishad (KSSP)[11] and the Delhi science Forum. Their dreams of democracy were still diffusionist. It was a dream of taking science to the villages. What was invented was the idea of the scientific temper, a pedagogic vision that a scientific worldview could be induced in a people. Unfortunately the radical scientist often visualized the scientific temper as an intellectual vaccine that could eliminate superstition, magic, religion fundamentalism. The left movements carried this same scientific view through science quizzes celebrating Newton, Bernal and Darwin. Here civil society and the progressive state shared a common vision of a positivist science. But in the later debates there came a split between the science policy of the state and the critiques of science by civil society. It was a science war, which emphasized that the citizenship provided by the new social contract was inadequate because it was a citizenship based on industrial time, which saw the citizen as a consumer and not an inventor of knowledge. It also realized that both science in India and the Indian constitution were disembedded knowledges.

The critique of science began as a human rights problem because projects on development either marginalized or cannibalized the culture of tribes, slums or the peasantry. The standard notion of human rights did not work because while it was adequate at the level of the individual it was unavailable at the level of the group. Secondly what one needed was a science that realized that nature was not just an object of an experiment or a resource but part of a way of life. As Tom Kocherry leader of the Kerala Fishers Forum claimed "70% of India depends on nature for its livelihood." Nature was thus not only a mode of production but a mode of thought. The movements realized that there were few life affirming notions of nature within science. The notion of wilderness used in American ecology was inadequate because for the American the wilderness was an unpopulated monument. One needed something beyond the American dialectic of wilderness and frontier or the British obsession with gardens. The worldview of the Bishnois,[12] the Chipko[13] movement came from their religious cosmology. It was not antiscience but only a critique of a statist science, which saw the pulp and paper industry as more eligible citizens than the tribal foraging for food and medicines. In the new model of development as an enclosure movement not only were tribals and marginal peasants displaced but rendered illegal. What was destroyed was not only the forest but a commons of knowledge about trees, fodder, forest produce, seeds, medicines, building. The commons was not merely a resource pool but a way of life that sustained a way of knowledge.

The movements were confronted with two facts. Firstly, the idea of rights was adequate for torture but helpless against science-induced displacement, obsolescence or even genocide. Secondly, they realized that in the battle called development the idea of nature itself had changed. It was not just farm, fish and fowl, it was also hybridized with technology. The citizen lives simultaneously in a natural, technological, biotechnological and information environment.[14] One had to meet these different hybridities simultaneously. One needs not only a new ethic for nature in science but a new ethic of technology. They also sensed the iatrogenic nature of science policy created particularly by the reductionist nature of scientific expertise. But the answer to it was not Luddism. The modern Luddite cannot smash the abstract machine. He can only rework the classifications behind abstract thought. He must turn futurologist, epistemologist and constitutionalist and he must also realize that the new politics of science is created by dissenting imaginations within and alternative imaginations without. A critique of science as an ongoing exercise cannot be located in fundamentalisms but in competing and reciprocal criticalities. For every Shiva or a Medha Patkar, there is a Chipko and a Narmada movement. Further there is no one construction of Chipko, Narmada or Balliapal anymore than there can be one master narrative of science. The power of the movements lay in the fact that they realized that politics is not just a protest against a dam or a forest bill. It must extend a challenge to official narratives of science and to the epistemologies that underlie it. Or to put it bluntly how do nonviolent movements search for a nonviolent science?

The movements realized that the Politics of Time was crucial at three levels of science. There was first the politics of the history of science. There was secondly the politics of memory and finally the politics of multiple time.

The history of science has always constructed itself as a rational cumulative continuous exercise. Science as an exoteric internalist narrative constructs itself in linear and progressive time. Science is conducted in victorious time, which has no place for defeated knowledges. While science deals with a diversity of times from mechanical, historical, evolutionary and quantum times (nanosecond), its own narratives are constructed in the impoverished time of unilinear narratives. For the movements, science fails as a narrative and as an act of story telling and yet they realize that it is the very unilinearity of time that provides its cognitive power. As Kuhn remarked the textbook as a reflection of a cognitive regime rewrites histories where defeated or alternative hypothesis have little place[15].

The politics of memory is a close corollary to the first because the progressive rhetoric of science is an amnesiacal one. It museumizes other forms of knowledge in the name of progress. It also renders obsolescent ways of life, which are abandoned because of the changing nature of technology. The innovations of science take place in standardized time. Science understands the grammar of progress but not the logic of obsolescence as a lived world. The paradigm as a monoparadigmatic space comes with an indifference about certain forms of time. Within the innovation chain, Socrates becomes a Schumpeterian idiot.

Democracy needs a multiplicity of times. A tribal performing shifting cultivation operates in a world of over twenty different kinds of time which emanates from the way he deals with soil, seed, seasons, rituals, fast, feast, rest, work, domestic and communal space. Farmers, women, patients, tribals live in a variety of times, which they need access to and which science denies them. It is within this context that ecology is as vital to science as quantum physics. What ecology smuggles into science is a notion of memory as a thesaurus of times. What the movements emphasized is that a democracy based on standard factory time is literally an oxymoron. At this point one must emphasize the difference and overlap between the different politics of memory. Ian Hacking talks of three forms of Foucaultian politics. The first dealt with the politics of the body, the second with the politics of populations. Hacking adds that the third form of hegemony is the politics of memory as an act of scienticization. But the politics of the second idea of memory deals with a liberation from history as the only form of memory with a plurality of times. The trouble with the official idea of sustainability is that it lacks such a repertoire of times.

Once the framework of multiple time is established, one challenges the abstractness of science. One begins with the great dualism of observer and observed, the subject-object dichotomy so deeply entrenched in science. Science as the movements and the dissenting academic suggest is not merely an object of production created through the optical gaze of the enlightenment but a subject of consumption and validation. Secondly the tacit division of labor between an expert who produces knowledge and a citizen who consumes it has to be rendered less asymmetrical by understanding the citizen as a person of knowledge. The worker, the peasant, the craftsmen are citizens of knowledge about science. Secondly this understanding cannot be devalued as "ethnoscience" while expert understanding is "philosophy of science." Such a hierarchy or devaluation creates the possibility of the museumization or appropriation of these other knowledges. Strangely even at a time when science is appropriating and patenting peasant knowledges, there is no epistemic acknowledgment of their status. Science begins a form of strip-mining where knowledge about local drugs, therapeutics, soils, and seeds is abstracted without considering the philosophies they are embedded in. Thirdly given the tendency of science and technological projects toward displacement, obsolescence, erasure, ecocide, the movements believe that the externalist idea of community involvement, participation and use of local materials is not enough. These are externalist measures. Even the subaltern emphasis on "Voice" is a trifle sentimental. These become mere epicycles that the scientific panopticon throws out to humanize itself. They do not touch the normal science of a discipline. The "movements" understand that participation does not constitute an epistemic challenge. It can merely add to the popularization of science or to an increasing awareness about decision-making. But the politics of an epistemic challenge requires a different set of constitutional or legislative guarantees. Probably the best way to explain it is by relating an anecdote.

A few months back representatives of what is now called denotified tribes—tribes which were once classified as criminal by the British—came to meet us. These tribes, which are over a few million strong, face a devastating medical situation. The face of the sickle cell anemia problem of which they have little understanding. They also face mental trauma from the everydayness of police torture and harassment. Gastroentitis, alcoholism, witchcraft are also widely prevalent. The community goes to the government primary health centers and to the herbal doctors. Their representatives had come with a suggestion. What they wanted was a dialogue, a seminar where patients, victims, medical practitioners of various persuasions, public health specialists, psychiatrists, human rights activists met and listened to their testimonies of health, illness and well-being. What they hoped would emerge from it was a new kind of health policy which they would present to the local legislature. They wanted a dialogic health policy which saw health and development together not in the language of expertise but in terms of a new notion of health politics.

What they wanted is not just participation. They wanted *presence* of two kinds. First, cognitive representation. They wished that when a policy was made the tribal doctor and the western health expert were both present and that the policy should represent the language of this dialogue. Such a presence goes beyond participation and ethnoscience to cognitive empowerment. The political economy and ecology of the tribal situation demands access to a variety of medical systems. Plurality was necessary for true access to a system because one can't have choice without alternatives. What one needs is the idea of cognitive justice, which is the constitutional right of different systems of knowledge to exist dialogue and debate. The idea of cognitive justice believes there is a link between survival and forms of knowledge. It includes not only the rights of dissenting scientists within a dominant paradigm but also the rights of alternative epistemologies and alternative sciences. The debate on alternative sciences cannot be exhausted by Lysenkoism and the racialism of Nazi science. Alternative sciences have existed long before as traditional agriculture or alternative medical systems. One does not need a Needham or a Feyerabend or a Charles Leslie to testify to it. A plea for cognitive justice also establishes the understanding that democracy within knowledge is crucial. The opposition of expert to laymen disguises to a certain extent the opposition between science and alternative sciences. One needs a parliament of epistemic debates but also the ecologies that would let these forms of knowledge survive and thrive not in a preservationist sense but as active practices. The idea of cognitive justice visualizes a commons of knowledge that citizens especially in subsistence cultures have access to both as consumers, critics, practitioners and philosophers. As a tactic it renders irrelevant the hierarchical contempt implicit in the notion of the ethnoscience as a lesser life form or model of being. The Levi-Straussian tricolor is not quite the scientist.

One now comes to the relation between the observer and the observed. The idea is based on what Arendt and Robert Romanyshyn elaborate as the astronautical self.[16] The astronautical self is disembedded self, which distances itself from the subject. The astronautical gaze grows, as Romanyshyn points out, into the anatomical gaze and two together create both objectivity and distance. But science as disembedded knowledge needs a different ecology as it travels from laboratory to the world. One understands the 'as if' heuristics of objectivity. But science in a grassroots sociology has, to use Habermas' terms, be both system and lifeworld. Objectivity and the astronautical self might be adequate for the heuristics of the system but life in a "lifeworld" has to be conceived as a ganglion of different connections. One needs a parallel methodology for a cognitive audit of science, a methodology, which is not after the fact as in cost benefit analysis, but one, which deals with epistemic connections in a wider sense. One believes that it is in this context that Linus Pauling, the great biochemist, once talked of an axiomatics of suffering for science beyond the felicific calculus of the Utilitarians. In search of such a frame the grassroots movement within a subsistence frame would talk not of an abstract relationship between observer and a silent observed. What is offered is a relationship of sensitivities, connections, and reciprocities between

System

Life world

Life — Livelihood

Life chances

Life style

Life cycle

Figure 16.2.

One also broadens simultaneously the idea of the observer to include a variety of "sightings" that go beyond the optical gaze of the observer. So the observer pluralizes beyond observer/observed into:

1. spectator
2. flaneur
3. consumer
4. victim/survivor/patient
5. witness
6. site

The above six forms are also what the critic John Berger would call "ways of looking." But the spectator unlike the observer watches the scientific game without paying it. The flaneur as the interested stranger consumes an object in delight but always floats away or around it. For the spectator and the flaneur, the object is still a spectacle. The consumer of science also participates in the scientific act and he is not just the individual citizen. The consuming body digesting science includes the state and the corporation. The consumer celebrates science but the victim has a more asymmetrical relation to it. The voice of the victim and his view might differ from the voice of the expert. The plurality of the new democratic ear/eye requires the viewpoint of the witness. The witness unlike the observer and the spectator lives out his testimony. The witness is not a neutral observer. He is marked by the lifeworld he has seen and must bear witness to. In interacting with the above roles, the scientist himself might realize that he has to be a witness. In fact he can become what one can dub a site, someone who acts and is acted upon. He experiments along with others and upon himself. Such a double embeddedness of science creates a different philosophical picture because the scientist no longer sees science as disembodied and disembedded. (See figure 16.3.)

Production/consumption, creativity/criticism come together to pluralize the intensity of the scientific gaze. When the iatrogeny of the scientific eye or the policy maker confronts the victim, witness or patient, a different kind of dialogicity is born in a heuristical, methodological and political sense. What the movements and their critique of science sought was not an anti-science but a plural vision which allowed for both the wisdom of normal

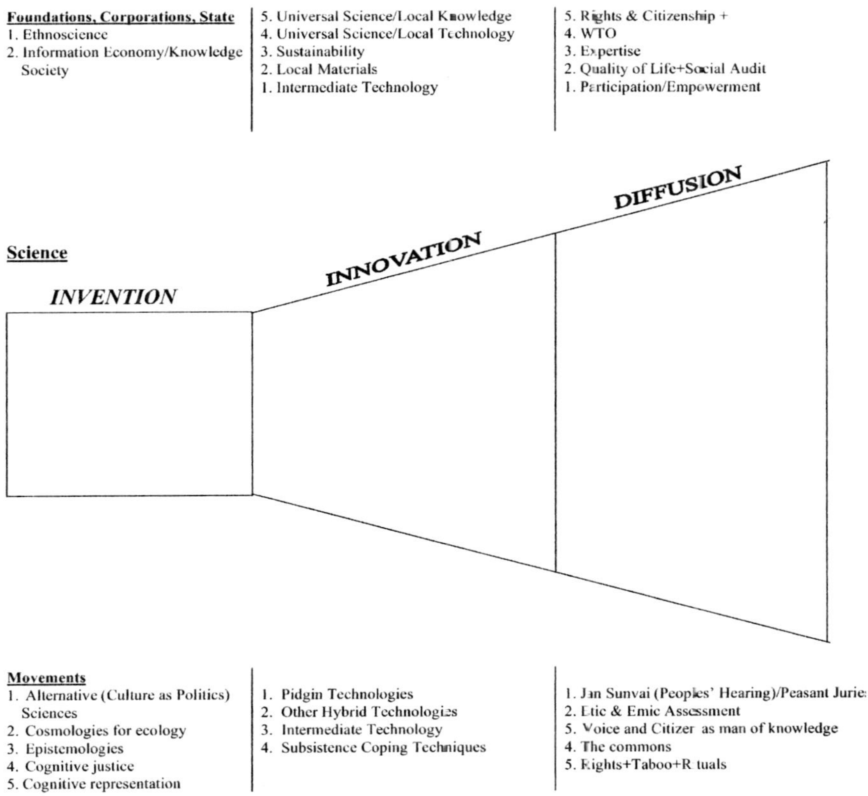

Figure 16.3.

science and the vision of the eccentric, the dissenter, the marginal, the vulnerable and the alternative worldviews. The playfulness, the new concepts and new reciprocity between science and social created through novelty by combination was extended by grassroots politics and philosophers to a wider domain. They forced the democratic imagination to contend with science by inventing new methodologies beyond boy-scout calls for participation and empowerment. A Boaventura Santos would be at home in this world. One wonders whether a Popper or Kuhn would have approved. Programs for open societies offer science as an image and model for democracy but they do little to add to the democratization of science or its imagination. A science *a la* James Conant that seeks only consumer or citizens approval is a disguised demagoguery that will work against the grassroots innovators.

The wisdom of the movements also forces them to be taxonomists who must anthropologically examine the classificatory structures of science. The hegemony of science is embodied in the classificatory act, which empowers the natural sciences over the social sciences, science over other knowledges and till recently, physics over the other sciences. Reductionism as a creative act produced the Green Revolution, the success of biophysics and genetic engineering celebrated by Monod in his *Chance and Necessity*.[17] Reductionism is both empowerment and disempowerment. One can have a science that celebrates both. But beyond this and/or logic, we need a different sense of part and whole. Science was based on a theory of levels where each level was more complex and superior to the one below it. But classification must go beyond the mere relation of the encompassing and the encompassed. One needs a new relation of parts to wholes within the science. One has to move beyond the dualism of reductionism/holism, or mechanism/vitalism. What follows is a different idea of parts and wholes within one science conceived as a set of relations. For example, information science could be a composite of the following relations where each contributes the whole but does not supplant the other. Let us put them in a tabular form.

The idea is to playfully open up a domain and visualize it not a social or natural science but a human science in a way similar or what Georgesceau-Roegen visualized for energy in the *Entropy Law and the Economic Process*.[18]

Information Sciences	
Cause-effect	Mechanical
Stimulus-response	Behaviorism
Sign-signal-feedback	Ethology / cybernetics
Sign-symptom	Medicine
Sign-symbol	Semiotic
Sign-symbol-meaning	Hermeneutics

VI

I realize that what I might have presented might appear like a catechism, but one hopes it is dialogic. The chapter is an attempt to show that the works of dissenting philosophers of science needs to be celebrated critically and not be subject to new forms of political correctness or to new kinds of McCarthyism. The dissenting academics and the movements are celebrations of knowledge. One has cited the American poet Robert Frost often in this chapter and so one would like to end by inviting the defenders of positivism to celebrate these new critiques not as revivalist discourses, which some could be, but as Frost would call " a lovers quarrel" with the world of modern western science.

NOTES

1. Ivan Illich, see particularly *In the Mirror of the Past,*1992, Lectures and Addresses, New York: Marion Boyers.

2. See Robert D. Romanyshyn, *Technology, Symptom and Dream,* 1989, New York: Routledge.

3. Hannah Arendt, *Eichman in Jerusalem,* 1965, Hardmondsworth: Penguin.

4. See Ashis Nandy, *Alternative Sciences,* 1980, New Delhi: Allied Publishers, and Vandana Shiva, *The Violence of the Green Revolution,* 1989, Dehradun: printed by author.

5. Boaventura de Sousa Santos, "A Discourse on the Sciences," *Review,* Vol. 15, No. 1. Winter 1992, pp. 9–48.

6. Shiv Visvanathan, "Transfer of Technology," *International Encyclopedia of Social and Behavioral Sciences,* 2002, Elseiver.

7. Steve Fuller, *Thomas Kuhn: A Philosophical History of our Times* 2000, Chicago: University of Chicago Press.

8. Shiv Visvanathan, *Organising for Science, The Making of an individual research laboratory,* 1984, New Delhi: Oxford University Press.

9. S. Parsuraman, and P. V. Unnikrishnan, *India Disaster Report: Toward a Policy Initiative,* 2000, Delhi: Oxford University Press.

10. See Shiva 1989.

11. The Kerala Sastra Sahitya Parishad (the Kerala literary and scientific association) was a forum established by Malyali science writers. It is today a movement within over 40,000 members dedicated to the popularization of science and to utilizing science for progressive ends.

12. The Bishnois are followers of the fifteenth-century saint, Guru Jambeshwar, who instructed his followers to protect plant and animal life; a cattle rearing and agricultural community, the Bishnois do not allow hunting or felling on their land.

13. The Chipko movement arose as a protest against logging abuses in the state of U.P. in India. The word "Chipko" literally means "embrace." It derives from the nonviolent practice of women who hugged trees, virtually interposing themselves between the tree and the contractors' uses.

14. See M. Whiteside, *Divided Natures,* 2002, Cambridge, MA: MIT Press.

15. Thomas Kuhn, *The Structure of Scientific Revolutions,* 1970, Chicago: University of Chicago Press.

16. Hannah Arendt, *The Human Condition,* 1958, London: University of Chicago Press.

17. Jacques Monod, *Chance and Necessity,* 1972, London: William Collin.

18. Nicholas Georgescu Roegen, *The Entropy Law and the Economic Process,* 1971, Cambridge, MA: Harvard University Press.

REFERENCES

Arendt, Hannah. 1958. *The Human Condition.* London: University of Chicago Press.

——. 1965. *Eichman in Jerusalem.* Hardmondsworth: Penguin.

Fuller, Steve. 2000. *Thomas Kuhn: A Philosophical History of our Times.* Chicago: University of Chicago Press.

Illich, Ivan. 1992. *In the Mirror of the Past.* New York: Marion Boyers.

Kuhn, Thomas. 1970. *The Structure of Scientific Revolutions.* Chicago: University of Chicago Press.

Monod, Jacques. 1972. *Chance and Necessity.* London: William Collin.

Nandy, Ashis. 1980. *Alternative Sciences.* New Delhi: Allied Publishers.

Parsuraman, S.; Unnikishnan, P. V. 2000. *India Disaster Report: Towards a Policy Initiative.* New Delhi: Oxford University Press.

Roegen, Nicholas Georgescu. 1971. *The Entropy Law and the Economic Process.* Cambridge, Massachusetts: Harvard University Press.

Romanyshyn, Robert D. 1989. *Technology, Symptom and Dream.* New York: Routledge.

Santos, Boaventura de Sousa. 1992. "A Discourse on the Sciences," Review, 15(1), Winter 1992, 9–48.

Shiva, Vandana. 1989. *The Violence of the Green Revolution.* Dehradun: Author Edition.

Visvanathan, Shiv. 1984. *Organising for Science, The Making of an individual research laboratory.* New Delhi: Oxford University Press.

——. 2002. "Transfer of Technology," International encyclopedia of Social and Behavioral Sciences. Elseiver.

Whiteside, M. 2002. *Divided Natures.* Cambridge: MIT Press.

17

Subjects or Objects of Knowledge?

International Consultancy and the Production of Knowledge

Maria Paula Meneses

INTRODUCTION

In "A Discourse on the Sciences" (Chapter 1), Santos offers a critique of the paradigm of modern science, envisaging the emergence of an alternative paradigm in which science forms part of a much broader constellation of knowledge, based on the assumption that all forms of knowledge, including the scientific, are context-bound.

The epistemological challenge provided by Santos, which identifies a universal constellation of knowledges based on the principle that each individual form of knowledge is incomplete in itself (1995, 340), is the basic presupposition used here to analyze and understand the problems facing the majority of scientific policies and practices in Mozambique. The case of consultancies is analyzed as an example of how the binary division between "North" and "South,"[1] donor and recipient, development and underdevelopment, and modern science and local knowledge is perpetuated from the point of view of the activities of a "new" category of social actors affecting the ways in which science is produced in the global South—the consultants.[2]

1. THE COLONIAL AND POSTCOLONIAL PRODUCTION OF UNEQUAL KNOWLEDGE

One of the "classic" dichotomies of modern science, especially in the social sciences, establishes a permanent opposition between "traditional" societies—which are considered remnants of the past—and "modernization"—the immediate source of progress—introduced by the colonial process. To create this dichotomy it was necessary to transform the past into a state of stagnation, reflected in underdevelopment, obscurantist practices, eternal tribal conflicts, etc. In short, the noncivilized Mozambican emerged, with colonial science, as part of the world of nature, which it was imperative to "cultivate" and transform.

The establishment of a new model of scientific rationality, as Santos states, was imposed as a "*totalitarian model, to the extent to which it denied the rational nature of all forms of knowledge that were not governed by its epistemological principles and its methodological rules*" (1999, 10). Indeed, for modern science it is important to spread scientific knowledge, assumed to be the culminating point in the process of "domesticating nature"; consequently, any other explanation is only tolerated as subaltern, local knowledge that can potentially be incorporated into this supreme and universal "corpus" of logic.

In destroying the local subject, it also annihilated its culture and its forms of knowledge, inculcating them with the "true" wisdom of science. This process took place in Africa with particular virulence from the eighteenth century onwards, a period during which the paradigm that separated nature from culture and society was starting to submit the former to the determinism of mathematical laws. Once transformed into a resource, nature (and the irrational, who came under the category of savages) had no other function than to be exploited to a point of exhaustion. Today, just as five centuries ago, the world capitalist system continues to strive for to transform people who possess other forms of knowledge into resources—now of a genetic nature—and as tools to be accessed through traditional knowledge.

Within the context of Africa, the question of the existence of other ways of understanding nature and society has to be seen in terms of the historical process of the negation of the knowledge of others or, in other words, colonialism. Colonial expansion was not only limited to economic and political areas. Nor did it cease with the end of colonial empires. It is therefore worth assessing as a whole how the "South" was, and still is, affected by this process of colonization, in order to establish the basis of a new scientific paradigm in which all the different forms of knowledge will have a place, all potentially interrelated and all legitimized by those who resort to them and ordain them as a form of power. The analysis of the configuration of various fields of knowledge in Mozambique is used in this chapter in order to detect the persistence of colonial subalternity in the field of knowledge production, a condition quite often assumed as a reproduction of "backwardness" in terms of development.

The Roots of the Modern and the Traditional

The population of Mozambique, which gained its independence in 1975, is almost entirely composed of native Africans, belonging to several ethnic and/or linguistic groups. Until the onset of Portuguese colonization toward the end of the nineteenth century, the various peoples that made up Mozambique did not live under a single political authority. They existed as independent entities, with different cosmologies and forms of social organization. The cultural mosaic that Mozambique represents is a clear indication of the wealth of forms of knowledge that the country possesses. However, nowadays when we talk about forms of knowledge, the prevailing viewpoint attributes a central role to science, with local or traditional knowledge described as alternative or peripheral.

The frontiers of civilization have been defined in terms of Western notions of the organization of society and of the body (as a social and physical order). Colonizing Africa was characterized as a mission to civilize a continent filled with suffering, inhabited by the savage, the "colonial animal," in which the natives symbolized filth, disorder, lawlessness and lack of knowledge (Fanon 1961, 54). By relocating the native into the realm of nature, colonialism enforced the disqualification of local knowledge, while promoting—quite often by force—modern science as a pillar of civilization. With the imposition of exogenous forms of knowledge that were not legitimate locally, the process of disqualifying, or even destroying, local forms of knowledge began to take place. Indeed, and without losing sight of the various discoveries and contributions made by modern science and the technologies associated with it, modern science also imposed a dependent status on other forms of knowledge, continually renegotiating the appropriation of the knowledge of the Other and only recognizing it as Other, subaltern and complementary to modern knowledge, but never functioning as forms of knowledge capable of interacting among each other in order to produce new knowledges.

The idea that traditional knowledge is practical, collective, has strong local roots, and reflects relevant experiences is central to the dichotomy of modern knowledge/traditional knowledge.[3] The absolutist or/or dichotomy underlying the scientific way of thinking is a product of the Cartesian model that divides the world between the subject who knows and the object that is to be studied. Yet, since all knowledge is socially constructed (Santos 1999, 44), any analysis should question the processes that legitimize the hierarchical organization of knowledge and power between local knowledge and global scientific knowledge.

An analysis of the postcolonial situation is one way of reflecting on the meaning of various forms of knowledge, this time from the point of view of those formerly colonized, as a means of attempting to redress the balance and to (re)assess the situation in terms of the contemporary "social environment." Yet we should also ask ourselves whether we continue to look at Africa in order to understand it on its own terms, or as a means by which the rest of the world can remain safely ensconced within its civilized status, which obliges us to accept the fact that colonialism has not ended with the ending of political relationships of direct subordination. Here, the argument that colonial difference is, in fact, a reflection of a localized epistemological construction to disqualify the knowledge of the Other becomes central. And since the "objectivity" of scientific knowledge (as with any other form of knowledge) is defined and limited by the context within which it is produced, this unequal relationship can only be overcome by breaking away from the paradigm that created it (Santos 1999, 35).

If, in the African continent during the colonial period, the main emphasis of the prevailing discourse lay in legitimizing modern science as a vehicle for progress, the emergence of nationalist movements at the start of the twentieth century refueled the debates on the nature and function of science. As in other areas of the world, the debates on science established themselves as debates on the politics of knowledge (see Visvanathan in this volume). The debates were not limited solely to discussing the use of knowledge for the development and emancipation of colonized peoples; they also aimed to provide the different forms of knowledge with the right to an existence without marginalization or subalternity in relation to official science, as defined and supported by the colonial state. In Mozambique, among the nationalist principles, liberty and equality were also to be applied to accessing and sharing the universe of knowledge. However, it is no less true that nationalist discourse, whilst challenging colonial domination, often also accepted the intellectual premises of modernity on which colonial domination was based.

In Mozambique, the goal of the newly state was to achieve the level of development presented by the West. The ethno-cultural differences were seen as politically dangerous, as a path to the advance of internal regional fractions; at the same time, many of the local African traditions of knowledge were regarded as backward, opposing the route toward development. During the first decade after independence, reference to other forms of knowledges—besides modern science—would be condemned as contributing to the spread of ethnic and regional divisions and rivalries in the country. Science was therefore seen from a positivist point of view, as the vehicle through which "development" or, in other words, the levels attained the northern countries, could be achieved. The policy was one of the transfer of technology, driven by the mechanisms of invention and innovation. At the same time, the dissemination of scientific and technological skills that had enabled other peoples to achieve substantial progress emerged as the very incarnation of democracy: scientific achievements should be made accessible to all. In term of public health policies, for instance, it was attempted to establish a health care system with a Western orientation, in which traditional medicine was only recognized as a means of accessing information about medicinal plants that could be use in biomedicine and biotechnology[4].

It is, however, worth considering in more detail the processes by which Mozambique, where the first minister for science and technology was only appointed at the start of the new millennium, became linked scientifically to the modern world.

After the civil war which escalated throughout the entire country, particularly in the 1980s, the peace process in the 1990s and the political opening up of Mozambique to a multiparty system and the construction of a democratic society cannot be dissociated from the influx of countless international organizations and, later on, the emergence and development of various national social organizations. In the same period, consultancy firms of varying types—entrepreneurial by nature—that carried out studies and consultancy work on a profit basis, were also established. It is important not to underestimate the impact of these firms, together with other NGOs (carrying out studies in specific areas such as justice, the environment, health, women and gender, democracy development and land, to give just a few examples) on the local scientific community. On the one hand, by offering employment opportunities to researchers from the public sector and to recent graduates—work that is usually well paid—they helped to create a "internal brain drain." In addition, since they had guaranteed funding for most of the work they undertook (although with strong limitations in terms of opportunities to gather information at grassroots level)—almost always from some external source—they represented one more form of competition for educational and research institutes, particularly the state ones (Silva et al. 2000). As a result, countless research projects and programs were carried out before a national system for science had been established in the country. The political changes that have taken place since the 1990s (and which represent a drawing back from the "socialist revolution") are reflected in the various neoliberal reforms introduced into the country, many of which imposed by international agencies, who nowadays consider technical support, information and the production and sharing of knowledge to be the main aspects of their involvement in the country (Mehta 2001). Science-as-commodity remains the central feature of the subordination of the "south" to the "north," through the metaphor of unidirectional development, a relationship which, in itself, reflects the imbalance between "north" and "south," namely that of donor/recipient, development/underdevelopment, knowledge/ignorance, teaching/learning, advising/following, planning/implementing.

The meaning behind this idea of development that has been applied to the country so far contains, in itself, the notion that, as in the past, it is the "north" which possesses knowledge and the "north" which ensures that science will supplant superstition and obscurantism and benefit the needy, namely those who lack any suitable environmental ethics and long-term sustainable planning, etc. A common element in these different forms of development discourse is the emphasis on the difference between specialist knowledge and local forms of knowledge. These discourses accentuate and deepen the contrast between rational/magical (religious), universal/particular, theoretical/practical, and modern/traditional.

These extremely powerful dichotomies influence the way in which arguments are constructed and favor one form of knowledge at the expense of another (or others). The "north" thinks, knows, disseminates, diagnoses, plans, organizes, executes and validates the research undertaken in and into the "south"; it advises and prepares models and lessons that have already been reviewed in other countries. The "north" produces, synthesizes and disseminates knowledge; the "south" produces data and information. As a result, in Mozambique, as in many other peripheral African countries, international aid operates within the premise that the "south" has problems and the "north" has the solutions to them. But can "sound" knowledge of the south only be found in the "north"? Or can such knowledge be produced exclusively in terms of references to the notions progress and development of the "north"?

As we shall discuss later in greater detail, in the "south" the consultancy culture or the culture of imported specialists[5] has served to reinforce the elitist and technocratic approaches of science and knowledge. The participation of communities and any form of consultation with them are seen merely as a concession to the principles of democracy, rather than understood as mechanisms necessary for the design and implementation of a coherent development policy. The separation between "the thinkers" and the "subjects" to whom this knowledge is applied and between those who propose reforms and those who implement them is therefore accentuated—both on a national and an international level. This approach reaffirms the tradition of locating the problems where the policies are to be implemented and never where the development policies are diagnosed, planned and formulated. Effective and sustainable policies do not only require sound knowledge, but also knowledge and a desire to implement them on the part of all the actors who are affected by them.

The political option followed in Mozambique never addressed the colonial relationship as an unequal power relationship within the fields of knowledge. For most Mozambican researchers—trained in the country or abroad and without a *locus* for self-reflection—the epistemological reference continues to be the universe of the "north," including the search for alternatives to the dominant paradigm of modernity and the colonization of knowledge. It is not surprising that in Mozambique the main emphasis is on conferring modern science with the hegemonic status of valid knowledge, reinforced through its status as "official science," reproduced and retransmitted through various mechanisms—centralized state policies, knowledge acquired through university departments, etc. The heavy cognitive and cultural barriers that have constructed the Other as a savage have prevented other forms of knowledge from flowing and interacting. The researchers remain imprisoned within their scientific armor, rejecting forms of knowledge that are familiar and closer to them.

The Eurocentric nature of modern science, as a type of globalized localism,[6] is very deeply rooted in the ability of the rhetoric of modern science to convincingly establish modern knowledge as the core and the all the other knowledges circulating in the world as the periphery. The perpetuation of the idea of underdevelopment reflects the inability of the north in recognizing the accumulated experience of the "south." The "south," lacking in experience, can only acquire it through the sources accumulated in the "north." This ability to reproduce the Other *ad eternum* through the cultural and epistemological dichotomy between scientific and alternative knowledge still functions nowadays under various guises: the consultancies, the lack of reflection on the meaning and uses of knowledges and the imported theoretical and methodological frameworks.

The universities and research units are increasingly being consecrated as the current centers of modern (civilized) political power in relation to the marginal communities (the "new savages"). In the specific case of Mozambique, it was mostly the members of this elite alone who took part in the preparation and discussion of the recent law on science and technology.[7] From the discussion that took place during the formulation of this policy, it transpired that, as in the past, for the majority of Mozambican researchers not only the science/traditional knowledge dichotomy but also the social sciences/natural sciences dichotomy remained unchanged. As in the past, the social sciences were once again designed to reproduce the Other as an object with "culture," who must be supplied with the forms of knowledge that only modernity possesses.

The construction of a new paradigm therefore involves various struggles, the first of which is to overcome the inferiority complex that exists in relation to "northern" knowledge (Houtondji 2001). This process encapsulates a double epistemological break, both with the inferiority complex as a reflection of the colonial nature of power and with the hegemonic power of modern science which continues to defend its status as the prime vehicle for promoting progress, now labeled "development." The next step is to recognize the incompleteness of each cultural universe, an incompleteness that is, however, only perceived when a given culture enters in contact with other cultures. This brief theoretical incursion enables us to highlight the central argument of this work, so eloquently formulated by Santos (1995)—that knowledge is only recognized as relative when it comes into contact with other forms of knowledge (either in the form of opposition, inequality or complementarity).

THE CONSULTANCIES—AN EXTREME CASE OF THE LOCALIZED GLOBALISM?

Far from wishing to deny the validity of the countless scientific projects undertaken in Mozambique, it is nevertheless important to provide a critical analysis of the consultancy work.

An understanding of the politics of cooperation and the trajectories of the studies on development aid enables us to obtain a more accurate idea of the persistence of the colonial relationship as a knowledge contract. Although the supposed objective of consultancy work is to improve project performance, alleviate poverty, etc., it has increasingly come to echo a strong feeling that the "knowledge" created by "aid" programs ultimately serves to reorganize the dependent and subaltern relationships of the "south" by reconstituting the colonial nature of power (Stirrat 2000).

The term "consultancy" is normally applied to research work whose unfolding—in terms of objectives, methodologies and approaches—is defined from the outset by the agency funding the work, through the "terms of reference." By resorting to a "consultant," specific specialist knowledge is being purchased in order to assess/supervise a particular project or research activity. In Mozambique, consultancy work ranges from providing support for the formulation of policies and assessments of the performances of the various sectors, to the planning monitoring and evaluation of individual projects. In short, the work of the consultant consists of helping the international agencies to promote the type of development they wish to advance.

African researchers are, in general, excluded from any effective participation in the construction of new theoretical and methodological tools. The very image of Africa remains heavily defined by exogenous theoretical references. At the same time, many research projects in the field of social sciences exist as such in formal terms only, since the researchers themselves spend little time in dialogue with their informants and are unable to determine whether their theoretical framework is appropriate for the situations they encounter and even less able, if it does prove inappropriate, to modify it. Given the lack of financial means to carry out independent research (and to survive), Mozambican social scientists resort to consultancy as an alternative and assume the risk of increasing intellectual proletarization. This explains their poor ability to function as autonomous agents of social transformation. Thus, by joining the "consultancy culture," Mozambican social scientists become links in the chain of production of knowledge about the "south" and for the "south."

Local Development or the Localized Local?

Santos argues that an epistemology of the "south" should be based on three approaches: learning that the south exists, learning to go to the south and learning through the south and with the south (1995, 508). But where is this "south" that we are encouraged to redeem? In criticizing the activities of the consultants, my starting point departs from Santos formulation (2003a, 253), when he affirms that the epistemological diversity in the world is potentially infinite. There is neither pure knowledge nor complete knowledge; there are only constellations of knowledges. It is therefore increasingly clear that claims for the universal nature of modern science are only one form of particularism, whose particularity lies in having the power to define all rival forms of knowledge as particular, local, context-bound and situation-bound.

Development in Mozambique is undeniably linked to progress in knowledge, which includes modern science. Yet how can we release the "new" local and traditional knowledge that the international agencies have been promoting lately as a vector of development? One of the features which most clearly confirms Santos' assertion is the fact that awareness of the strength of local knowledge is entering into the existing landscape of knowledge in the country, having been almost (re)discovered by the agents of the core financial institutions in the capitalist system, such as the World Bank.

Since Mozambique is a country in which science, as previously mentioned, was imposed as part of the colonial process and later emerged through its subordinate position in the world system, effectively crushing and disqualifying local knowledges, how can we construct a network of knowledges that is so vital to the country's growth? How can we build up a knowledge characterized by solidarity, mediated by opportunities for participation, by a sense of cultural incompleteness, of epistemological restlessness?

An increasing number of voices have begun to argue that an understanding of the world demands a much deeper understanding of "development discourse," claiming that the ways in which we know and construct the "developing" world are a precondition for changing that world. Yet, despite this persistent challenging of the uniformity of the "Third World," attempts to place Africa within the global critical panorama of the problems of the continent have proved difficult, since the trend remains that of celebrating a universal image, lacking in specific and analytical critical perspectives. Most of the knowledge produced on and for Mozambique in terms of perspectives and alternatives to development is derived from generalizations and partial observations gathered through consultancy work.

Despite great diversity, some common points exist which enable us to outline a profile of the consultant: it is an individual who works in a multidisciplinary team and whose function is to help the development agencies provide support for the "less developed." The consultancy teams normally include a social scientist, an economist and various other specialists, in accordance with the particular objectives of the project. These teams are usually contracted by companies that specialize in consultancy work; in other cases it is the agencies themselves that contract the consultants on an individual basis and are responsible for the organization of the team.

An analysis of the terms of reference reveals what each development agency sees as relevant in their appraisal of any given project. What is disturbing and at the same time surprising in these documents is that they declare as universally valid or relevant what in fact is a specific vision of the world, in which particular aspects considered representative of the whole.[9]

With the image of Mozambique "elaborated," as in the past, on the basis of the opposition between the modern (synonymous with the "north") and the traditional (synonymous with the ethnic, local), subalternity is maintained, with the country, like the continent itself, described as synonymous with an inability to produce "wealth" and to develop. The definition of essential areas for development distils the problems of the country and the region. The assumptions underlying the idea that these themes remain relevant and exclusive are normally shared or rarely questioned by those who work in the consultancy teams. The first of these assumptions is derived from a positivist concept of the world and of knowledge. The world is seen as composed of facts that can be empirically verified/tested (how many people "are poor," how many plants contain "medicinal properties," which are most important ethnic characteristics, etc.). It is equally relevant to note the importance attributed to quantitative data in this type of work: calculations represent the essence of knowledge and, therefore, of power and control over what takes place. The second assumption, which follows on from this, presumes that all the elements that exist in the world are interrelated and it is the task of the consultant to systematically decipher these associations. As a result, it is possible to anticipate what will take place in the moment or situation that follows any form of change and, consequently, to plan actions and interventions that will have predictable *a priori* results. In short, consultancy work is aimed at changing the world in a predetermined manner, in order to improve it. This way of viewing the world is a reflection of a rational and determinist perspective on the future of the world which Santos has termed proleptic reasoning.[10]

In the world of the consultancy, "culture" and "ethnicity" are associated with something which only "others" have, that is local and frequently considered illogical and irrational. The task of the researcher as consultant is to "decode" this universe, to "localize" these histories and to overcome this kind of irrationality. This "intrinsically cultural" irrationality is portrayed in the reports in various forms: it may be agro-economic—for example the fact that rural populations also depend for their survival on harvests and so turn their backs on the direct market laws—or it may be epistemological, involving recourse to "traditional medicine," etc.

Consequently, the consultancy reports have a very similar structure. The doubts are absent from the text and tables and quantitative statistical information are by far the main methods of supporting statements. Another interesting element is the impersonal tone of the text—the authors are not named and their impressions, doubts and queries are absent, just as the objects of their work, the "agents of local culture," are not present. In short, in the work of the consultant, it transpires that any limitations in terms of development lie in the cultural difference between the "south" in relation to the "north." The role of the consultant is therefore to seek to break down these barriers by proposing alternatives to overcome these cultural limitations. However, as Santos (1999, 58) argues, any knowledge only exists as self-knowledge or as a moment of recognition of the incompleteness of the knowledge of a given culture when it comes into contact with another. It therefore has to be recognized that a lot of Western concepts relating to development and knowledge are themselves cultural artifacts and the ways in which people use, modify and manipulate development initiatives must be understood in this way as incomplete knowledge, unable, due to the powers of metonymic reasoning, to recognize its own inability to make constructive and equal contact with other knowledges. This new way of imposing the hegemonic logic of knowledge is a central feature in many of the consultancy reports. As the development programs are implemented, we can expect to witness the global homogeneity, that is, a particular globalized localism imposes itself, to the point where the rationale underlying the development program is transformed into a universal rationale, due to the support and the actions emanating from the consultants. They are, therefore, the levers of modernity operating in the field.

A contemporary vision of the old dilemma about how to conceive of the "other" is reflected in the current perspective on "indigenous local knowledge." The (re)discovery of the local appears to imply a rejection of the negative effects of modern practices—for example, a preference for peasant knowledge rather than that of the foreign agricultural specialist. The reality, however, is quite different. An arbitrary selection is made of what the consultants consider "usable" or "disposable" knowledge. Knowledge that can help the general and centrally formulated program for development is considered valid and beneficial. Other knowledges which cannot be understood in this way and which seem to have no rational use and are therefore part of the "culture" (or "superstition," as is frequently the case with traditional medicine), are disqualified and do not even feature in the consultant's report. In other words, within the context of the consultancy work carried out by the major international agencies, local knowledge is only valid when it serves the projects of modernity.

Participation by Communities or Their Continuing Invisibility?

The principal methodology employed in much of the consultancy work—the so-called participatory process—is another insidious way of determining and implementing projects from above. From this point of view, the term "development" embodies various qualitative objectives; reinforcing popular participation, enhancing local forms of democracy, improving living conditions and welfare and consolidating community efforts in the development program. The objectives of "equitable growth" and "improving the living conditions of all residents" are designed to be achieved through specific interventions in various sectors that make use of methodologies that combine processes introduced from above with increased community participation. These objectives and methodologies constitute both the means and the targets to be achieved through "development" initiatives. This discourse is part of a new orthodoxy of development used, since the last decade, both by NGOs or the World Bank.

The main idea of the participation process is that local knowledge and the skills of the target populations should be considered central to the formulation of programs. It is therefore up to local actors to define the objectives and main needs. However, when the techniques employed are analyzed—maps, calendars, diagrams, consultations, etc.—it once again becomes clear that these processes are firmly anchored in the cultural tradition of Western rationality (Meneses 2000). The subaltern relationship is aggravated by the fact that the methods defined as essential in overcoming the stagnant underdevelopment of the communities—the building of main roads, economic infrastructures, etc.—are always proposed from above, from outside the community (Meneses 2003). The ultimate objective of this program of participation, supposedly led from below, is therefore firmly based on the rationalist tradition. Far from truly attempting to construct a multicultural democratic and participatory project based on the community, this approach, involving a selective choice of participants, reflects a way of thinking that aims to empower the individual citizen but implies the negation of the community.

In the final analysis, the causal relationship between participation and the increased power of the communities is based on establishing a false relationship between knowledge and subjects/actors, subjects whose power is supposedly reinforced on a conscious level but within a vacuum, divorced from social and political action. This approach gives priority to participation in terms of the production of knowledge rather than in terms of communities managing and making decisions about the project itself. In defending this position, the approach that is evident in most of the consultancy work that has been carried out in recent years in Mozambique constructs the communities—the targets of development interventions—as passive agents awaiting the arrival of an emancipatory "intervention" brought messianically by the organizations and agencies who manage the development programs.

These distinctive forms of appraising the "culture" of the "other" contain in themselves many explicit signs of a colonial perspective on power. In the past, as today, "local culture" was only comprehensible when understood and absorbed into the dominant paradigm of modern knowledge. In the specific case of Mozambique, most of the ideas underlying the theoretical corpus of these programs are based on "truths" produced about the African continent and Mozambique that date back to the nineteenth century and the greater part of the twentieth century, that is, during the colonial era. As before, landscapes are constructed in which the juxtaposition of "us" and the "other," the past and the present, tradition and modernity and culture and rationality are reproduced.

The methodologies and theoretical frameworks present in much of the research work carried out in Mozambique ultimately mark the discourse and analysis of the actual researchers, meaning that, apparently unconsciously, this discourse remains partial and heavily defined by neocolonial perspectives, the intrusions of the hegemonic global perspectives of the dominant society. This does not mean that the agencies themselves and the actual consultants favor neocolonial oppositions. The important point is to identify the extent to which colonialism, as the regulator of science, hampers the construction of an emancipatory knowledge in which different knowledges may coexist as a reflection of the various ways of being that exist in the world. The continued imposition of exogenous terms and the refusal to accept that Western knowledge—in whatever form it takes—is a globalized form of local knowledge, are in fact the great stumbling blocks to development in the region.

In this new guise that the consultancy reports represent, the local dissolves in investigation, in discussions. The consultancies are, in fact, new devices for reconstructing the local by reinforcing inferior difference, the local as once more localized and made exotic, without recognizing its right to be linked—through equivalence—to similar realities in other places. The incorporation of African researchers in the research is used to show that this form of reconstructing the object of the research is not racially or culturally driven. However, the fact that the social scientists are African does not necessarily guarantee that the final result of the research is legitimate and representative of the aspirations of the people. This is perhaps one of the greatest lessons to be learned from consultancy work.

It is this situation that has led African researchers to affirm, when assessing their own performance, that their role nowadays is similar to that of the "illiterate" informants of the colonial era; just as the interpreter of yesterday was obliged to provide the information required on the colonial questionnaires, so today's researchers have to complete the information required for the consultancy's terms of reference (Hountondji 2001). In most cases, the consultancies do not represent a dialogue of knowledges but a monologue from within the field of Western science. In this way, the African researcher, consciously or unconsciously, serves whoever investigates Africa; he remains a face without a voice through the investigation, an object to help reach other objects, rather than the subject of a possible other/new discourse on types of knowledge and futures (Mudimbe 1988). Even if individualized, in the majority of cases his work serves to enhance the information base of the "north." The "north" continues to recognize information about Africa, now under the mask of "traditional knowledge," essentially when it falls within the "ethnic/exotic" domain.

The consultant is frequently obliged to write what his client wants to know or actually wants to be written. Because the consultant is normally disengaged from the program after the assessment or monitoring that he/she has been contracted to undertake has been completed, there is no mechanism (apart from professional ethics) to regulate the consultant's responsibilities in terms of the results of his activities.

The time has come for a critical analysis of the social sciences from within, from Africa, rather than in terms of the exotic and the ethnographic. A new approach to the social sciences will demand a more profound consideration of the subject of ethnicity, given the ductile and dynamic nature of the cultural representations it involves. Ethnicity is closely linked to practices of territorialization and classification that date from the colonial administration; in this sense the ethnic and ethnicity are no longer "essential" categories, but rather practices and utterances that have been socially constructed. Ethnicity, tradition, the idea of Africa itself are, without doubt, invented social structures (Ranger 1983; Biaya and Bibeau 1998) which accurately reflect their existence as the products of a particular set of social and political dynamics. The "invention" of ethnicity, of the tribe and tribalism does not mean that these realities do not exist or that those who reclaim them have a distorted view of reality. Instead, it is important to understand the context of the politics of identity in which these realities are publicly activated and why, at a given moment, certain characteristics of identity superimpose themselves over others. The question of identity must therefore be seen as apart of a much wider social and political processes in which the state must be analyzed as the privileged arena for the definition and recomposition of identities (Santos 2003b). Ethnicity—in its various forms—therefore emerges as a means of access, of inclusion in the social contract that the state represents, in terms of the distribution of the material and symbolic resources embodied in the state.

In sum, most of the problems that the consultancies extract as a form of production of knowledge are the results of extreme simplification and generalization of the data that have been gathered. Moreover, there is a lack of comparison between case studies, a weak theoretical and methodological framework, an absence of conceptual rigor and a mechanical transition from the results of these studies to the formulation of policies and, in a more general sense, the use and abuse of these results in order to legitimize the policies recommended. Nowadays, whether we like it or not, only the results produced by consultancies that have been validated by the global institutions have the necessary credibility to be used in the formulation of national policies. The opaque relationship between the validation of knowledge and the power of the agency is a critical factor, which is rarely mentioned in the reports. Many of the ideas and trends that become dominant do so not necessarily because they are the result of careful investigation whose merits are recognized and whose effectiveness in explaining or transforming situations is proven, but due to the (ideological, political or financial) power that lies behind the research.

AGAINST THE TYRANNY OF THE "CORE SCIENCE": WHAT ALTERNATIVES?

As this analysis shows, the implantation of modern science in the former African colonies was an essential aspect of the process of regulating relations between the "civilized" and the "savages." Today the debate between science and alternative forms of knowledge in fact constitutes a residual localism originating in this ideological conflict.

Yet how can differences be described without emphasizing the exotic and the local? How can emancipatory knowledge be constructed without also constructing barriers that once again limit and disqualify other forms of knowledge? Following the critique formulated by various authors (Santos 1995; Escobar 1995; Visvananthan 1997; Stirrat 2000; Diawara 2000), it is important to resist against the domination of the core and propose analytical alternatives that are capable of functioning in two directions: one in the sense of combating the dominant notion of development conceived in terms of hegemonic core perspectives and another which aims to understand—at the grassroots level—how social groups interact with the impositions that have been inflicted on them and what types of resistance they have mobilized against them.

Despite the ceaseless attention paid by researchers to the contact zone between the First and the Third World, many end up losing sight of what is perhaps the central relationship in this "zone," the relationship between knowledge legitimized by the fact that it belongs to privileged research institutions and the vast mass of humanity who, by various means, are denied any chance of participating in any dialogue with the science of the "north." In Santos' words, "the two contact zones constitutive of western modernity are the epistemological zone, where modern science and common knowledge confront each other, and the colonial zone, where the colonizer and the colonized confront each other. These two zones are characterized by the extreme disparity among the realities in contact and by the extreme inequality of the power relations among them" (2003a, 268).

The contemporary world is composed of plural communities of knowledges in which multiple subjects coexist. Historically speaking, all these centers of knowledge are legitimate and should be equally recognized as such. This multi-situational proposal is neither radical nor new, but it does demand that modern science, produced by the core, recognizes the need to question the legitimacy of its status and its limits as universal knowledge. Such an attitude requires core science (and its defenders in various peripheral locations) to develop a self-reflexive awareness that can alert it to the variety of competing paradigms, temporarily marginalized, that coexist alongside the paradigms of the dominant Western knowledge. Even more importantly, it must accept that there are different relationships between distinct knowledges.

At the same time, local knowledge should start to be understood not as authentic information that is unique and exclusive to a particular group, but as a knowledge that is advanced through dialogue and resistance among global and local knowledges. This knowledge, which is local because it is situated, produced and shared within a particular group, contains practical and theoretical wisdom, including representations and social practices that are related to a particular technical knowledge. Local knowledge—like scientific knowledge, is not—contrary to what the essentialists and the romantics claim—distributed homogenously and uniformly within the group which uses and (re)produces it. It varies from group to group, according to social status, ethnicity, gender, etc. The ownership, transmission and negotiation of this knowledge is therefore intrinsically linked to social differences and power relationships. Far from being a static grouping, this knowledge is the object of endless interpretations, depending on the various situations and interests involved.

From the situation and point of view of an anthropologist working in an underdeveloped country such as Mozambique, what are the implications, both for social scientists and for their "subjects," of the processes of globalization in its current phase? Scientific engagement extends far beyond the simple production of scientific knowledge relating to this "traditional" world, or reflections about oneself through the other. This engagement is essentially an interactive dialogue with an emphasis on the search for a means of communication between the different communities and movements in particular places and regions. Perhaps that is why the question of written ethnography sometimes becomes irrelevant. Research what? For whom?

Speakers of Portuguese and English—the main languages used in the consultancy reports—are a minority group in the country. This is probably one of the reasons why these reports hardly ever reach those about whom they are written. When they do arrive, they are just a mass of papers written in *"a language people don't understand,"* full of formulae and concepts elaborated in an impenetrable style. The images consist mainly of official maps, and photographs of animals, buildings or plants. Just as in colonial landscapes, the actors melt away, creating spaces that have no history and can later be filled by other powers. In short, the public in Mozambique for whom the work is destined represents a very much-reduced sphere within the private world that promotes development. Yet statements like *"these whites are really clever. They know there's something important here"* show that the "objects" of the consultancy work are aware of the fact that knowledge is an arena within which power can be affirmed.

These issues have forced me to reflect on my own position and my role as an intermediary in the process of (re)fabricating the other, and of affirming difference without the right to equality. How do we exit from this labyrinth of questions that have no answers? The feeling I have nowadays is that most of the information and reports that have been produced have no application. They are a tragic reflection of an information-producing system that does not know how to mediate between the so-called traditional and the modern or between the "I" and the "Other." The "other" part of "my self," a partial link to other universes and voyages into identity, does not fit this framework. The choice of a university degree by the consultancy institutions as a means of legitimizing specialist authority, and the emphasis on using the title before my name when being introduced to the community we were working with was always a painful experience that highlighted our "other identity" and made us feel part of another universe whose interests were opposed or estranged from those of the community. In this sense, since I am a specialist consultant and the owner of superior knowledge, my authority can never be seen as a contribution to knowledge and can never have credibility. It will always be simply a reproduction of the violence and political power that places a researcher in a position where they can proclaim authority over knowledge without truly possessing it.

In this current phase, perhaps what anthropology has to offer us is a means of understanding ourselves, in the midst of our incompleteness. It is possibly one of the contact zones in which diatopical hermeneutics can best be practiced. Whilst the body is dispersed throughout the world of diversity, reason is forced to let itself be seduced by complexity and the search for stories that are not normally detected. From this point of view, it is the stories, which find us, and the people, and situations that chose us to be their interlocutors.

This practice of research begs one more question—what is to be done about the inevitable distance created between the subject and the object of the research? Anthropological facts are a hybrid product resulting from the interaction between the anthropologist and the subjects of his study. These facts should therefore never be considered the direct, lived experience of the subjects who have been analyzed, but as the aseptic written translation of their experiences. Participatory observation, with its requirement to submerge oneself in alterity, shapes the researcher much more than his "object." The researcher must be aware that there is no merit whatsoever in using participatory observation if it only contributes toward increasing the (dis)information of the core in relation to the peripheral "south." Gradually we can extract ourselves from the indifferent figure of the classical anthropologist by drawing a dividing line between the disinterested researcher—who is only concerned with discovering an exotic other so that he can see himself reflected in the mirror as modern—and the researcher who is interested in the huge experience that human culture represents.

What similar phenomena can be found in other "locales" that are autonomous in relation to each other, but subject to the same forces of hegemonic globalization? For Santos the answer would be comparative translation and links between the local.[11] Only in this way will it be possible to identify the complex relationships between the local and global and to construct, according to Santos' proposal, "grammars of translation" that enable us to highlight the knowledges that have up to now been silenced or banished by global knowledge. The solution lies increasingly in our growing awareness of incompleteness and our recognition of the other as equal and as a reciprocal part of a project to broaden the vast mosaic of diversity that is unique only as a whole, but is not represented by its parts. Each locale will therefore always be a point of reference in global discourse and vice versa. In this sense, in order to transform the practice of social research in Africa, it is vital to create translation bridges that can enable researchers to break out of the work in which they are confined and seek out dialogue with new social agents and new social practices operating in an equally new context that involves articulation between local, national and global struggles.

The key question is therefore the practice of social science as a project for emancipatory action and, by definition, the influence of context on practice. Although political context has greatly influenced the debate on the role of the social sciences in the "south," little attention has been paid to the central issue of the destiny of social research. Now may be the time to stress the transgressory aspects of the social sciences and to shift from observing to becoming more interventionist, by redefining the logic and practices of development.

A new social science demands equal status. The Mozambican social scientist cannot therefore continue to be seen as the person who discovers new facts; on the contrary, we should be acting as the translators of this reality and this routine and thus apparently not very relevant or visible, knowledge. The identification of visible or invisible areas of knowledge depends on how theory is constructed and on the relevance given to it. This is why I agree with Santos' appeal to learn from the south and with the south (1995, 508). It is vital to reverse the current process which limits the production of science in the name of development to a pathology that involves listing all the evils which, viewed through a morbid exoticism, highlight the issue of wars seen from a tribal perspective, the phenomena of corruption and the environmental destruction associated with extreme poverty (which includes the drama of HIV-AIDS) as the main features of African political societies today.

As Santos (2001) states, it is necessary to create opportunities for translating intellectual desires for "representation" into political practice, to radically alter the social and political situation which divides the world between the modern and the traditional, the civilized and the savage and the developed ands the underdeveloped. A reflective perspective underlies this desire to implement innovative projects that actually serve the needs and interests of the recipients (Mbilinyi 2002). Or, to put it another way, the aim of this other social science is to identify the utility of the projects: who they are for and why?

In opposition to the monism that refuses to acknowledge the variety of "truths," the relativism that denies the importance of truth and the nihilism that denies the existence of truth, it is vital to accept a system of interacting knowledges that revive and uphold contact with what is really important and true. The alternative should be formulated as a project for interrelated knowledge, a landscape of multiple knowledges that recognizes, on equal terms, the ontological robustness of both small and large theories and forms of knowledge. Turning to Santos' proposals (2003a) once again, it is vital to promote the translation of concepts and practices, perhaps one of the greatest epistemological challenges we have to face. The knowledge necessary for growth and development has to be anchored in multiple realities, as a recognition of the plurality of knowledge that must be respected and interconnected. Only this type of knowledge will be truly authentic, since it is based on and owned by those who need it, as part of a process in which various forms of information act simultaneously. Only in this way is it possible to absorb knowledge of traditional medicine whilst also seeking out biomedicine and to absorb the power of science when we lock into the mirror and see ourselves from the other side, knowing why . . .

Differences in the concept and meaning of knowledge have to be negotiated and new effective epistemological partnerships have to be built. Only in this way the authoritarian and normative taxonomies that divide the world up between the natural and the social and the traditional and the scientific will be destroyed and reconfigured. It will then be possible to reassess the categories that have been imposed from outside, in which any predetermined order imposed on things is rejected and in which a form of politics is constructed which respects difference. The challenge is therefore to fight the monoculture of knowledge (Santos 2003a), not only in theory but also in the way in which we practice what we have studied, in our actions.

In a world that offers an infinite map of knowledges, it is difficult to create a single meaning for the meaning of knowledge. It is this consensus that needs to be discussed, in addition to how we can create a taut and constructive dialogue within it. This is why Santos appeals for the visibility of the other, the need to create space for other knowledges and an openness to realities that are absent from modern reasoning and invisible to the eyes of so many (1995, 2003a). It is essential to expand the present and to make "other" physical and imaginary locales more visible, to make space for knowledges and recover experiences. It is also time to disengage all the geocultural categories that seek to preserve their hegemonic status and organize space and time in epistemological terms, by making knowledge a perpetual process and the main foundation of a more just and valid society.

NOTES

1. "South" does not only refer to a geographical location; it also "expresses all forms of subordination brought about by the capitalist world system" (Santos 1995, 507).

2. Most of the experiences referred to here are the result of my involvement, over the last ten years, in various consultancy projects essentially concerned with "natural resources" and "historical and social patrimony," supported by various development agencies. I have also benefited from the fact that I participated as a member of the steering committee in the process of formulating the policy for science and technology in Mozambique.

3. Whilst arguing for the need to make visible the types of knowledge that Western modernity has transformed and termed traditional, as opposed to modernity and progress, I would also like to draw attention to the extreme case of reactionary essentialism in which difference—recognized as such—is distilled and reproduced as the traditional, unchanged in time and space.

4. However, today, Mozambique has around 1,000 biomedics, whilst practitioners of traditional medicine—finally recognized as part of the health system in 2003—number over 10,000.

5. Several warnings have been sent during the last decade about the implications of the expansion of the international consultancies, particularly in Africa. There is, however, an important different between living in a country and merely visiting it for professional reasons. Overseas consultants may leave behind ideas, articles, documents and recommendations but it is those who live it h region, the country or the community who have to carry out the tasks and live with the consequences of a project.

6. Santos defines "globalized localism" as "the process by which a particular local phenomenon is successfully globalized" (2001, 71).

7. The simple fact that the discussion documents were written in Portuguese restricted the opportunities for many "local, lower class" Mozambicans to participate, since only 30 percent of the population does speak this language.

8. For Santos, "localized globalism" refers to the specific impact of the hegemonic globalization on the local (2001, 71)

9. This latter point has been theorized by Santos as "metonymic reasoning," a kind of reason that claims to be the only form of rationality and therefore does not exert itself to discover other kinds of rationality or, if it does, it only does so to turn them into raw material (2003a, 240).

10. Santos has proposed to concept of "proleptic reasoning," a kind of reason that does not exert itself in thinking the future because it believes it knows all about the future and conceives of it as linear, automatic, and infinite overcoming of the present (2003a, 240).

11. For Santos, the proposed new paradigm of knowledge requires continuous translation in order to create mutual understanding between the potential and available experiences (2003a).

REFERENCES

Biaya, T. K., and G. Bibeau. 1998. "Afrique revisitée." *Anthropologie et Sociétés* 22 (1).

Diawara, M. 2000. "Globalization, Development Politics and Local Knowledge" *International Sociology* 15 (2):361–71.

Escobar, A. 1995. *Encountering Development: The Making and Unmaking of the Third World*. Princeton, NJ: Princeton University Press.

Fanon, F. 1961. *Les damnés de la terre*. Paris Maspero.

Gardner, K., and D. Lewis 1996. *Anthropology, Development and the Post-modern Challenge*. London: Pluto Press.

Hountondji, P. J. 2001. "Tempting Traditions: Internal Debate Needed in Traditional Cultures." *Compas Magazine*, March:12–13.

Mbilinyi, M. 2002. *Alternative Views on Development and Poverty Reduction: Tanzanian Grassroots perspectives about Food, Land and Democracy*. Paper presented to the Conference on Africa and the development challenges of the new millennium. Accra, CODESRIA & Third World Network.

Mehta, L. 2001. "The World Bank and Its Emerging Knowledge Empire." *Human Organization* 60 (2):189–96.

Meneses, M. P. 2000. *Sobre mapas e fronteiras*. Work presented to the Seminar of on Women and Gender. Maputo, Mozambique: CEA-UEM.

———. 2003. "Os 'outros' e 'nós': a questão do acesso, uso e gestão dos recursos naturais em Licuáti." In *Conflito e transformação social em Moçambique*, ed. Boaventua Sousa Santos and João Carlos Trindade. Porto: Afrontamento.

Mondlane, E. 1969. *Struggle for Mozambique*. London: Harmondsworth.

Mudimbe, V. Y. 1988. *The Invention of Africa*. Bloomington: Indiana University Press.

Ranger, Terence. 1983. "The invention of Tradition in Colonial Africa." In *The Invention of Tradition*, ed. E. Hobsbawm and T. Ranger. Cambridge: Cambridge University Press, 211–62.

Santos, B. S. 1995. *Toward a New Common Sense*. New York: Routledge.

——. 1999. *Um discurso sobre as ciências*. Porto: Afrontamento.

——. 2001. "Os Processos da Globalização." In *Globalização: fatalidade ou utopia?* ed. B. S. Santos. Porto, Portugal: Afrontamento, 31–106.

——. 2003a. "Para uma sociologia das ausências e uma sociologia das emergências" *Revista Crítica de Ciências Sociais* no. 63:237–80.

——. 2003b. "O Estado heterogéneo e o pluralismo jurídico." In *Conflito e transformação social em Moçambique*, ed. B.S. Santos and J.C. Trindade. Porto, Portugal: Afrontamento, 47–95.

Silva, R., T. Cruz e Silva, J. Barreto, C. Bungala, and F. Mbofana. 2002. *Estudo sobre os sistemas de investigação científica em Moçambique*. Maputo: MESCT.

Stirrat, R. L. 2000. "Cultures of consultancy." *Critique of Anthropology* 20 (1): 31–46.

Visvananthan, S. 1997. *A Carnival for Science*. Delhi: Oxford University Press.

18

The Splendors and Miseries of "Science"

Coloniality, Geopolitics of Knowledge, and Epistemic Pluriversality

Walter Mignolo

INTRODUCTION

In the English version of his provocative piece on "Discourse on the Sciences," Boaventura de Sousa Santos makes two crucial statements that I would like to use as a springboard for my own argument:

a) Modern science, which emerged in the sixteenth century out of the scientific revolutions of Copernicus and Galileo, and later built on Newton's work, was abandoning the esoteric speculations of its founders to become the ferment of an unprecedented social and technological transformation ([1987] 2002, 11).
b) The new scientific rationality, being a global model, was also a totalitarian model, inasmuch as *it denied rationality to all forms of knowledge that did not abide by its own epistemological principles and its own methodological rules* (2002a, 13; italics mine).

The deeply engrained belief of "science" as a point of arrival in the history of humankind is at stake in the previous two quotations. I would like to pursue them further here, inviting the reader to imagine how to look at "science" from the perspective of forms of knowledge to which rationality was denied precisely in the name of "science." That is, "science" is not only a practice, but it goes together with a metadiscourse that—on the one hand defines certain practices as sciences, attributes value to it and, on the other, evaluates and disqualifies knowledges that do not fit the criteria established by the metadiscourse. The metadiscourse is constructed and transformed by scientists themselves who define their own identities, and by philosophers who created a discipline called philosophy of science precisely to define, evaluate, and police both the scientific practices and its conceptualization.

In the sixteenth century—when the origin of science as we know it today can be located—affirmation of scientific and denial of nonscientific forms of knowledges were related to literacy. A certain conceptualization of knowledge, from theology to science going through (secular) philosophy, and from the sixteenth century on, became the standard of both emancipation and regulation used in the double and complementary movement Sousa Santos has been articulating in several of his works. But above all, the epistemic standards established in the name of theology, philosophy, and science made possible the denial of rationality to all other forms of knowledge. In this precise sense, the complicity between modernity and knowledge, self-defined as a planetary point of arrival, was at the same time coloniality as a planetary epistemic denial. Therefore if science and knowledge could be, as Sousa Santos has it for legal discourses, emancipating or regulatory, one particular kind of emancipation is decolonization or decoloniality (as some would prefer to avoid misunderstanding with the term "decolonization" as used during the Cold War to refer to political decolonization at the level of the state). Decolonization is being used today in the sense of decolonization of knowledge. Decolonization in this sense is akin also to the concept of "liberation" used in theology and philosophy of liberation in Latin America, and their critique of science as a justification for developing projects that indeed are imperial designs. Thus, science shows one of its faces when looked at from top down and a quite different one when seen from bottom up. Thus, an epistemic shift is necessary in order to liberate and decolonize knowledge. Shifting the geography of knowledge (let's say, thinking in and from Harvard or from la Universidad Andina Simon Bolivar in Ecuador) makes you wonder about the universality of scientific knowledge. From top down, the dominant paradigm is that of modernity, that is, of newness. From bottom up, the dissenting paradigm is that of coexistence with a dominant one, and it is spatially, rather than temporally, located. The paradigm of newness imposed itself on two moves, and both happened simultaneously during the sixteenth century. On the one hand, the very conception of the Renaissance meant the colonization of time and therefore the invention of the European Middle Age. On the other hand, the "discovery" of America allowed for the colonization of space and the invention of the European difference, which will be imperial (with the Muslim world and the Ottoman empire) and colonial (with the Indigenous people in Indias Occidentales and later on, America).

My argument here has three parts. The first part looks into the crisis of rationality within the "internal" history of European modernity and Western civilization. An important development in this sphere, which I cannot explore here in detail, is "eminist epistemology," critique of the patriarchal foundations of scientific rationality (Alcoff 1993). The second part explores the possibilities of decolonizing knowledge and draws a map of "epistemic diversity" within the emerging paradigms of coexistence. I insist that if "colonialism" may be taken as something of the past, "coloniality" is alive and well, and it is coloniality, in the sense of the colonial matrix of power (and in this case, of coloniality of knowledge), that was never relinquished and it is the necessary conditions of the "advances" toward the completion of the incomplete project of modernity. The third part explores the epistemic colonial matrix of power through efforts under way of decolonizing being and knowledge. Scientific totalitarianism is indeed today one aspect of "global coloniality" (the complementary side of neoliberal concepts of globalization) that takes in appropriation of land and exploitation of labor, militarization and control of authority, and control of gender and sexuality (as we witnessed in the United States in 2004); but epistemic colonization engenders more and more not only resistance but process of conceptual delinking; that is, process of thinking otherwise and imagining a world in which many world, could coexist, beyond the principles of neoliberal civilization that infects the entire spectrum of the social fabric and, of course, the reproduction of coloniality. Thus, one line of my argument is the shift from the paradigm of newness (modernity, postmodernity, neoliberalism) to the paradigm of coexistence (decoloniality of knowledge and of being, spatial coevalness coexisting with temporal and linear march toward a point of arrival, leaving behind what is not modern, postmodern and neoliberal).

THE HUBRIS OF THE ZERO-POINT: FROM EPISTEMIC TOTALITARIANISM TO THE "PARADIGM OF A PRUDENT KNOWLEDGE FOR A DECENT LIFE"

At stake here is the "hubris of the zero-point"—an expression coined by Colombian philosopher Santiago Castro-Gomez (2002a, 2002b). The idea, however, has been around for a while. Sylvia Wynter (1992, 2003), a Jamaican radical intellectual and cultural critique, has identified two radical shifts in the geopolitics of knowledge. The first was toward the centralization of universal principles and on the colonization of time (the invention of the Middle Age), and the other the colonization of space (the invention of America). The colonization of time and space needed a point of reference and, historically, it was self-defined as the Renaissance. The shift moved from a polycentric world to a unicentric one governed by the hubris of the zero-point: a universal point of observation marked by the figure of God and articulated in and by Christian theology. Science, in the modern sense of the world, emerged at that time with Copernicus. The World Map or *Orbis Universalis Terrarum*, popularized by Abraham Ortelius toward 1570 gave to the epistemic shift a visual image: all coexistent territorialities were subsumed under the world map that is still guiding our conception of continental divides and locations of continents on the earth. In philosophy, Rene Descartes provided to the future generation the secularization of the zero-point of observation by detaching the mind from the body. Secular philosophy and science established themselves in the nineteenth century, while England, France and Germany engaged in a new wave of colonial expansion across the globe. Consequently, the scientific revolution of the sixteenth century could be looked in three interrelated spheres of the modern shift of the geography of knowledge:

a) Theology provided the tools and the impulses to the epistemic shift that disqualified, in Europe, nontheological knowledges that complemented the physical expulsion of the Moors and the Jews from the Iberian Peninsula, and provided the criteria for understanding, evaluating and classifying the Indian population of the Americas in the theological frame of European Christianity. All this constitutes the *spatial epistemic shift*. In the first case, the spatial epistemic shift confronted Latin and Greek (and European imperial languages), considered the languages that encapsulated knowledge with the rest of languages of the world (Mandarin, Russian, Arabic, Hebrew, Aymara, Quechua, Nahuatl), that were taken as deficient and lacking categories of knowledge to understand the mysteries of the faith and the complexity of the world. But there was also a temporal one. Man of the middle of the fifteenth and sixteenth century self-fashioned an image of his times as Renaissance and invented the Middle Age. This was the *temporal epistemic shift*. The temporal shift occurred within the internal tradition of knowledge, that is, within the Greek and Latin epistemic legacies marked by the translation of those legacies into vernacular languages. Vernacular languages, in Europe, however were the empire's companions, as Elio Antonio de Nebrija, a philologist who wrote the first grammar of a vernacular language (Spanish), explained to Queen Isabella when she asked him what was a grammar good for.
b) The double colonization, of time and space, ended up by suppressing the possibilities of a pluriversal world that became more and more, universal. Science provided not just a method of inquiry, but a philosophy, a politics and ethics of knowledge. That is, every way or form of knowledge that did not conform to the philosophical, ethical and political standards of sciences were dismissed or marginalized in time (e.g., tradition, primitives) and space (cultures outside the realm of European present where science emerged and established itself from Copernicus to Newton).

Hegemonic epistemic totalitarianism was theological and Christian during the sixteenth and seventeenth centuries, but the hegemony began to switch toward ego-logical and secular totalitarianism with the pioneering works of Francis Bacon and Rene Descartes. Now this paradigmatic change or paradigmatic transition was, as I said before, internal to European history. Both epistemologies, theological and ego-logical, belong to the same paradigm, the "paradigm of newness" that characterizes the very concept of the modern and of modernity. The question is that epistemic totalitarianism brought together methodological innovations, ethical imperatives and economic interests motored by the role of science in a capitalist economy; that is, an economy oriented toward accumulation of wealth that more often than not put scientific "discoveries" (e.g., pharmaceutical products) to the service of personal accumulation in detriment to public health. What is needed today is to uncouple the technical and methodological aspects of science from the ethical imperatives, economic interests and political projects. The most innovative scientific discoveries, in other words, do not carry embedded in themselves an ethics and a politics, and even less a blind assumption that economic discoveries will be beneficial by itself to humanity at large. But if an ethics and a politics are "managed" from above, it will be certainly an ethics and a politics that physically and institutionally that will be separated from science but that will operate within the same totalitarian frame provided by Theo and the ego-politics of knowledge. Thus, if uncoupling the methodological and technological from the ethical and political is necessary, it may not be possible to do it within the same paradigm of newness. We need another paradigm, a paradigm of coexistence rather than of newness.

THE "DISCOVERY" OF COLONIALITY AND THE EMERGENCE OF "AN OTHER PARADIGM"

"Coloniality," in the social sciences and the humanities, was recently discovered. The verb here is appropriate because coloniality was always there although invisible. It was invisible because the rhetoric of modernity that looked for and justified newness, progress, and development hid the fact that these goals were achieved by a small portion of humanity in detriment of the majority—the beneficiary celebrated newness, progress and development without mentioning the costs in human lives and human conditions that were carrying the burden of modernity. Sousa Santos' expression, "a prudent knowledge for a decent life" acquires its full and weighty meaning when looked at from the perspective of coloniality. Coloniality refers to the darker side of modernity, the miseries that the discourse of modernity has to hide (and often ignore out of blindness or of bad consciousness) in order to celebrate its splendors. It is well known in scientific circles that Norbert Wienner published in 1950 *The Human Use of Human Beings: Cybernetics and Society*, only two years after the publication of his path-breaking *Cybernetics, or Control and Communication in the Animal and the Machine*. Wienner was always recognized as one of the great scientists of the twentieth century, but he was bad-mouthed as "eccentric" and "exotic" and "not serious" when in the second book he pointed to the excesses of sciences in the exploitation, management and control of human beings. A society of fellows who, willingly or not, take advantage of the current complicity between science, ethics, politics and economics, needed to bring down his critical opinion. The critique of Wienner's critical social consciousness shows involuntarily that great scientific discoveries do not carry in themselves an ethics and a politics. From the perspective of coloniality it could be added that science and technology in complicity with capitalism carry the seed of coloniality, the control *of* all living organisms; and that is what the scientific community was unable or unwilling to recognize.

Let's review the modern concept of science and knowledge. The historical changes in the "modern" concept of knowledge (since the Renaissance) could be understood semantically in the changes of meaning of "scientia" from Latin into the vernacular "science" (either in French or English) or "wissenschaft" in German. The word "science" itself is simply the Latin word for knowledge: *scientia*. Until the 1840s what we now call science was "natural philosophy," so that even Isaac Newton's great book on motion and gravity, published in 1687, was *The Mathematical Principles of Natural Philosophy* (*Principia Mathematica Philosophiae Naturalis*). Newton was, to himself and his contemporaries, a "philosopher." Something new was happening in natural philosophy, however, and it was called nova scientia, "new" knowledge. It began with Mikolaj Kopernik (1473–1543), whose Polish name was Latinized to Nicolaus Copernicus. Perhaps an indication that knowledge was being racialized, Polish names and language were outside the domain and authority of Latin as the language of "scientia." The change in question is already perceived at the turn of the sixteenth century when, as discussed previously, you look at Acosta's *Historia Moral y Natural de las Indias* and Bacon's *Novum Organum*. The entire philosophical paradigm that Acosta inherited from Antiquity (the Bible, Greek and Latin authors—with the remarkable silence of the Arabic contribution to science and mathematics [Mignolo 2002b]), which he contrasted and challenged from his own experience in the "Indies" (e.g., the New World), was displaced by Bacon's "Novum Organum," that is, a new method for the acquisition of knowledge.

By the end of the eighteenth century the success of the "new scientia" was such that Immanuel Kant would have liked to have an organization of society that would follow the law of the cosmos (as described by Newton), which Kant conceived as "cosmopolis," and began to spill over all the very "universal principles" of knowledge. Thus, the point in question is not the "nova scientia" as a techné to organize information and to structure argumentation per se, but the ideology built upon and in complicity with new social conditions: the Glorious Revolution in England in 1688, the American Revolution in 1776, the French Revolution in 1789 and the Industrial Revolution that unfolded since the end of the eighteenth century. All of it combined, as Sousa Santos puts it, in a new set of social conditions:

> At the social level, this was also the cognitive horizon most consonant with the interests of the rising bourgeoisie, who considered the society they were beginning to control as the final stage in the evolution of mankind (Compte's positive state, Spencer's industrial society, Durkheim's organic solidarity). Hence, Newton's prestige and the simple laws to which he reduced all the complexities of the cosmic order readily turned modern science into the hegemonic model of rationality that then spilled over from the study of nature into the study of society (2002a, 19).

Theology, secular philosophy and secular science (both lodged in the house of ego-logy) are internal transformations of Western epistemology, paradigmatic changes, if you wish, within the same cosmology. Certainly arguments could be built in favor of stressing the epistemic break of the Enlightenment and even of the "scientific revolution." However, and beyond the structure of scientific theories and of the scientific method, the ideological matrix of Christian Theology, secular philosophy, and science is obviously the same. There are no interferences of Mandarin language and knowledge or/of Arabic language and knowledge or/of Aymara language and knowledge. The three configurations have been expulsed and constructed as the exterior of modernity. Obviously, the exterior is only *ontological* from the perspective of modernity. From the perspective of *coloniality* the exterior is the necessary frontier of *modernity* defined from the very perspective of modernity itself. That is, there is no objective point of view in which *modernity and its exteriority are being seen*. The "seer" sees from the perspective of modernity. And it was the establishment and the self-legitimization of this ideological matrix that was responsible for denying "rationality to all forms of knowledge that did not abide by its own epistemological principles or its own methodological rules," as Sousa Santos puts it (2002a,13). This was the foundation of epistemic totalitarianism. Thus, the discourse of sciences is a regional discourse, the discourse of the regional history of European thought. That particular history, however, is double. On the one hand, it is the "history of European modernity" but it is also the "silenced history of European coloniality." While the first consists of a history of self-affirmation and celebration of intellectual and scientific achievements, the other is a history of denials and deferrals of all other forms of rationality. The double discourse of self-affirmation and of deferrals is the double history of modernity/coloniality from the European Renaissance to the postmodern North Atlantic.

III. SCIENCE AND THE RACIALIZATION OF THE "THIRD WORLD"

Indian scientist and activist Vandana Shiva (1993, 1994) has made and repeated in various places and events a series of crucial points to question the splendors of science and to show its miseries from a Third World perspective, as she puts it, or from the perspective of coloniality, as I prefer to avoid the risk of essentially linking people with geographical areas at the same time without forgetting that coloniality of power implies being attentive to the geopolitics of knowledge and geo-historical sensibilities. Modern colonialism operated on the basis of classification not only of people by color, religion and languages (distinct from European colors, religions and languages), but also by racializing continental divides (that is, Europe, Asia, Africa as a Christian trilogy that was modified by the "discovery" by Christians of the fourth continent, America, and converting the trilogy into the Christian geopolitical tetragon). Science was born and flourished in one of the continents. The others had adopted it, rejected it or suffered its consequences. Shiva has exhibited how a series of conceptual misgivings had dangerous results. Let me give you one example of the consequences of the marriage between scientific knowledge, colonialism and market ideology: to produce more to sell more, reducing costs by selling more, and benefiting everyone by selling more even if reducing cost means to cut jobs. "Scientific local-knowledge" (advertised as universal) overruled "experience-based knowledge" in the name of efficiency. In the forest and in the fields a new category appeared: "weed." Weed (like Indians and Blacks) was a "scientific" classification that eclipsed and erased (or used when relevant) the "experience-based" classification of persons who did not learn from the books of science but in the daily and yearly (through several generations) dealing and living with and in the forest. "Weed" follows the same logic of the categories of "Indians" and "Blacks": the imposition of a classification of what shall be discarded. However, discarding Indians, and Blacks and weeds implies also discarding the knowledge that "Indians" and "Blacks" have and, on the other, discarding the knowledge that people in India who had lived in the country and in the forest for centuries have about the complexity of what the ignorance of the Western "expert" reduces to weed, hiding what is his or her own ignorance. At that junction, the principle of a prudent knowledge for a decent life is overruled by the greediness of timber companies and their lust for wood—nothing better than declare primitives or ignorant all those who know what the civilized expert ignores.

Shiva's work combine two spheres of standpoint epistemologies, feminism and geo-historical location. The first operates at the level of the biopolitics of knowledge (e.g., the line that links knowledge with the biography of the scientist), while the second at the level of the geopolitics of knowledge (the line that links the scientist with his or her location in the colonial geo-historical order of the modern world—e.g., Third World, Middle east, Europe, etc). If the reader at this point protests that biography and geopolitics do not have anything to do with science, which is practice detached from biographical configurations and geo-historical locations, we are right in business because it is precisely that ideological assumption (that scientific knowledge is so detached and universal) that I am calling here into question. The two spheres (biographic and geo-historical) cross each other at the level of race and racism, since being a woman of color establishes a shift within feminism itself: solidarities in the critique of sciences at the level of gender and of race had to be negotiated differently, and, consequently, they have to be attentive to the geographic locations in the politics of knowledge. The critique of the theoretical principles and belief system of science (either theological or ego-logical, or both, have different motivations) and the hidden geographic and biographic subjective understanding (European sciences were invented and sustained by white European men, anchor in specific languages, countries and families).[1]

Shiva's geopolitics of knowledge enacts a critique of science geo-historically located in the First World from the "exteriority" of modernity and the process of racialization to which Third World "nature" and "people" have been subjected.[2] While Shiva (based in India) criticizes the discourse of science from a subjective understanding of women of color and Third World history, Dana Haraway (based in the United States) (1991a, 1991b) critiques it from a subjective understanding of white women and First World history. "Point of view" of course is not a guarantee of anything, and people who suspect political correctness in the defense of "point of view" prefer "the criteria of excellence," as if "excellence" would be exempt from point of view or if one assumes that the hubris of the zero-point is not A point of view because it is THE point of view.

So far so good, but there are still issues to explore. Meera Nanda, an Indian scientist based in the United States has raised an important concern and called our attention to the danger of postmodern critiques of science. Nanda's argument is that postmodern critiques of sciences coincide (and was taken up by) Hindu Nationalists who, in a second step, developed a philosophical justification of Vedic Science; calling for the legitimacy of Vedic knowledge as scientific. I would go into the fact that this type of argument plays right into the hands of Western fundamentalism, replicating it by changing its content but not its logic. How do we get out of this entanglement? Part of Nanda's argument, critiquing Vandana Shiva and Ashis Nandy whose critique of sciences plays in the hands of Hindu Nationalists, was to support her argument in Amartya Sen's and Marta Naussbaum's. Nanda's argument will help us in coming back to Sousa Santos' dictum "a prudent knowledge for a decent life."

One of Nanda's strong arguments against Shiva and Nandy, and the Indian Left, is that their critique of Western science played—as I said—in the hands of Hindu Nationalism and their defense of Vedic Science. I would like to stress that Shiva and Nandy are not supporters of Hindu Nationalism, but that their critique to Western imperialism *runs parallel* to Hindu Nationalists' critique Westernization of India. One among many examples from Nandy:

> The literature of the ultra-Hindu nationalist group, for instance, is replete with exhortations to improve the Hindus socially, morally, and psychologically. This improvement is invariably defined according to the values thrown up by the experience of the "Hindu defeat" in the hands of the non-Hindus: hard discipline, masculinity, impulse control, power, and nationalism geared to *realpolitick. Even when the authoritarian Indian takes a revivalist position, thereby trying to return to the uterine warmth of an idealized past and an idealized motherland, what he projects into the past and into the concept of the nation are nothing other than conventional Western middle-class values.*[3]

As for Vandana Shiva one could say that even if some of Nanda's criticism are justified, Shiva could not be reduced to her position in "eco-feminism." In other works such as *The Monoculture of the Mind* (1993) and *Stolen Harvest: The Hijacking of the Global Food Supply* (2000), Shiva has been consistently denouncing and revealing—based on the particular experience of neoliberalism in India—the rhetoric of modernity under which the logic of coloniality continues to increase the "wretches of the earth," as Frantz Fanon had it. Now let's assume that *in spite of* (and not *because of*), some of Nandy's and Shiva's argument could play in the hands of Hindu Nationalist arguments. If we choose that line of argumentation, it would be easy to show that some of Nussbaum's and Sen's arguments play right into the hands of neoliberal arguments. What do I mean by neoliberalism? Briefly, the idea that the market should be the point of reference for the organization of society, to value human capabilities and make many human life dispensable, that trade union should be reduced and citizens *diminished in their social* role and life and *curbed toward their function as human resources and labor providers.* These are some of the principles of neoliberal civilizing ideals. They presuppose precisely the *universality of human capabilities.* More specifically, some of the basic neoliberal principles (that we can compare with the Hindu Nationalism), are based on the *rhetoric of modernity* that, in the sixteenth and seventeenth centuries were mainly based on conversion as salvation; since the nineteenth century, salvation was articulated as the benefit of civilization and, increasingly, of progress. Thus, according to the neoliberal creed, sustained economic growth is the best way to ensure *human progress*. The question is: why shall we assume that *progress* is necessary for *human happiness and good life*? In order to achieve these goals, the market in *neoliberal global designs* shall be free and without interferences from the State in order to ensure that *economic globalization would be beneficial to everyone*. Accordingly, governments should mainly function to provide the infrastructure to advance the rule of law with respect to property rights and contracts. In other words, and in the words that Nanda derives from Nussbaum and Sen, "believe that is both possible and desirable to move all *people*, every where above the threshold for a good life defined in terms of capabilities." Today it is neoliberalism the main agent of moving all people in and out of the benefit of globalization. Instead of "moving all people everywhere above the threshold of for a good life," which is the *rhetoric of modernity actualized by neoliberal principles*, what we have, as Vandana Shiva argued[4], are increasing war, increasing marginalization and poverty and increasing genocides in Africa, and a return to totalitarian regimes forced by the brutal entry of neoliberal religion of the market in Russia and the increasing violence of the *clash of civilization* announced by Samuel Huntington in 1995. Thus, the question raised by Nanda is not only relevant for India. It is a question valid for the entire world as there is an increasing *complicity* (forced or voluntary) between neoliberal agents in Europe and the United States and local supporters and defenders in the rest of the world. On the other hand, there are *increasing* mixed reactions and oppositions to the *universal creed* that progress lifts poverty and to the belief that war is necessary to bring freedom around the world. These "moves" move in two directions. On the one hand, religious fundamentalisms react to the impositions of market fundamentalism, and imperial State terrorism enacts the movement of people supported by the legality of the State itself backed up by military forces. On the other hand, progressive intellectuals like Shiva, Nandy, Nussbaum and Sen (and an increasing number of social movements), all with different agendas, for sure, can make it "possible and desirable" to produce fruitful dialogues and to coordinated and compatible moves. This should be possible once the *pretense of a right universal truth* good for everyone, gives rise *to pluriversal needs and interests, political projects, and ethical commitment to work toward a world in which many worlds would coexist, rather than toward one world in which many world would subsist.*

Of course I have not been implying, as the last sentence makes clear, that Nussbaum and Sen *are neoliberals* in the same way I am not saying that Shiva and Nandy *are Hindu nationalists*. I am saying that Nussbaum and Sen on the one hand and Shiva and Nandy on the other may have *complementary* agendas between them as progressive intellectuals in different locales of the geopolitical distribution of intellectual labor. All of them can be criticized for their positions' support of either the universal totalitarian beliefs that support neoliberal agendas and the nativist totalitarian beliefs that support Hindu Nationalist agendas. Thus, what I am trying to argue is that while Shiva and Nandy are helpful in critiquing neoliberal principles under the feet of Nussbaum and Sen, Nussbaum and Sen are helpful in that they can be used for critiquing the Hindu Nationalist principles under the feet of Shiva and Nandy. A pluritopic critique is of the essence if one wants to move away from the *cage created by modern epistemology and the naturalization of invented dichotomies* which, today, are ending in opposing fundamentalism: the global fundamentalism of the market (e.g., neoliberal civilization) that is being confronted by a variegated emergence of religious fundamentalism (Islamic, Hindu), the dissenting agendas *across* political projects grounded in different geo-historical locations and biographical experiences (such as Nussbaum, Sen, Shiva and Nandy, among others). Geo-historical locations and biographical experiences are not taken here as multicultural manifestations of some universal principle but, quite on the contrary, as the geo-historical distribution of knowledge that the logic of coloniality enacted and naturalized. Colonization of knowledge is a process that to be operative and effective, it needs to create the colonial difference in the domain of knowledge and in the domain of being: "I am not thinking, therefore I do not exist" is a formula of imperial designs instilled among the dominated population through universities, media and all other forms of communication.

If a dialogue between multiple projects moving toward another globalization can take place in order to shift the geo- and biopolitics of knowledge, to decolonize knowledge and being, it would be under the assumptions that many people around the world (for example, Shiva and Nandy in India, Shariati in Iran, Yasinne in Morocco, Dussel in Mexico, Haraway and Hardy, Sen and Nussbaum in the United States, Sousa Santos in Portugal, etc.) are questioning science and neoliberalism in their complicity with imperial global designs. Science cannot be measured and evaluated *only* by the conceptual structure of scientific theories and the *application* across the divides created by imperial discourse (race, gender, sexuality, developed and developing countries, First World and Third World, etc.). Decolonization of knowledge and of being is imperative to shift the geo- and biography of knowledge and transform the colonial epistemic difference (that produces such divides). One has to think only of the fact that among the many issues, science explores to advance and improve human life; it *never explores domination, exploitation, marginalization.* Curiously enough, scientific knowledge's goals are to improve human life (and human knowledge and understanding, as the McArthur Foundation has it) without touching the logic that produces domination and exploitation. Decolonization of knowledge and of being in all its different geo- and biopolitical aspects as illustrated by the previous examples, implies a critique of modernity and modern sciences from the perspective of coloniality. To claim the universality of certain geo- and biopolitical grounded statement in favor of another, as Nanda does using Nussbaum and Sen against Shiva and Nandy, means to fall back into the assumption that certain *abstract universals* (liberals, Christians, Marxists, Islamics) are better than others. And such claims, it doesn't matter who makes them and from what geo-historical and biographical location, have always been the seed of a totalitarian destiny.[5] There are no safe places that could be justified in the name of abstract universals, be them science, Christianity, Marxism, Islamism, etc. Ethics is the bottom line of "a prudent" knowledge for a decent life.

Let's take, for example, two basic principles that Nanda derives from Sen and Nussbaum *to judge whether a society is regressing or progressing*:

> Two levels of human capabilities serve as the objective criteria for judging whether or not any society can be said to be *developing* or *regressing*. The first level sets the ground floor for a life that is suitable for humans as a species distinct from other animals and includes bodily capacities like avoiding hunger and thirst and includes the distinctively human traits like humor and play and the ability to reason and make moral judgments. The next level describes capabilities that make a human life a *good* life. These capabilities include, for instance, the ability to "imagine, to think and to reason" and the ability for critical reflection and crucially for women, the ability to "live one's own life and nobody else's." (Nanda, 231; italics mine, last sentence quotation from Nussbaum)

First of all, what constitutes *good* life and whether society is *developing or regressing* are not only particular points of view, based on the criteria that humans distinguish themselves from nature but, above all, are particular statements that do not pause to evaluate that their own particularity is being assumed as universal. If, at this point in history, one assumes criteria such as the one Nanda derives from Nussbaum and Sen as universal, one is indeed *regressing* in the sense that it is assumed that there is only one set of criteria that names the universal and establishes what is progress, what is stagnation and what is a good life.

Some of Nanda's criticisms of Shiva and Mies on eco-feminism are well taken. In the case of Shiva, however, it would be necessary to distinguish, on the one hand, Shiva's sometimes excessive links between earth and womanhood from where she derives many of her conclusions. However, there are other criticisms that deserve more scrutiny because they reveal some of the basic Nanda' assumptions that are not only different but mainly contrary to Shiva's. For example, Nanda takes on Shiva's assumption that "subsistence economies were the 'original affluent societies,'" in the sense that subsistence economies took care of "basic vital needs of their members" (Nanda, 231). The problem is that for Nanda, Shiva defines the nature of the basic needs and the level at which they could be satisfied by "cultural norms alone" (Nanda, 231):

> By so relativizing needs, Shiva is able to gloss over some grim facts behind her much-celebrated "affluence" of non-Western agricultural systems. The fact [is], for example, that subsistence economies did not (and still do not) supply all the nutritional needs of all their members at level that is biologically adequate for maintaining basic capabilities. (Nanda, 231)

Nanda counters such "cultural relativization of needs" with the challenges she finds in Amartya Sen and Martha Nussbaum in their "capabilities approach to human functioning." The "capabilities approach" is based on a tree-diagram that from a universal principle derives and spreads cultural differences. Thus,

> The capabilities ethics is based upon a *"critical universalism"* that holds a set of basic human capabilities as *intrinsically* worthwhile for a flourishing human life, while admitting that these capabilities may be expressed differently in different cultures and different historical epochs. (Nanda, 231)

The logic of "critical universalism" is the logic that needs and sustains "multiculturalism." That is, multiculturalism is accepted as a matter of fact, on the assumption that there are *some universal principles* from where *cultural differences* emerge in variegated numbers of geographical locations and historical epochs. But, really, who can say, beyond God [any God] or a group of human beings who pretend to take the place of God in a secular form, what "universal principles" are? My own position starts from the premises of an ethics based on *critical cosmopolitanism*[6]; and critical cosmopolitanism begins from the recognition of the "epistemic colonial difference" against which Shiva and Nandy have been struggling and continue to do so; while for Sen and Nussbaum the epistemic colonial difference is out of their sight; that is precisely what "critical universalism" means, blindness to the colonial difference and, consequently, the possibility of conceiving critical cosmopolitanism to erase the colonial epistemic difference. But what does the epistemic colonial difference mean? It means that since the sixteenth century until today, all knowledge that have not been cast in some of the modern European imperial languages (Italian, Spanish and Portuguese, in the first wave of colonial expansion; French, English and German in the second wave, since the eighteenth century), grounded in Greek and Latin, were not and are not sustainable knowledges. Therefore, even if one recognizes that sometimes Shiva or Nandy run the risk of supporting dubious nationalist positions, one shall not fail to recognize that they are struggling in the violent epistemic borders created by the colonial epistemic difference, that is, the difference that makes the difference in the modern/colonial world and that *moves* the system—that is, the hegemonic concept of knowledge (e.g., political theory, political economy, expertise in agriculture and deforestation, in pharmaceutical products, etc.). On the other hand, Sen's and Nussbaum's blindness to the colonial epistemic difference (of which "critical universalism" is a consequence) puts them at risk of supporting fundamentalist positions in power, today, under the name of neoliberal philosophy and civilization for which "critical universalism" is far from being a challenge and even less a threat.

The question is that there is a geopolitical space of knowledge that Nanda overlooks in her resetting of the dichotomy "modernity vs. nativism." That being said, I take her outstanding contribution to be the links she establishes between postmodern and feminist critiques of science, in the United States (Haraway, Harding, etc.) and the matching of this critique with those enacted by Hindu nationalists. However, in between, there is the colonial epistemic difference that is overlooked. Or, if you do not want to reify it, let's say that in my vocabulary that colonial difference(s) and colonial differencial(s) are instances of the same phenomenon that organizes the modern/colonial world. If modernity is left alone, without *its constitutive dark side*, coloniality, then there is no other alternative that conceives society and history in the linear time of *progress*. That is why, for Nanda, the "prophets are looking backward" in which "back" (which is spatial) has the temporal connotations that "primitive" has, for example, acquired in the eighteenth century. If we bring coloniality into the picture and underline the *simultaneity* of histories around the world, then "modernity" and "nativism" are opposed contents of *the same logic*: the *logic of coloniality*.

My endorsement of Shiva's critique of science shall not be interpreted as a move away from one kind of totalitarianism in order to embrace a different one—one from the Third World and women of color perspective instead of from the First World and masculine perspective. A Third World perspective is a claim for epistemic and political rights of existence, simply, that have been denied by First World institutions like the IMF or the Genomic Research Institute or by institutionalized knowledge in sites of high learning in France, Germany or the United States, but it is far from being enough. There are several possibilities open to a feminist of a Third World view. The method is always to see what is being done under the banner of one or another perspective. What we should retain from Haraway is that "objectivity in science" from a feminist perspective cannot be measured by some method that inspects the "correspondence" between the scientific law and scientific description, but the "subjective understanding" from where any scientific claim is made. In the first case ("correspondence"), the "point of view" has been neutralized as if there were only one and that only one was the warranty of objectivity by "science" itself and replaced by "scientific rigor," "objectivity" and, lately, by "excellence." A feminist critique of science denounces the fact that the proclaimed scientific objectivity hides the masculine "subjective understanding" upon which science as a practice was created and claims an "objective understanding" to which all other subjective understanding shall submit. I would like to add that, historically, the idea that objectivity implies a "neutral point of view" is anchored in the zero-point of observation I described at the beginning of my argument, a transference and translation from the "eye of God" as the ultimate warranty to the "eye of Reason" in a secular world. The "discovery" of coloniality has opened up two different doors: one, analytical, leads to the geopolitical and biopolitical territory hidden under the criteria of scientific objectivity (and translated also into technological efficiency). That is to say, "subjective understanding" is unavoidable and if it may not intervene in the construction of a mathematical formula or the particular way in which a lab experiment is conducted, there is no question that subjective understanding will intervene in the why of a mathematical formula or of lab experiment, in the first place, and, in second place, in the ways in which a given research project intersects with social conditions and personal story. Unawareness of such crossing will bring back to the blindness of bad intentions of those who critiqued Wienner for being a superb scientist but a little eccentric in dealing with social issues that were outside his range of expertise! As if to understand social issues you have to be a sociologist and not just an intelligent and critical member of that society!

Shiva, by underlining the Third World instead of the feminist perspective, reveals the fact that science and scientific discourse not only hide their gender and sexuality subjective understanding, but also the fact that science is geopolitically marked and, therefore, enters into a structure of power differentials in which all knowledges that do not fit the mold of what has been self-defined as "science" is disavowed as traditional, unsustainable, folkloric, etc. That is, since the idea of the Third World implied a geopolitical racialization of the interstate system (already implied in the Christian division of the three continents attributed to the hierarchy established among Noah's son (Japheth, Shem and Ham), "science" in the First World offers the necessary "knowledge" to take advantage of the "natural resources" of the Third World where there is no "knowledge" but "culture" and "nature" (Pletstch 1981; Coronil 1997; Escobar 2000). In the larger part of the world, systems of knowledge have been built around the sustainability and the pleasures of life, and not with the primary goal of obtaining economic profit. Imperial expansion, from the sixteenth century to today, was predicated, among other things, about replacing local knowledges IN the colonies with the local knowledge FROM the metropolis. The more "scientific" knowledge embraced the "perspective" and needs of capital "development" (e.g., accumulation), more it disavowed "nonscientific" forms of knowledge:

> In the "scientific" system which splits forestry from agriculture and reduces forestry to timber and wood supply, food is no longer a category related to forestry. *The cognitive space that related forestry to food production,* either directly or through fertility links, *is therefore erased with the split. Knowledges systems which have emerged from the food giving capacities of the forest are therefore eclipsed and finally destroyed, both through neglect and aggression.* (Shiva 1993, 14)

Neither feminist nor Third World perspectives by themselves guarantee anything, since there is no safe place. But feminist and Third World perspectives are absolutely necessary to reveal the narrow and limited scope of "science" in their masculine and First World perspective. Of course no one can claim victory just jumping around with the feminist and Third World flag. It has to be shown that knowledge is not just a conceptual apparatus floating in the spirit of the humanity, but is located in a geopolitic of imperial and epistemic knowledge and in the structure of gender divisions in the Christian and capitalist West.

CONCLUDING REMARKS

Knowledge as founded in alphabetic writing, Christian theology based on the classic tradition and on mathematics and astronomy, became the reference point and the measuring stick for people who were traveling around the world (only Europeans traveled all over the world since the end of the sixteenth century) and classifying all other types of knowledge. If Copernicus, for instance, did not have much to do, directly, with colonization, he greatly contributed to set up the perspective of European modernity and to eclipse coloniality.

Bacon wrote:

> For first, I propose a natural history that does not so much charm with its variety or gratify by the immediate fruit of experiments, as provide light for the discovery of causes and supply the first mother's milk to the infant philosophy. (Bacon, 25)

> For man is only the servant and interpreter of Nature and he only does and understands so much as he shall have observed, in fact or in thought, of the course of Nature. . . . No force whatever can unfasten or break the chain of causes, and Nature is only overcome by obeying her. So it is that those two objects of mankind, *Knowledge and Power*, come in fact to the same thing; and the failure of works derives mostly from ignorance of causes. (Bacon, 29)

The unsaid, in Bacon's dictum, is that what he calls *Knowledge and Power* is only the *Modern* perspective, that is, the perspective Bacon is advancing as one key figure of *modernity*. Hidden from view is the coloniality of power and knowledge. That is, everything that the *Modernity of Power and Knowledge* disqualifies by asserting itself as one perspective that becomes "better" and "universal." For, what was wrong with cosmologies different from Bacon's beyond the fact that other cosmologies needed to be wrong to make *Modernity of Power and Knowledge* stand out as true. Bacon's strategy by asserting a secular conception of *Modernity of Power and Knowledge* operated under the same logic implemented by Christian Theology in the previous centuries and, above all, in the sixteenth century, when it had to account for an unknown part of the world and an unknown variety of people. *Modernity/Coloniality* came into the picture together with the construction of the *epistemic colonial difference*.

Since there is no safe place, the decolonization of knowledge and of being need to start by uncoupling science from ethics and politics (which of course have to be the space where the balance between science and economy takes place). "A prudent knowledge for a decent life" requires that science be at the service of strengthening and improving *human relations* (not just a minority of human lives in detriment of the other), and not to improving personal fortunes. The paradigm of newness encourages competition, war, fight for privileges and individual "success." The paradigm of coexistence puts social life over individual success and accumulation of wealth; it works toward a pluriversal (rather than a universal) world (a world in which many worlds could coexist). The critique of science attempted here shall not be confused with the benefit humanity can draw from "scientific" invention, creativity and explorations. The critique of science was its complicity, in the modern/colonial world, with a capitalistic ethics of success and a political theory that enforces and reinforces the complementary relations between accumulation of knowledge and accumulation of life and that only rhetorically will accept that what we need is a prudent knowledge for a decent life; we do not need celebration of achievements that contribute to the increasing marginalization and expendability of human lives.

NOTES

1. Geopolitics has always been hidden in the assumptions, and some time claims, that European scientist and philosopher made about knowledge. Geopolitics of knowledge, as an analytical tool, allows for unveiling the veiled and localizing the universal. A case in point, Charles Bambach's study of Heidegger's philosophy, particularly chapter 3 "The Geo-Politics of Heidegger's Mitteleeuropa" (Bambach 2003, 112–79).

2. Long time ago politics scientist Karl Pletsch (Pletsch 1981) made the point clearly mapping the distribution of scientific knowledge during the Cold War: First World have objective science, together with democracy; Second World have science, but not democracy and nonobjective (that is, ideologically driven); while the Third World has culture, but not science. Pletsch, like Wienner, was silenced in the field of political (and social sciences) in general.

3. "Adorno in India. Revisiting the Psychology of Fascism" in *Exiled at Home*; Delhi: Oxford University Press, 1998, 104–5, italics mine.

4. See Vandana Shiva, "Globalization and Its Fall Out," http://www.zmag.org/sustainers/content/2003-04/02shiva.cfm.

5. See Susan George, "A brief history of neoliberalism," http://www.globalexchange.org/campaigns/econ101/neoliberalism.html.

6. "The Many Faces of Cosmopolis: Border Thinking and Critical Cosmopolitanism," Public Culture 12.3, 2000.

REFERENCES

Alcoff, Linda. 1993. *Feminist Epistemologies*. Ed. Linda Alcoff and Elizabeth Potter. New York: Routledge.

Bambach, Charles. 2003. *Heidegger's Roots. Nietzsche, National Socialism and the Greeks*. Ithaca, NY, and London: Cornell University Press.

Castro-Gomez, Santiago. 2002a."The Social Sciences, Epistemic Violence, and the Problem of the 'Invention of the Other." *Critical Conjunctions· Foundations of Colony and Formations of Modernity*. Special Issue of *Nepantla: Views from South*, 3 (2):269–86.

———. 2002b. "Not Longer Broad but Still Alien Is the World: The End of Modernity and the Transformation of Culture in the Times of Globalization." *Latin American Perspectives on Globalization: Ethics, Politics and Alternative Visions*. Ed. Mario Sáenz. Foreword by Linda Alcoff. Boulder, CO: Rowman and Littlefield Publishers, 25–39.

Coronil, Fernando. 1997. *The Magical State: Nature, Money and Modernity in Venezuela*. Chicago: The University of Chicago Press.

Crombie, A. C. 1979. *Augustine to Galileo*, vol. 1. "Sciences in the Middle Ages," 5th to 13th Century. Cambridge, MA: Harvard University Press.

Escobar, Arturo. 1999. *El final del salvaje. Naturaleza, cultura y política en la antropología contemporánea*. Bogota: CEREC-Instituto de Antropología.

Eze, Emmanuel Chukwudi. 1997. "The Color of Reason: The Idea of 'Race' in Kant's Anthropology." *Postcolonial African Philosophy*. London: Blackwell, 123–70.

———. 2002. *Achieving Our Humanity: The Idea of the Postracial Future*. New York: London.

Gibson-Graham, J. K. 1996. *The End of Capitalism (As We Knew It). A Feminist Critique of Political Economy*. London: Blackwell.

Ginzberg, Ruth. 1989. "Uncovering Gynocentric Science." *Feminism and Science*, ed. Nancy Tuana. Bloomington: Indiana Univesity Press, 69–85.

Gould, Stephen Jay. 1987. *Time's Arrow, Time's Cycle: Myth and Metaphor in the Discovery of Geological Time*. Cambridge, MA: Harvard University Press.

Grosfoguel, Ramon, and Ana Margarita Cervantes-Rodriguez, ed. 2002. *The Modern/Colonial/Capitalist World-System in the Twentieth Century: Global Processes, Antisystemic Movements and the Geopolitics of Knowledge.* Westport, CT: Greenwood Press.

Haraway, Donna J. 1991a. "A Cyborg Manifesto: Science, Technology, and Socialist-Feminism in the Late Twentieth Century." In *Simians, Cyborgs, and Women: The Reinvention of Nature.* New York: Routledge, 127–48.

———. 1991b. "Situated Knowledges: The Science Question in Feminism and the Privilege of Partial Perspective." In *Simians, Cyborgs, and Women: The Reinvention of Nature.* New York: Routledge, 183–202,

Harding, Sandra. 1986. *The Science Question in Feminism.* Ithaca, NY: Cornell University Press.

———. 1989. *Is Science Multi-Cultural? Postcolonialism, Feminisms, and Epistemologies.* Bloomington: Indiana University Press.

Hegel, George W. F. 1991. *The Philosophy of History.* Trans. J. Sibbree. New York: Prometheus Books.

Kant, Immanuel. 1763/1960. *Observations on the Feeling of the Beautiful and Sublime.* Trans. John T. Goldthwait. Berkeley: University of California Press.

Lander, Edgardo. 2002a. "Los derechos de propiedad intelectual en la geopolítica del saber de la sociedad global." In *Indisciplinar las ciencias sociales. Geopolíticas del conocimiento y colonialidad del poder*, ed. Catherine Walsh, Freya Schiwy, and Santiago Castro-Gomez. Quito: Universidad Andina y Abya-Yala, 73–102.

———. 2002b. "Eurocentrism, Modern Knowledges, and the 'Natural' Order of Global Capital." In *Critical Conjunctions: Foundations of Colony and Formations of Modernity*, ed. Saurabh Dube, Ishita Banerjee Dube, and Edgardo Lander. Special Issue of *Nepantla: Views from South*, 3/2, 245–68.

Maldonado-Torres, Nelson. 2002. "Postimperial Reflections on Crisis, Knowledge and Utopia. Transgresstopic Critical Hermeneutics and the 'Death of European Man.'" Special issue on "Utopian Thinking," ed. Roman Grosfoguel, *Review. Historical Systems and Civilizations, A Journal of the Fernand Braudel Center for the Study of Economies* 25 (3):277–317.

———. 2003a. *Thinking from the Limits of Being: Levinas, Fanon, Dussel.* New York: Routledge.

Mignolo, Walter D. 1995. *The Darker Side of the Renaissance: Literacy, Territoriality and Colonization.* Ann Arbor: The University of Michigan Press.

———. 2002a. "The Geopolitics of Knowledge and the Colonial Difference." *South Atlantic Quarterly* 101 (1):57–96.

———. 2002b. "José de Acosta's *Historia natural y moral de las Indias*: Occidentalism, the Modern/Colonial World, and the Colonial Difference." In *Natural and Moral History of the Indies*, ed. Jane E. Mangan. Trans. Frances Lopez-Morillas. Durham, NC: Duke University Press, 451–518.

———. 2002c. "The Zapatistas's Theoretical Revolution: Its Historical, Ethical, Political and Epistemological Consequences." Special issue on "Utopian Thinking," ed. Roman Grosfoguel, *Review. Historical Systems and Civilizations, A Journal of the Fernand Braudel Center for the Study of Economies* 25 (3):245–77.

———. 2003. "The Enduring Enchantment (Or the Epistemic Privilege of Modernity and Where to Go from Here)." *SAQ* 101 (4):928–64.

———. 2003a. "'Un paradigma otro': colonialidad global, pensamiento fronterizo y cosmopolitanismo critico." *Historias Locales/Deseños Globales: colonialidad, conocimientos subalternos y pensamiento fronterizo*. Madrid: Ediciones Akal.

———. 2003b. "Globalization and the Geopolitics of Knowledge: The Role of the Humanities in the Corporate University." *Nepantla: Views from South* 4.1.

———. *Local Histories/Global Designs: Coloniality, Subaltern Knowledges and Border Thinking*. Princeton, NJ: Princeton University Press.

Nanda, Meera. 2003. *Prophets Facing Backward. Postmodern Critiques of Science and Hundu Nationalism in India*. New Burnswick, NJ: Rutgers University Press.

Nandy, Ashis. 1994. *The Illegitimacy of Nationalism. Rabindranath Tagore, and the Politics of Self*. New Delhi: Oxford Press.

Philip, Kavita. 2001. "Seeds of Neo-Colonialism? Reflections on Ecological Politics in the New World Order." *CNS* 12.2: 4–47.

Pletsch, Carl E. 1981. "The Three Worlds, of the Division of Scientific Labor, circa 1950–1975." *Comparative Studies in Society and History* 23 (4):565–90.

Prigogine, Ilya. "Science, Civilization and Democracy: Values, Systems, Structures and Affinities." *Futures* 18 (4):493–507.

Readings, Bill. 1996. *The University in Ruin*. Cambridge, MA: Harvard University Press.

Santos, Boaventura de Sousa. 1995. *Toward a New Common Sense: Law, Science and Politics in the Paradigmatic Transition*. London: Routledge.

———. 1998. "De la idea de la universidad a la universidad de las ideas." *De la mano de Alicia. Lo social y lo político en la postmodernidad*. Bogota: Universidad de los Andes.

———. 2001. "Toward an Epistemology of Blindness: Why the New Forms of 'Ceremonial Adequacy' neither Regulate nor Emancipate." *European Journal of Social Theory* 4 (3):251–79.

———. 2002a. *Um Discourse sobre las Ciencias*. Portugal: Edicoes Afrontamento. (Orig. pub. 1987.)

Shiva, Vandana, ed. 1993. *Monocultures of the Mind: Perspectives on Biodiveristy and Biotechnology*. London: Zed Books, and Malaysia: Third World Network.

———. 1994. *Biodiveristy Conservation. Whose Resource? Whose Knowledge?* New Delhi: Indian National Trust and Cultural Heritage.

Wynter, Sylvia. 1991. "1492: A New World View," in *Race, Discourse and the Origin of the Americas: A New World View*. Ed. Vera Lawrence Hyatt and Rex Nettleford. Washington: Smithsonian Institution, 5–57.

———. 2003. "Unsettling the Coloniality of Being/Power/Truth/Freedom. Towards the Human, After Man. Its Overrepresentation. An Argument." *New Centennial Review* 3 (3):257–337.

19

Beyond Eurocentrism: Systematic Knowledge in a Tropical Context. A Manifesto

Orlando Fals-Borda and Luis E. Mora-Osejo

On January 4, 2001, in Bogota, Colombia, the authors issued a self-described "manifesto"[1] to Colombian scholars inviting them to rescue local values of self-esteem and creativity and to resist intellectual colonialism by European and North American colleagues. After months of discussion, they decided to revise the document and publish an expanded text. Eyebrows were raised not only because this is an interdisciplinary document (signed by a biologist and by a sociologist without any visible signs of disagreement on such sensible matters) but also because both are known to have been trained and given doctorates during the 1950s at the University of Florida and Mainz Johannes-Gutenberg University, respectively.

TOWARD A CONTEXTUAL HYPOTHESIS

We both have admitted and still believe that we received the best possible training from our professors, who were world-recognized authorities. We assimilated their teachings eagerly and transferred their frames of reference, methodological and theoretical orientations into Colombia. Without being fully conscious of it, we also transmitted their thematic preferences, naturally derived from the actual circumstances of their background, their times and their institutions. This somewhat spontaneous transfer of knowledge did serve a good purpose in helping to awaken to Colombian conditions.

In Colombia in the 1950s universities had been ignored by governments and left without sufficient resources. Moreover, traditional elites were inclined to imitate and adopt foreign traits and cultures. We both were given good opportunities to serve our society, and we were left mostly to ourselves in regard to the nature and content of the subjects we taught. Our positive educational direction became strained after about a decade, when we noticed that some central concepts, principles and theories learned abroad did not find commensurable replication or confirmation in the conditions in which we were living and working. Social, economic and political factors were so resistant that they virtually obliterated much of the respected academic rule of the knowledge accumulated abroad.

In the social sciences, we were challenged by structural and open violence, an intractable monster still at large; in the natural sciences, especially biology, the challenge came from observing and dealing with the unique characteristics of the tropical milieu in the Amazonian and Andean regions, all of which contradicted well-known frames of reference. These discrepancies and negativities were of such nature and scope that we felt we had to relearn much of our disciplines. We discovered the obvious fact that contexts mattered for meaningful observation and abstraction, and to deal in a satisfactory way with natural and social phenomena.

After much deliberation we concluded that we had to fly alone. At the time of our Manifesto in 2001, we agreed to frame our concern for contextualization in the following manner:

> As human products, scientific frames of reference receive inspiration and concreteness from cultural, historical, and geographical contexts. This is a universal process with different modalities, which are justified by the search for a fullness of life and spiritual as well as material satisfaction for those who intervene and share in the creativity/inquiry process and in its diffusion. (2001, 7)

This is nothing new, of course, and during our university training there had been a nebulous understanding of what was meant with "context." We tried to combine the contextual hypothesis with the more contemporary concept of endogenesis, as expressed in our Manifesto, where *endogenesis* is a biological term referring to "growth from within."

With the advantage of hindsight, after fifty years of experience and study in tropical and violent areas in Colombia, we realize that we could have done better had we reread some of our Euroamerican university textbooks and manuals more critically. As a belated exercise, we did so with unsatisfactory results. We understood context better through contact with, or by reading from, local respected elders and sages in our base or ethnic communities, or by reflecting on our own socialization at home.

As a theoretical construct, contextualization could be seen as a widespread principle of life, almost common sense. Although it had not been a clear-cut heritage from our university training in Europe and the United States, it could be useful nevertheless for our scientific and disciplinary purposes in Colombia. From the standpoint of sociology, one could find passing references to "context" in the revered essays by W. F. Ogburn (1912–1956) (1957), and we could relate it to the wellknown principle of "definition of the situation" by W. I. Thomas and F. Znaniecki (1958, 43), which included recognition of the role of church and family. But the same idea is of course older, if we remember Blaise Pascal's dictum that "what is truth on one side of the Pyrenees is error on the other."

Somewhat clearer in the broader sense of the term "context" that we use in the Manifesto, where the idea is considered as "perspective/vision" or "Weltanschauung" as in the sociology of knowledge when examining different paradigms. Karl Mannheim (1936, 89) stated that "no human thought is immune to the ideologizing influences of its social context," including common sense knowledge. A similar approach is found in Berger and Luckmann when they state that "specific agglomerations of reality and knowledge pertain to specific social contexts, and these relationships will have to be included in an adequate sociological analysis of these contexts" (1966, 13). Yet context is still limited to the social, and seen mostly as a latent or constant factor.

From the standpoint of the natural sciences, we discovered considerable assistance in the work of philosophers of biology like Ernst Mayr (1988) who had been fighting mechanistic and deterministic interpretations. Mayr appealed to the concept of "living systems" as "complex open systems" and defined them as follows: "The complexity of living systems exists at every hierarchical level from the nucleus to the cell, to any organ system, to the individual, the species, the ecosystem, the society" (p. 14). This definition broadened the scope of the life sciences and social sciences. It was flexible enough to include our idea of contextualization as a dynamic, open phenomenon that implies a respect for, and concern with, meanings, symbols, discourses, values and norms connected with a complex time-and-space dimension that is ecological, social and cultural.

LAGS AND PARADIGMS

Once we adopted an open and systemic contextual hypothesis as an instrument to determine an ethos or collective vision of a "mother society" with living beings, observers, and creators, we could move on to criticize Eurocentric paradigms and Eurocentric teachings by locating them where they belong: within their own time/space realities and milieux. This observation can induce independence among intellectuals in poorer nations, as inside observers and actors who are more fully in charge of their own ways and means, including the mysterious patterns of our unique tropical world. It would have been expected that the idea of Eurocentrism, as a formal concept, would not be in the curricula of European and/or American universities, and this was not the case when we were students there. Yet there have been some self-critical attitudes among radical intellectuals, like Sartre and Rolland. References can be found in the works of Edward Said, Immanuel Wallerstein, André Gunder Frank.

The first systematic treatise on the subject of Eurocentrism comes from the Egyptian economist Samir Amin (1985). For Amin, Eurocentrism is a culturalist expression of ongoing capitalist world expansion. Like its better-known political cousin, Development, Eurocentrism proposes the western mode of life, economy and culture as a model to be adopted by the rest of the world, as the only solution to the challenges of our times. It fits well with the ideologies of globalization. However, as European and North American scientific paradigms have been conceived in the contexts of temperate zones and their historical, cultural and material development, they are similarly conditioned by those contexts in the determination of collective thinking and action. These paradigms have spread to the rest of the world creating a breach between their societies and ours in the South. Advanced countries have been able to develop a strong knowledge-building capacity, in part with inputs from resources and riches of our southern nations. Some of those paradigms became dominant in many societies but the resulting imbalance has produced not only a powerful economic base in the North supported by sophisticated technologies but also a one-sided worldwide political and economic system posed to favor the richer countries where southern realities and facts may be unknown, disregarded, or unilaterally exploited.

At least on the basis of the contextual hypothesis, it appears to us that those ex novo facts do not make Euroamerican scientific paradigms—notably Cartesian positivism, Newtonian mechanism, Marxian determinism, and Parsonian functionalism—any better, superior or more pertinent for local purposes, than those generated in other parts of the world. They are conditioned by history, culture and environment. Descartes' thinking was molded by the insecurities of the Thirty Years War and his secular dualism and belief in control of nature were understandable rebellions and alternative quests. Today what may best be recovered from him may be his revolt against the status quo including traditional universities. Newton, like Talcott Parsons, was overtaken by the relativity theory and Heisenberg's uncertainty principle, although some of Newton's camouflaged insights taken from medieval magicians could still be worthy of attention today. And Marx and his European followers should be understandably forgiven for their lack of knowledge of Latin American history and culture.

Paradigms derived from the reflections of those great men are now like still pictures on a movie camera, often just parroted and being presented as dominant patterns of thinking for contemporary problems in the so-called Third World and elsewhere. We forget that they are all social constructs and are subject to revision and interpretation. It is therefore understandable that if a scientific frame of reference is not well rooted in its milieu, theoretical and practical lags would occur with consequent dysfunctional implications for sociopolitical and other systems. We also state in our Manifesto that such situations are worsened when frames of reference are copied or badly adapted from extraneous paradigmatic sources, like those implied by Eurocentrism and Development. Their imitations often become a source of local disorganization and anomie. For example, the Cartesian rationality of the "Green Revolution" of the 1950s was bent against those it purported to serve, the poor peasants; advanced machinery created difficult lag problems when imported to retarded economies; and functional American institutions found strong cultural resistance when tried elsewhere. Such has been the fate of much of Development in all these years since it was invented in 1949 by the American government.

More endogenous paradigms rooted in our own realities and circumstances would likely have contributed more effectively to our progress. Our experience has shown that such alternative paradigms, more open than those imported from advanced countries, could offer constructive ways out of our problems, as suggested next.

COMPLEXITY IN THE TROPICS

Life conditions in the Andean and Amazonian tropical regions are unique and diverse. They need to have adequate and proper explanations, management techniques, and governmental institutions according to local exigencies. We detail these processes in our Manifesto. From the scientist's point of view, the knowledge of local realities turns out to be as rich and useful as made possible by his/her personal involvement, that is, through life experience or "vivencia." Scientific insight and authority come from this involvement with real life. We have learned that endogenesis of this kind opens the way for useful discoveries and initiatives apt to alleviate social problems within the local world. We as insiders to the tropics are in a privileged position to produce, analyze and systematize this knowledge with the help and contribution of autochthonous peoples.

Fortunately awareness of these abilities is increasing, as witnessed by local inventions by Andean farmers (improved seeds and plants, potato sorters, and the like) and city dwellers (in community improvement programs, dance and music) and in civic movements like peace programs, land takeovers from *latifundists*[2] and water distribution systems.

We know that environmental factors in the tropics are complex and clearly distinct from those of other world zones, notably the temperate ones. Yet we have received from the North, often with good intentions, economic, social, cultural paradigmatic and developmental advice that many of us have considered final and sufficient. This limiting procedure has had the additional effect of nullifying or undervaluing the intellectual production of our people, which may be more pertinent to our needs. Intellectual colonialism on our part is one negative result, as shown in many published works. Such extraneous anel/or incompatible formulae are precisely the ones that in our zones have had negative environmental impact. For example, our forests utilize subtle ways to gain mineral nutrients from poor soils, as is the case with leaf decomposition induced by rain, then captured by mychorrhizal fungi, which then return the nutrients to the trees. This is a continuous growth cycle that our aboriginal and peasant communities understand well. In fact they have created or discovered for centuries many varieties of plants useful for mankind, and they have developed basic behaviors and social organs congruent with those processes. Yet dominant paradigms constructed in temperate zones usually cannot accommodate such indigenous, ancient wisdoms. They are closed to them and lead to tragic mistakes like those caused by transferring neat, homogeneous planting systems from North America to Amazon pluvial bushes. It is known that the highest indices of organic diversity are found in our territories, as for instance in savannah lands, coral reefs, and deep-sea waters, besides humid forests. As suggested in the Manifesto, we need to enrich traditional knowledge about these systems with academic knowledge about their potential, and also to develop effective procedures for their sustainable use.

Steps are being taken in this direction. Similar challenges are met in customs, values and patterns of social organization that we have given ourselves. This is important because it is also among us where there is the greatest loss in biodiversity, and where there are the biggest threats for the survival of life not only here but also for the entire world.

NEED FOR ENDOGENESIS

Explanatory and reproductive endogenesis is necessary. It is required by the infinite contextual reality of our milieu, a task that is not adequately anticipated by Eurocentric paradigms, techniques and institutions. We are more fully aware of the marked differences of our local world, especially in fragility and complexity in regard to climate, soil and ecosystems as compared with those of other zones. This, in turn, conditions human behavior and enriches cultural patterns. The harmonious reconstruction of the relationship of people and nature in our country implies a rediscovery of the peculiarities of our daily living and our socialization processes. Independent scientific research focused on the intricacies of our medium and on our specific social, cultural and historical development is proving to be useful for our needs, as described above. Let us remember that the tropical climate is characterized by the Circadian seasonal rhythm—summer in daytime, winter at night. This condition is accentuated as one goes higher into the mountains. It is also characterized by intermittent oscillations of radiation, relative humidity and temperature during the daily light cycle, notwithstanding the stability shown by monthly averages. Moreover, even in reduced tropical forests there are hundreds of trees, but each species has just a small number of individuals in place. The megafauna is also found in small numbers. The habitat appears then to be structured like a fine mesh of specific niches. Such is the way in which the immense biodiversity of the tropical ecosystems is formed, in the lowlands as well as on the mountains. This process also gives rise to ecological patterns of thought, feeling and action in our cultural and ethnic groups.

It may be seen that with these dynamic contextual flows effective solutions can be gained which cannot be transposed from foreign, authoritative or closed constructs. It is increasingly important for academics and practical leaders to exercise our investigative self-discipline in observing realities and making inferences accordingly.

CREATIVITY AND KNOWLEDGE ACCUMULATION

Our Manifesto ends with an invitation. Although it is possible, logical and convenient to develop scientific paradigms and technical frames of reference with our own intellectual and practical means, we cannot do this ignoring what is foreign. However, intellectual colonialism and development cannot continue as we form reference groups of our own. This task is best done when it is interdisciplinary and open, and when it anticipates possible universal interest. To this end the capacity of our local human element has been amply confirmed for centuries. We have been able to accede readily to the imbedded factors of our human and natural environments, to be socially and culturally creative with universal recognition and to produce and invent in material fields with traditional and modern know-how. We have done this up to the present, albeit hampered by poverty; ignorance and disease; political discrimination and rigid seigniorial structures; economic dependence and exploitation; violent upheavals; and mental and spiritual captivity. Of course we have tried not to become xenophobic nor isolated from the intellectual world. But we have tried to comply with the need to accumulate diverse patterns of thought and action congruent with our specific modalities of growth and progress—a process defined as "knowledge summation" or "convergence" in the Participatory Action Research (PAR) school that we developed in the so-called Third World.

We cannot overlook the type of intellectual accumulation that has occurred in the North with its technical superiority. Yet such northern vision and know-how can be more useful and pertinent to the South if they become horizontal and symmetric with enough respect for what we in the South have learned and discovered on our own with PAR and other schools, and with the concourse of the common peoples.

The twenty-first century has opened with a battery of new critical tools inspired in diverse consolidated systems of thought, yet they can converge. Such may be the course with PAR. This may be fortunate, as such convergent systems can help all of us—in the North as well as in the South—to understand those complex, irregular, multilinear and fractal dimensions of our tropical social and natural structures. In this manner we can work together more constructively. For example, European complexity and open systems theories like those by P. B. Checkland, Robert Flood or Ernst Mayr can be enriched with the findings of Chile's Humberto Maturana or of the Colombian Desana Indians' "circuit of the biosphere"; the theory of chaos (Mandelbrot, Prigogine) can be refreshed with daily-living studies like those of Venezuela's Jeanette Abuabara; Peter Reason's "participatory worldview" can be contextualized with Father Camilo Torres' "participant utopia"; and holism (cf. Gregory Bateson, Fritjof Capra) can find support in aboriginal and oriental thinkers.

An alliance of North and South colleagues sympathetic to these intellectual, social and political developments may be formed by all of us who are interested in similar problems and motivated by convergent interests—an alliance between peers that could everywhere face up to the structural injustices and global defects of the modern world (cf. Fals-Borda 1996). With such objectives in mind, we can stimulate combined research-and-teaching attitudes and practices within and outside educational institutions which are able to overcome discriminatory distinctions, such as those between the academic and the popular, between the scientific and the political, and to stimulate self-esteem among our peoples and in our academic communities. This is indispensable for satisfactory living in countries as privileged as Colombia.

NOTES

1. The full text of the original manifesto can be seen in *Globalisation, Education and Society,* 1 (1) (March 2003).
2. A *latifundista* is the owner of a large estate.

REFERENCES

Amin, S. 1985. *Eurocentrism: Critique of an Ideology*. New York: Monthly Review Press.

Berger, P., and T. Luckmann. 1966. *The Social Construction of Reality*. New York: Doubleday.

Fals-Borda, O. 1996. "A North-South Convergence on the Quest for Meaning." *Collaborative Inquiry* 2 (1):76–87.

Fals-Borda, O., and L. E. Mora-Osejo. 2003. "Manifesto." *Globalisation, Education and Society* 1(1).

Mannheim, K. 1936. *Ideology and Utopia*. London: Routledge and Kegan Paul.

Mayr, E. 1988. *Toward a New Philosophy of Biology*. Cambridge, MA: Harvard University Press.

Ogburn, William F. 1957. *On Culture and Social Challenge*. Chicago: University of Chicago Press.

Thomas, W. I., and F. Znaniecki. 1958. *The Polish Peasant in Europe and America (1918–1920)*, vol. 1. New York: Dover.

20

From an Epistemology of Blindness to an Epistemology of Seeing

Boaventura de Sousa Santos

INTRODUCTION

In his celebrated essay of 1898, Thorstein Veblen criticizes classical economics for promoting an impoverished, tautological or circular relation between facts and theory, a relation that he designated as "ceremonial adequacy" (1898, 382). Once the laws of the normal and of the natural are formulated "according to a preconception regarding the ends to which, in the nature of things, all things bend" (1898, 382), the facts either corroborate such concept of normality and the propensity to predefined ends, and are established as relevant, or they don't, in which case they are discarded as abnormal, marginal, irrelevant. Veblen's plea was for the replacement of this normative and illusory adequacy for a real one, the abandonment of a "metaphysics of normality and controlling principles" for the observation of the real economic life process made of real economic actions by real economic agents.

With this plea, Veblen launched a debate in economics which has been with us ever since in all the social sciences and indeed in science as a whole. The debate can be formulated in the following terms: what counts as representation, if representation counts at all? and what are the consequences of misrepresentation? The most intriguing features of this debate are, on the one hand, that it is by far easier to establish the limits of a given representation than to formulate a general coherent representation of limits and, on the other hand, that the consequences of misrepresentation tend to be different from those predicted, thus confirming, if nothing else, the misrepresentation of consequences. In other words, it has been much easier to criticize ceremonial adequacy than to create a credible alternative to it. Veblen illustrates this condition very well. At the outset of his article he mentions approvingly and as an example to follow the "eminent anthropologist" M. G. de Lapouge, whose work is given as a symbol of the evolutionary revolution going on in other sciences (1898, 373). If, however, we read the article by Lapouge and note the scientific results accepted by Veblen we are confronted with a delirious racial anthropology in which the binary of dolichocephalic-blond and brachycephalic types account for such laws as the law of the distribution of wealth, the law of attitudes, the law of urban indices, the law of emigration, the law of marriages, the law of the concentration of dolichoids, the law of urban elimination, the laws of stratification, the law of intellectual classes and the law of epochs.

The dilemma of this harsh evaluation of Lapouge's evolutionary science and the way Veblen draws from it lies precisely in that it is almost self-evident. The blindness of others, particularly of those in the past, is both recurrent and easy to establish. But if that is the case, whatever we say today about the blindness of others will probably be seen in the future as evidence of our own blindness. The dilemma can be formulated thus: if we are blind, why is it so difficult to accept our own blindness? And, if that is the case, what is the point of seeing at all? My contention is that the consciousness of our own blindness, which we are forced to exercise while unveiling the blindness of others, should be at the core of a new epistemological stance which calls for a plurality of knowledges and practices, since no knowledge or practice in isolation provides reliable guidance; and for an edifying, socially responsible, rather than technical, application of science, since the consequences of scientific actions tend to be less scientific than the actions themselves.

In this chapter I address the issue of the resilience of ceremoniality in our scientific management of adequacy. Accordingly, I concentrate on the two steepest slopes of the debate: the issue of the representation of limits and the issue of the misrepresentation of consequences. Concerning the first issue, the representation of limits, I will argue that the most intractable difficulty lies in that for modernity and for modern science there are indeed no insurmountable limits. Accordingly, the representation of limits is as provisional as the limits it represents. Concerning the second issue, the misrepresentation of consequences, I will argue that the project of modernity anticipated two mutually constituted consequences of modern science: social regulation and social emancipation. However, to the extent that the possibilities of modernity were reduced to those of capitalism, the two consequences were torn apart: some knowledges and social practices, by far the dominant ones, took social regulation as the primordial consequence of their endeavors, while subordinate knowledges and social practices took social emancipation as their privileged consequence. The problem, however, is that, in this paradigm, the regulation that does not emancipate does not even regulate and, vice versa, the emancipation that does not regulate does not even emancipate.

KNOWLEDGE-AS-REGULATION AND KNOWLEDGE-AS-EMANCIPATION

Before I address the two issues—the representation of limits and the misrepresentation of consequences—I must situate them in a broader epistemological landscape, the project of western modernity as an epistemological paradigm.

The project of western modernity is organized around a discrepancy between social experience and social expectations, herein lying its utmost novelty. For the first time in western history, experience does not have to and indeed should not coincide with expectations. Seen from the perspective of social experiences, social expectations are excessive and, vice versa, seen from the perspective of social expectations, social experiences are deficient. The normality and symmetry of this disjuncture is rendered by the twin pillars upon which western modernity is based: the pillar of social regulation and the pillar of social emancipation. Each one of them is constituted by three principles or logics.[1] The pillar of regulation is constituted by the principle of the state, formulated most prominently by Hobbes; the principle of the market, developed by Locke and Adam Smith in particular; and the principle of the community, which presides over Rousseau's political theory. The pillar of emancipation is constituted by three logics of rationality as identified by Weber: the aesthetic-expressive rationality of literature and the arts; the cognitive-instrumental rationality of science and technology; and the moral-practical rationality of ethics and law. The tensions between the two pillars, which derive from the play of excesses and deficits in the discrepancies between experience and expectations, are to be managed by the exercise of reason as a normal state of affairs designated as progress. The project of modernity thus aims at a harmonious and reciprocal development of both the pillar of regulation and the pillar of emancipation whereby the harmonization of potentially incompatible social values, such as justice and autonomy, solidarity and identity, equality and freedom, will be ensured.

The epistemological dimension of the paradigm of modernity matches the scope and the structure of the double binding of social regulation and social emancipation. We know that any form of knowledge implies a trajectory or progress from point A, designated as ignorance, to point B, designated as knowing. Forms of knowledge are distinguished by the way they characterize both the two points and the trajectory that connects them. There is, therefore, neither ignorance in general nor knowing in general.

Each form of knowledge recognizes itself in a certain kind of knowing to which it opposes a certain kind of ignorance, which in its turn is recognized as such only in contrast with that kind of knowing. All knowing is knowing of a certain ignorance, as all ignorance is ignorance of a certain knowing.

The paradigm of modernity comprises two main forms of knowledge: knowledge-as-emancipation and knowledge-as-regulation. Knowledge-as-emancipation entails a trajectory between a state of ignorance, which I call *colonialism* and a state of knowing, which I call *solidarity*. Knowledge-as-regulation entails a trajectory between a state of ignorance which I call *chaos* and a state of knowing which I call *order*. While the former form of knowledge progresses from colonialism toward solidarity, the latter progresses from chaos toward order. In the terms of the paradigm, the mutual binding between the pillar of regulation and the pillar of emancipation implies that these two forms of knowledge balance each other in a dynamic way. What this means is that the knowing power of order feeds the knowing power of solidarity, and vice versa. Knowledge-as-emancipation derives its dynamics from the excesses of order, while knowledge-as-regulation derives its dynamics from the excesses of solidarity (Santos 1995, 25).

My argument is that this social and epistemological paradigm suffered a historical accident. Something happened that only retrospectively can be conceived of as inscribed in an evolutionary necessity: from mid-nineteenth century onwards the possibilities for the implementation of the paradigm of modernity were reduced to those made available by world capitalism. This accident produced an enormous turbulence between social regulation and social emancipation, which eventually led to the cannibalization of social emancipation by social regulation; from being the *other* of social regulation, social emancipation was transformed into the double of social regulation. But since social regulation does not sustain itself without its other, its cannibalizing social emancipation led to a double crisis of regulation and emancipation, each one feeding on the other. This is the situation in which we now find ourselves. At the epistemological level this historical process led to the total primacy of knowledge-as-regulation over knowledge-as-emancipation: order became the hegemonic way of knowing while chaos became the hegemonic form of ignorance. Such imbalance allowed knowledge-as-regulation to recodify knowledge-as-emancipation in its own terms. Thus, knowing in knowledge-as-emancipation became ignorance in knowledge-as-regulation (solidarity was recodified as chaos), and, conversely, ignorance in knowledge-as-emancipation became knowing in knowledge-as-regulation (colonialism was recodified as order). My argument is that the persistence of ceremonial adequacy and its problems, both as concerns the representation of limits and the misrepresentation of consequences, have much to do with the conversion of order as a way of knowing into colonialist knowledge and with the concomitant conversion of solidarity, as a way of knowing, into chaotic ignorance. In my view, the way out of this situation in a context of paradigmatic transition consists in reassessing knowledge-as-emancipation, granting it primacy over knowledge-as-regulation. This implies, on the one hand, that solidarity be turned into the hegemonic form of knowing, and, on the other, that a certain degree of chaos be taken in as a consequence of the relative negligence of knowledge-as-regulation.

THE REPRESENTATION OF LIMITS

In the study of the representation of limits in economics and in social sciences in general it will be helpful to consider the case of sciences which have faced the issues of both representation and limits most dramatically, either because of the nature of the objects they study, or because of the type of technical capabilities they have been designed to develop. I mean archeology, involved in the study of objects and behaviors that are very distant in time; astronomy, involved in the study of objects that are very distant in space; cartography, concerned with the representation of spaces through maps; and photography, concerned with representation as "reproduction." It will still be useful to consider an artistic activity, painting, which, at least since the Renaissance, has been haunted particularly by the question of representation.

Drawing freely on the procedures and strategies that these knowledges and practices have designed to overcome the dilemmas and fallacies of representation, I want to show, first, that such procedures, strategies, dilemmas and fallacies are at the core of modern scientific knowledge as a whole and, second, that within the range of alternatives made possible by such procedures and strategies, social sciences in general and mainstream economics in particular have chosen the alternatives least suited to promote solidarity as a form of knowing. Underlying my argument is the idea that such procedures and strategies are the meta-technologies that allow the scientist to produce recognizable and convincing knowledge and that such meta-technologies, which are internal to the scientific process, are as partisan as the technological interventions of science in social life. The key concepts in my analysis are the following: scale, perspective, resolution, and signature. All of them have been developed by the above-mentioned disciplines as they have confronted, in closest contact, the limits of representation and addressed the dilemmas emerging therefrom.

THE DETERMINATION OF RELEVANCE

The first limit of representation concerns the question: what is relevant? The question of relevance lies in that the relevance of a given object of analysis does not lie in the object itself but in the objectives of the analysis. Different objectives produce different criteria of relevance. If we would submit the choice of objectives to the open and potentially infinite scientific discussion that characterizes the analysis of scientific objects, we would never be able to establish a coherent criterion of relevance and carry out any intelligible scientific work. As long as we discuss objectives we cannot agree on objects. Since the discussion is potentially infinite, the only way to make science possible at all is to postulate the equivalence or fungibility of alternative objectives. It is, therefore, by denying or hiding the hierarchy of relevance among objectives that modern science establishes the hierarchy of relevance among objects. The distortion is thus imminent and indeed unavoidable. The established relevance is a sociological, or better, political economy fact disguised as an epistemological evidence. The invisibility of the disguise is premised upon the credibility of the distortion. The distortion is made credible by creating, in a systematic way, credible illusions of correspondence with whatever is to be analyzed. Two procedures are used to produce such illusions: *scale* and *perspective*.

We don't observe phenomena. We observe scales of phenomena. Though scales are important to all the disciplines I am drawing from, it is in cartography that scales are most central. Indeed, the main structural feature of maps is that in order to fulfill their function of representation and orientation they inevitably distort reality. Jorge Luis Borges told us the story of the emperor who ordered the production of an exact map of his empire. He insisted that the map should be exact to the most minute detail. The best cartographers of the time were engaged in this important project. Eventually, they produced the map and, in fact, it could not possibly be more exact, as it coincided point by point with the empire. However, to their frustration, it was not a very practical map, since it was of the same size as the empire (Borges 1974, 90).

To be practical, a map cannot coincide point by point with reality. However, the distortion of reality thus produced will not be considered inaccurate if the mechanisms by which the distortion of reality is accomplished are known and can be controlled. Maps distort reality through three specific mechanisms, which, since they are used systematically, become intrinsic or structural attributes of any map. Such mechanisms are: *scale*, *projection*, and *symbolization*. For the purposes of this chapter I limit myself to scales.[2]

Scale, as Monmonier has defined it, "is the ratio of distance on the map to the corresponding distance on the ground" (1981, 4). Scale involves, then, a decision on more or less detail. "Since large-scale maps represent less land on a given size sheet of paper than do small-scale maps, large-scale maps can present more detail" (1981, 4). Since maps are "a miniaturized version" of reality (Keates 1982, 73), mapmaking involves the filtering of details, "the selection of both meaningful details and relevant features"

(Monmonier 1981, 4). As Muehrcke puts it, "what makes a map so useful is its genius of omission. It is reality uncluttered, pared down to its essence, stripped of all but the essentials" (1986, 10).[3] One easily understands that the decision on scale conditions the decision on the use of the map, and vice versa: "Small-scale maps are not intended to permit accurate measurements of the width of roads, streams, etc., but rather to show with reasonable accuracy the relative positions of these and other features" (Monmonier 1981, 4).

Geography, which shares with cartography the concern for spaces and spatial relations, has also contributed important insights on scales, both scales of analysis and scales of action. As to the former, there are phenomena that can only be represented on a small scale, such as climate, while others, like erosion, for instance, can only be represented on a large scale. This means that the differences in scale are not simply quantitative but also qualitative. A given phenomenon can only be represented on a given scale. To change the scale implies to change the phenomenon. Each scale reveals a phenomenon and distorts or hides others.[4] As in nuclear physics, the scale creates the phenomenon. Some of the fallacious correlations in geography derive from the superimposition of phenomena created and analyzed on different scales. The scale is "a coherent forgetting" (Racine 1982, 126) that must be carried out coherently. Mediating between intention and action, scale applies also to social action. Urban planners as well as military chiefs, administrators, business executives, legislators, judges and lawyers define strategies on a small scale and decide day-to-day tactics on a large scale. Power represents social and physical reality on a scale chosen for its capacity to create phenomena that maximize the conditions for the reproduction of power. The distortion and concealment of reality is thus a presupposition of the exercise of power.

Different scales of analysis create different patterns of regulation and promote different action packages. As regards *regulation patterns*, it must be borne in mind that representation and orientation are two antagonistic modes of imagining and constituting reality, one geared to identifying position, the other geared to identifying movement. Large-scale regulation is rich in details and features; describes behavior and attitudes vividly; contextualizes them in their immediate surroundings; is sensitive to distinctions (and complex relations) between inside and outside, high and low, just and unjust. Large-scale regulation invites a pattern of regulation based on (and geared to) representation and position. On the contrary, small-scale regulation is poor in details and features; skeletonizes behavior and attitudes, and reduces them to general types of action; provides sense of direction and schemes for shortcuts. In sum, small-scale regulation favors a pattern of regulation based on (and geared to) orientation and movement.

Besides having different regulation patterns, different scales of analysis also condition different *action packages*. An action package is a connected sequence of actions structurally determined by predefined boundaries. I identify two kinds of boundaries: those defined by range and those defined by ethics. According to range, we can distinguish two ideal-types of action packages: the *tactical* and the *strategic action package*. According to ethics, we can also distinguish two ideal-types of action packages: the *edifying* and the *instrumental action package*. In the light of the previous examples, I would suggest that large-scale analysis and regulation invites tactical and edifying action packages, while small-scale analysis and regulation invites strategic and instrumental action packages. Social groups or classes that are predominantly socialized in one of these forms of analysis and representation tend to be specifically competent in the type of action package associated with it. In a situation in which large-scale and small-scale analysis and regulation intersect, the large-scale action package tends to be defensive and to regulate normal, routine interaction or, at the most, molecular struggles, while the small-scale action package tends to be aggressive and to regulate critical, exceptional situations, triggered by molar struggles. These tendencies may hold true irrespective of the class nature of the social groups involved in the specific action package.

Of all the social sciences, mainstream economics has been the most focused on orientation, with greatest involvement in science-based intervention in social life. For that reason it has favored small-scale analysis, of which mathematical modeling is the most characteristic illustration. Small-scale analysis has been as prevalent in macro as in microeconomics. As in the case of maps, small-scale analysis privileges a pattern of regulation geared to orientation and movement and an action package based on strategic and instrumental actions. The efficacy of the orientation is premised upon the vagueness of representation, that is, upon the negligence of details and contrasts, dismissal of submerged meanings and practices, disregard for the different durations of both declining qualities and emergent qualities. In such a mode of representation the uncertainty of the position is made irrelevant by the dynamics of the movement. Based on such a representation, the efficacy of the orientation depends upon one condition: it must be sustained by extrascientific political forces powerful enough to promote movement to cover for the social costs of the negligence of position.

This means that the preference for small-scale and thus for orientation over representation is an epistemological decision which, rather than sustaining itself, is grounded on a sociological, political economy fiat. The definition of the relevant features of a given course of action is determined by the regulation objectives, and not the other way around. Different objectives and thus different interests create different relevant facts.

This may be illustrated with the analysis of a given labor conflict in a factory producing for a transnational corporation through franchising or subcontracting. The factory code, that is, the production law of the workplace, as a form of local legality, regulates the relations in production in great detail, in order to maintain workplace discipline, prevent labor conflicts, reduce their scope whenever they occur, and eventually settle them. The labor conflict is the nuclear object of the factory code, because it confirms, *a contrario*, the continuity of the relations in production, which are the *raison d'être* of the factory code. In the wider context of national state labor law, the labor conflict is only a dimension, however important, of industrial relations. It is part of a broader network of social, political, and economic facts in which we easily identify, among others, political stability, inflation rate, income policy, and power relations among labor unions, business and government. In the still wider context of the transnational regulation of international franchising or subcontracting, the labor conflict becomes a minute detail in international economic relations, hardly worth mentioning.

Thus, the different regulatory orders operating on different scales translate the same social objects into different relevant objects. However, since in real life, the different regulatory scales do not exist in isolation but rather interact in different ways—in our example the regulatory purposes of the three scales converge in the same social event—there may be the illusion that the three regulatory objects overlap point by point. In fact, they do not coincide at all. Workers and sometimes the employer tend to have a large-scale view of the conflict, with all its details and relevant features, a concept molded by local regulation. Union leaders and sometimes the employer tend to see the conflict as a crisis in a process of continuous industrial relations. Their view is predominantly molded by national state regulation; consequently, their actions in the conflict aim at a compromise between the medium-scale and the large-scale view of the conflict. For the transnational corporation, the labor conflict is a tiny accident in a globally designed investment and production system; if not promptly overcome, it can be easily circumvented by moving the production to another country.

Transnational corporations favor the small-scale view of the conflict because this is the scale at which they organize their global operations. Together with the multilateral financial institutions they are the small-scale actors *par excellence*, covering vast parts of the globe and most drastically reducing the amount of detail or contrast as a condition of operational efficacy. Mainstream economics tends also to favor the small-scale view of the conflict. The fact that this view converges with the view of the transnational corporations is, in epistemological terms, a coincidence, and, in political economy terms, the coverup for a combination of interests. Mainstream economics creates the reality that maximizes the efficacy of the regulation it propounds.

THE DETERMINATION OF DEGREES OF RELEVANCE

Once relevance is established a further question must be asked: how relevant? In Western modernity and in modern science the degrees of relevance are established by another procedure operating in tandem with scale: *perspective*. Leon Battista Alberti (1404–1472) is considered the founder of one-point perspective in Renaissance painting, even though the mathematical laws of perspective were formulated for the first time by the Florentine architect Fillipo Brunelleschi (1377–1446). In his treatise *De Pictura* of 1435 Alberti compares the painted picture to an open window: "a picture, in his view, should be made to seem as if it were a pane of transparent glass through which we look into an imaginary space extending in depth" (Andrews 1995, 1). In order to achieve that, he devises a method for drawing a mathematically correct representation of space in which the relative size of objects at different distances and the apparent convergence of parallel lines will be as convincing to the eye in art as they are in nature (Gilman 1978, 17). As Gilman says, "[i]n the fifteenth and early sixteenth century . . . perspective arises out of and gives expression to a sense of certainty about man's place in the world and his ability to understand that world" (Gilman 1978, 29). The system of proportions between the objects to be painted and their images and between the distance of the observer's eye and the painting creates an intelligible world organized around the viewpoint of the spectator. The credibility of this "illusionistic" art (Gilman 1978, 23) lies in the mathematical precision of the individual's point of view. Renaissance's perspective is both a show of confidence in human knowledge and the artistic counterpart of individualism.

However, this precision and this confidence are obtained at a very high price: the absolute immobility of the eye. The illusion is real on the condition that the painting be seen from a predetermined and rigidly fixed point of view.[5] If the spectator changes his place the illusion of reality vanishes. Gilman is thus right when he says that "[t]he very fulness and definition of perspective space implies the radical incompleteness of our vision, and the point of view becomes a drastic limitation, a set of blinders, as well as an epistemological privilege" (Gilman 1978, 31).

The imaginative structure of proper perspective underlies, as I said, both modern art and modern science. It is also through perspective that degrees and proportions of scientific relevance can be established. There is, however, an important difference in the operation of perspective in art and in science. In modern art the painter conceives of the spectator as his radical other. The painter paints *for* the ideal spectator. The painter imagines the spectator's gaze in order to deceive it effectively. The painter is the only one with access to reality and both he and his spectator know that. The illusion of reality develops in tandem with the reality of the illusion. On the contrary, the modern scientist sees himself as the ideal spectator; he puts himself at the center of the privileged point of view to observe the reality fully revealed to his gaze. Even though he does other things besides being a mere spectator—otherwise no scientific work would get done—these other things, other than having the spectator in mind, are the product of the spectator's mind. In other words, they are the spectator at work. As the creator is absorbed by the spectator, the reality of the illusion is cannibalized by the illusion of reality and, as a consequence, the latter becomes the reality of reality. Accordingly, the modern scientist believes in the illusions he creates to an extent that the painter does not. Nor would the scientist be as comfortable with the epithet of "illusionistic science" to characterize his work as the painter was with that of "illusionistic art" to characterize his.

This conflation of the creator with the spectator in modern science had a crucial consequence. Because he always externalized the spectator, the painter could make a distinction between the ideal spectator, the one eye of the viewer, and the significant spectator, his patron or *mecenas*. On the contrary, the scientist could not make such a distinction, the ideal and the significant spectator being both himself. This made it impossible to ask for and to question the significant spectator for whom the scientist, as a creator, worked. The negative consequences of such unquestioning grew with the conversion of science into a productive force and thus with the growing impact or even interference of the significant spectator on the work of the scientist.

Of all social sciences, mainstream economics has been the one in which the choice among alternative significant spectators has been most drastically reduced to a single one, the capitalist entrepreneur. As the latter's impact on scientific work grew, the invisibility of the reality of illusion allowed for the illusion of reality to become the entrepreneur's reality. The latter's preferences and limitations, rather than being blinders, became epistemological privileges. Consequently, a political economy fiat could be credibly smuggled into the scientist's epistemological claims. The efficacy of the orientation made possible by small-scale analysis was reinforced by the monopolistic appropriation of the significant perspective.

THE DETERMINATION OF IDENTIFICATION

I have so far dealt with the first limit of representation, the determination of relevance. The second limit of representation deals with the question:

how to identify it? Once the relevant level of observation and analysis has been established, it is necessary to identify the relevant phenomena. Identification consists of two major démarches: *detection* and *recognition*. Detection has to do with the definition of the traits or features of a given phenomenon. Recognition consists in the definition of the parameters according to which the detected phenomena will be classified as a distinct element of a system of explanation or of a system of interpretation. The procedure underlying both detection and recognition is *resolution*.

Resolution refers to the quality and details of a given identified phenomenon, be it a social behavior or an image. Resolution is central to both photography, remote sensing technologies, and archeology. In photography, resolution or resolving power is the capability to image spatial detail. This capability may be referred either to the film or to the lens. The resolution of the film is determined by the size distribution of its silver halide grains (the larger the grains the poorer the resolution). The resolution of the lens is determined by its optical properties and size. The number of line-pairs per millimeter defines the level of resolution (Avery and Berlin 1992, 36). In remote sensing technologies the most important type of resolution for my purposes here is spatial resolution: "it is the measure of the smallest object that can be resolved by the sensor or the area on the ground represented by each pixel. The finer the resolution, the lower the number" (ERDAS 1997, 15).[6] In archeology the resolution refers to the homogeneity of events and behavior and their relation to the archeological record (Gamble 1989, 23).

There are many degrees of resolution but they are usually reduced to two: *coarse-grain* and *fine-grain resolution*. For instance, in photography high-speed films operate with minimal lighting conditions but only incorporate large-diameter grains and for that reason have a lower resolution than low speed films (Avery and Berlin 1992, 38). In archeology, a coarse-grained assemblage is one where at any one location the correspondence between an event and the archeological record it generated is poor; and, conversely, a fine-grained assemblage is one where the materials deposited reflect more precisely the activities that were carried out at those locations and in relation to the immediate environment (Gamble 1989, 23, 24). For my purposes here it is important to note that whenever a system of resolution is constituted by more than one component, the resolution level of the system is determined by the component with lowest resolution. For instance, in photography the resolution system is constituted by two components, the film and the lens. If the two don't have the same level of resolution, the resolution level of the photography will be determined by the lowest-rated component (Avery and Berlin 1992, 37).

In my view, resolution, just as scale and perspective, is at the core of modern science and operates at two different levels: the level of methodology and the level of theory. Both methods and theories are present in the scientific identification of objects to be analyzed; but methods predominate in the process of detection while theories predominate in the process of recognition. The quality of the scientific identification is thus determined by a system of resolution comprised of two components: methods and theories. It is commonly observed that the development of research methods has outpaced the development of theories particularly in the social sciences. For that reason it is not surprising that it is still common to go back to the nineteenth-century founding fathers to look for theoretical guidance, whereas the research methods and the data-gathering techniques we use today are extremely more sophisticated than those available in the nineteenth century. This means that the resolution level of our methods is higher than the resolution level of our theories and, consequently, that while the quality of scientific detection tends to be fine-grain, the quality of scientific recognition tends to be coarse-grain. In other words, our detection capabilities by far exceed our recognition capabilities.

Even though this discrepancy is inherent to all social sciences, for reasons that deserve to be elucidated, mainstream economics is the one in which the gap between detection resolution levels and recognition resolution levels is widest. And probably for the same reason, it is also the one in which the very existence of the gap has been most fiercely denied. As a result, because the level of resolution of identification is determined by the lowest-rated component, that is, by theory and thus by recognition resolution, mainstream economics operates and intervenes in social life in a coarse-grain mode but manages to legitimize its operation and intervention as if it were of fine-grain resolution quality.

The consequences of economics' interventions in society cannot but betray the excess of this claim. Among such consequences, the most negative can be designated as the *fallacy of exogeneity*. It consists in defining as relations among exogenous entities the internal transformation such entities undergo as their mutual endogeneity develops. Sam Bowles has recently exposed this fallacy in his analysis of market preferences (Bowles 1998). As Bowles emphasizes, mainstream economics has cherished, as one of its fundamental axioms, the axiom of the exogenous preferences, the celebrated minimalist conception of an undersocialized *homo economicus*, an individual actor with exclusively self-regarding and outcome-based preferences (Bowles 1998, 103). Against this vision he convincingly argues in favor of the endogeneity of preferences, that is, the extent to which the markets affect the preferences that are supposed to impact on it as external forces. In particular, he focuses on a group of preferences which he calls "nice traits"—"these are behaviors which in social interactions confer benefits on others" (Bowles 1998, 92)—and shows how the markets may block or discourage the development of such traits.

In my view, it is not surprising that the fallacy of exogeneity should occur most specifically in markets. Contacts in markets are ephemeral and impersonal. Given the high resolution of methods, mainstream economics can detect, as individual and separate, entities or factors that keep minimal distances among themselves. The meaning of such distances, that is, the understanding of what might be separating entities or, on the contrary, uniting them can only be provided by theory and recognition resolution, and since the latter is coarse-grain it is unable to discriminate contexts, networks, interpenetrations, embeddedness. This explains why the endogeneity of preferences does not emerge clearly and is accordingly discarded.

THE IMPOSSIBILITY OF DURATION

The third limit of representation blocking the road to unceremonial adequacy is the limit of time and time perception. Once relevance has been determined and the object identified, it is necessary to determine the object's temporal location. All objects exist in time-spaces and therefore neither their relevance nor their identification can be considered completed before the time-spaces are determined. This determination is most difficult because, while in scales, resolution, and perspective the distinction between subject and object operates unproblematically, in the determination of time-space the subject and the object both exist in time-space. To solve this difficulty, modern science has tried to neutralize differences by hypostacizing the most elusive frame: the *hic et nunc*, the here and now, presence and simultaneity. Modern perspective has made possible such simultaneity between subject and object, between painter and spectator. Through perspective, simultaneity is attained scientifically, since once the viewer is immobilized by the logic of the system, the space is totally unified. "[S]imultaneity in perceiving a picture, also requires a synchronization of what is represented; by grasping the picture spatially as a unit we also assume the depicted events to be simultaneous" (Andrews 1988, 35). Disregarding time differences is thus a condition of analytical confidence. However operational, this present-orientation and simultaneity are totally arbitrary and vulnerable to the *fallacy of false contemporaneity*. This fallacy consists in assuming that there is only one way of being contemporary and that it applies indifferently to all participants in a given event or behavior. When World Bank officials meet with African peasants (a more hypothetical than real occurrence) it is assumed that the simultaneity of the encounter determines that both groups are contemporary or coeavel in the same way. The fact that this is an unequal encounter in which two contemporaneities clash—the peasants' present reality is conceived by them as a past present and by the World Bank as a present past—however crucial, gets thereby obscured and goes uncontrolled. In this context there is no room to account for the different contemporaneities meshed in the simultaneous.

Of all social sciences, mainstream economics is the most prone to navigate in the fallacy of false contemporaneity. This is linked to the features of determination of relevance and identification characteristic of mainstream economics. Starting with relevance, the privilege granted to small-scale analysis means also that orientation and movement are privileged to the detriment of representation and position. The compression of time is thus particularly drastic; duration cannot be grasped and residues become indistinguishable from emergent qualities. To the extent that residues and emergences are still distinguishable, the orientation bias of small scale tends to be overzealous in the identification of obstacles to movements and consequently to exaggerate the identification of observed features as residues. While archeology excels in finding residues in order to explain the evolution of behavior patterns, mainstream economics excels in finding them and in discarding them as trash. It is ironic that much of what the archeologists of the twenty-second century will know about us will be revealed by the trash we left behind.[7] This should alert us for the situatedness of our findings and the relevance we ascribe to them. The epistemology of trash cannot be as easily discarded as the trash it refers to.

Turning now to the determination of the degrees of relevance, I would like to show how the use of perspective by mainstream economics prevents the identification of durations, rhythms, sequences, tempos, synchronies and dissynchronies. As I mentioned above, what is characteristic of mainstream economics in this regard is the monopolistic appropriation of the significant spectator by the capitalist entrepreneur. The dramatic intensification thus produced of the significant other, smuggled in as the self, has two main consequences: a hyperspatialization of past times and fast-speed interventions.

The lessons from archeology are particularly pertinent in this regard. The temporal construction of the archeological records can occur in two ways. The first, extremely rare, can be called the *Pompeii mode* (Binford 1981). It occurs whenever it is possible to determine rigorously the dates in which different events and objects enter simultaneously in the archeological record. Hiroshima will be the Pompeii of the archeologists of the future. The second mode, much more common, can be called the *palimpsest mode*. It refers to the situations in which the same archeological layers comprise objects and residues from very different periods and times not susceptible of exact dating.

The hyperspatialization of past time in mainstream economics consists in an inherent bias in favor of the Pompeii mode which, given its extreme rarity (e.g., global oil shock; a world war; a global financial crash, etc.), implies the systematic misrepresentation of social palimpsests as social Pompeiis.

This bias derives from the pressure made intense by the intensification of the significant spectator, to privilege clearly delimited, highly homogeneous and simultaneous findings.

The second consequence is fast-speed intervention. Highly spatialized simultaneous social fields call for fast-speed interventions, the ones that maximize the orientation and movement preferences of small scale. Fast-speed interventions, like fast-speed films, require very little exposure and can operate in virtually all conditions; but, also like fast-speed films, they have a very low resolution level; they are coarse-grain interventions. Their speed, together with the coarseness of their resolutions, makes such interventions both highly intrusive, highly fallible, and highly destructive. The Rapid Rural Appraisals by the World Bank economists throughout Africa and Asia are a good example of fast-speed interventions[8].

This type of interventions, which indeed, irrespective of the names they bear, are much more common than what we may imagine, symbolizes the destructive side of scientific research. Since the very beginning modern science has assumed a posture, which Schumpeter was to attribute later on to capitalism: the capacity for creative destruction. In epistemological terms, such posture resides in the very idea of scientific revolution conceived of as a radical break with and a departure from all previous knowledges. Bachelard has formulated it better than anyone else with his concept of *rupture epistemologique* (1972). By discarding all the alternative knowledges, modern science has revealed itself as a waste maker, a condition that we, the few privileged inhabitants of consumer society, share as well. This is, by the way, another dimension of the above-mentioned epistemology of trash and, indeed, another aspect of a political economy of waste making in modern science. Two questions must be asked in this regard: how much waste do we have to make in order to produce scientific consequences? Who suffers most with the pollution we thereby produce?

Of all the social sciences mainstream economics has been most involved in fast-speed intervention. For that reason it is most directly confronted with the dilemma, which I will call the *excavation dilemma*. Excavation is the core procedure of archeological research. It is through excavation that one has access to the archeological record. The excavation site is a well delimited area where the systematic search for residues deposited underground takes place, a search that when successful is the only way to identify behavioral patterns and adaptive strategies in our most ancient past. The dilemma, however, is that once the excavation is conducted and the residues are collected, the archeological work destroys the archeological site forever, making it impossible to start all over again: once taken out of the depositional formations in which they were integrated, the collected objects cannot be put back in. The dilemma resides therefore in that an eventual advancement in knowledge necessarily entails a definitive and irreversible destruction: the destruction of the relations among objects and, with it, the elimination of any possible alternative knowledge about them.

This dilemma has been fully acknowledged by archeologists and strategies around it have been designed. For instance, according to Sharer and Ashmore, "[S]ince the excavation process itself destroys an archeological site, it should be confined whenever possible to situations in which adequate planning, time and money are available to ensure the maximum useful knowledge about the past is recovered" (1987, 564). Similarly, Robert Dunnell considers that excavation "is expensive, destructive to the record and at best yields great detail about a few widely separate sites . . . Excavation, once the hallmark of archeology, will [in the next fifty years] be employed only when all other means of data acquisition have been exhausted" (1989, 65).[9]

In mainstream economics, on the contrary, this dilemma has never been acknowledged, in spite of the fact that it is dramatically present in most of the scientific interventions and, above all, in fast-speed interventions. As a consequence, and contrary to what happens in archeology, no alternative research strategies have been designed. The blindness vis-à-vis this dilemma increases the possibility that the creative destruction of mainstream economics becomes just destructive destruction.

THE DETERMINATION OF INTERPRETATION AND EVALUATION

The final limit of representation has to do with *interpretation* and *evaluation*. It is through interpretation and evaluation that our research objects are integrated in the wider contexts of politics and culture at which level science-based transformation occurs. Such integration is made possible by establishing links between social action and patterns of political and cultural formation. Because of the nature of the scientific object, archeology is probably the science in which establishing such link is the most central task. The term used by some archeologists to designate such link is *signature*. In my view this concept has heuristic capabilities far beyond archeology. In archeology, signature describes the link between behavior and distinctive patterns of residue formation (Gamble 1989, 22). Signature is thus about authorship, intelligibility, and purposefulness. This means that interpretation and evaluation is premised upon the knowledge of the agents involved (authorship), their knowledge practices (intelligibility), and their projects (purposefulness).

This is a domain in which the limits of representation already dealt with converge to make the signature of reality in social sciences in general and in mainstream economics in particular highly deficient. Concerning agents, the smaller the scale of analysis the stronger the emphasis on orientation and movement. The representation of agents tends to privilege those that move and need orientation, that is, *docile bodies*. The smallest the scale, the highest the docility of the bodies. The one-point perspective reinforces this effect. The immobility of the spectator's eye, which is particularly intense in mainstream economics, can only guarantee the illusion of reality to the extent that mathematical proportions are strictly kept. The represented bodies have to be kept in a cage, be it an iron cage or a rubber cage. Outside the cages there are no agents, whether friends or enemies. At the most, there are *strangers*, indifferent bodies. Docile bodies and strangers are thus the two possible categories of agents, hardly a fine- grain resolution of social agency.

The impact of perspective on the representation of knowledge practices is equally constraining. As Gilman reminds us, the intelligibility of the world made possible by Renaissance perspective was obtained at an exacting price, the immobility of the eye and the blinders necessary to create the single view (Gilman 1978, 31). This single view is what best characterizes modern science and its epistemological break both with common sense and all other alternative knowledges. The other side of the strength of the single view is its incapacity to recognize alternative views. Social practices are knowledge practices but they can only be recognized as such to the extent that they are the mirror image of scientific knowledge. Whatever knowledge does not fit the image, is discarded as a form of ignorance. The single view, rather than being a natural phenomenon is the *ur*-product of the creative destruction of modern science. The epistemological privilege that modern science grants to itself is thus the result of the destruction of all alternative knowledges that could eventually question such privilege. It is, in other words, a product of *epistemicide*. The destruction of knowledge is not an epistemological artifact without consequences. It involves the destruction of the social practices and the disqualification of the social agents that operate according to such knowledges. In mainstream economics the particular intensity of the significant spectator has imposed a specially arrogant single view, and, as a consequence, the epistemicide has been broader and deeper.

Finally, the purposefulness in social action, that is, the agents' projects, is the domain in which the scientific signature of reality is most deficient. Projects are an anticipation of reality and as such imply a distance from current experience. This anticipation and distance has a specific temporality, the temporality of a bridge among noncontemporaneous or differently contemporary courses of action through aspiration and desire. The fallacy of false contemporaneity analyzed above makes such bridge a useless device, thus turning aspiration into conformism and desire into the desire of conformism. Moreover, the type of coarse-grain identification characteristic of modern science creates, as I mentioned above, a bias in favor of the proliferation of residues to the detriment of emergent qualities, a condition that leads to disqualifying as retrospective all the emergent qualities that don't fit the qualities of the project legitimated by science. The narrower the project, the wider the retrospective.

The limits of signature, be it of authorship, intelligibility or purposefulness, are therefore strict and, of course, the possibilities of interpretation and evaluation cannot exceed them. The result is an imaginative structure consisting of docile bodies and strangers, victims of successive epistemicides, navigating in a sea of residues "swept along into the future that others have laid out for them" as the temporally poor described by Rifkin (1987, 166).

This signature of social practice is highly selective and therefore the link it establishes between agents and patterns of behavior is at best speculative. As I indicated, at each stage of the signature process many alternatives are left out: alternative types of agents other than docile bodies and strangers; alternative knowledges other than scientific knowledge; alternative projects other than the project of the significant spectator. Dealing with discarded alternatives means to deal with nonexistent entities. There are at least two ways in which nonexistent entities may "occur" and, accordingly, two ways of trashing alternatives. First, there are alternatives that never occurred because they were prevented from emerging. Second, there are alternatives that did occur but the types of scale, perspective, resolution, time compression, and signature used by science did not recognize them at all, or took them for residues. Only a *sociology of absences* will be able to elucidate us about the limits of representation at work in each situation. While in the first situation, the alternatives that didn't occur, we are dealing with silences and unpronounceable aspirations, in the second situation, the alternatives that did occur, we are dealing with silencings, epistemicides, trashing campaigns.

Possible alternatives are, in epistemological terms, the missing links, the incomplete records, black holes, voids. Modern science suffers in general from *horror vaccui* and whenever possible discards alternatives in order to eliminate epistemological disturbances. The objectivity and the rigor of scientific knowledge is indeed a by-product of *horror vaccui*. Mainstream economics is, of all the social sciences, the most haunted by *horror vaccui*. The specific way it has dealt with the limits of relevance, identification, duration and interpretation and evaluation makes *horror vaccui* look particularly threatening and destabilizing. On the other pole of the spectrum we could locate archeology, which, however sharing with all the other social sciences the same *horror vaccui*, takes a much more relaxed attitude vis-à-vis it, trying to domesticate it rather than eliminating it. Glenn Stone, for instance, speaks of negative evidence in these terms: "negative evidence is a form of data. 'Data' are taken to be observations made of archeological phenomena, as opposed to the phenomena themselves. . . ." "Negative evidence refers to the failure to observe a given phenomenon (or lacunae in data sets)" (1981, 42). And he proposes that the interpretation of such absences be an integral part of the archeological analysis.

The sociology of absences is a daunting task.[10] As we shall see, it requires an epistemology of absences. Only an epistemology of absences will prevent interpretation and evaluation from being based on very blurred, coarse-grain signatures of social life. Otherwise, rather than signatures, we end up with wandering names looking for docile bodies and strangers.

TOWARD AN EPISTEMOLOGY OF SEEING

An insight into the consequences of the epistemology of blindness is not, in itself, an insight into the epistemology of seeing. Therefore, I will start from the consequences of the epistemology of blindness and move later to delineate an epistemology of seeing.

The consequences of blindness manifest themselves as the misrepresentation of consequences. Such misrepresentation must be analyzed at two levels: capacity to regulate and capacity to emancipate. In general, modern science has represented the phenomena in ways that fit its regulatory imagination. In the case of mainstream economics, this seems to be particularly true. The specific social construction of agents, as both docile bodies and strangers, is, in fact, geared to making social regulation particularly easy. Docile bodies and strangers or indifferent people are about the easiest possible targets of social regulation. One can even say that the undersocialized *homo economicus* looks like a hero when compared with either docile bodies or strangers, the two versions of the oversocialized *homo sociologicus*. However, as I hope to have shown, the oversocialized *homo sociologicus* is not the opposite of the undersocialized *homo economicus*; it is rather its double. The *homo sociologicus* is the *homo economicus* in action.

The facility of regulation is merely apparent for the following two reasons, one having to do with agents and the other with actions. First, I have claimed that the tension between experience and expectations is one of the most distinctive characteristics of modern regulation. The agents constructed by mainstream social sciences and particularly by mainstream economics are incapable of living through that tension. Docile bodies experience but do not expect or, which is the same, their expectations mirror their experiences one to one. On the other hand, strangers are indifferent both to experience and expectations, they can live both separately and without any tension. In either case, the tension between experience and expectations is lost. Once this occurs, order, which is the point of knowing in the knowledge-as-regulation, conflates with colonialism, the point of ignorance in knowledge-as-emancipation. In other words, it becomes the colonialist order, the degree zero of social emancipation. At the degree zero of emancipation, however, modern regulation cannot sustain itself, since it is the tension between regulation and emancipation that keeps both alive and credible.

The facility of regulation is also only apparent for another reason: because of the types of social actions constructed by science. While modern regulation is based on the tension between experience and expectation, it is also based on the symmetry between action and consequences. Modern science has been entrusted with the task of producing and reproducing this symmetry. Indeed, what makes a given action scientific is the control it exerts over the consequences stemming from it.

It is today well established that this symmetry, if ever it existed at all, has vanished forever. Our common experience is rather that of a growing asymmetry between the scientific capacity to act, which has increased exponentially, and the scientific capacity to predict consequences, which at best has stagnated. Accordingly, the actual consequences of a given scientific action tend to be far less scientific than the action itself.

The notion that consequences are therefore excessive in relation to scientific action is probably the manifestation of another fallacy of exogeneity, the exogeneity between actions and consequences. Having in mind my previous analysis of the limits of representation particularly as regards mainstream economics, the picture of scientific action that emerges is one constructed by (1) very small scale determination of relevance combined with a single-view perspective in which the significant spectator carries a heavy weight; (2) a coarse-grain identification resolution based on an imbalance between detection methods and recognition theories; (3) a gross distortion of sequences and temporalities by imposing Pompeii premises on social palimpsests and false contemporaneity on noncontemporaneous social layers; (4) a poor capacity to decipher the signatures of social practices, both concerning agents but also concerning knowledge practices and projects. A scientific action thus constructed bears the imprints of the consequences, which the fallacy of exogeneity attributes then to external nonscientific causes. The "less-than-scientific" character of the consequences is inscribed in the very "scientific" character of the actions from which they derive. A scientific form of social regulation that cannot control the consequences of its operation cannot by any standard be considered a reasonable or reliable form of regulation.

Modern science has become the privileged form of knowledge-as-regulation, in spite of the fact that, as I have shown, social regulation cautioned by it is neither reliable nor sustainable. On the other hand, modern science has totally deserted the other possibility of knowledge inscribed on the paradigm of modernity: knowledge-as-emancipation. Mainstream economics is also in this case the extreme version of a syndrome that involves modern science as a whole. The solution that mainstream economics has given to the problems confronting the limits of representation converged, as we saw above, on a view of social reality fit to be regulated by a type of order close to colonialism, that is, a type of order that transforms the other into a manipulable and fungible object. This is, as I have suggested, the degree zero, the point of ignorance of knowledge-as-emancipation. In this form of knowledge, as we know, the point of knowing is solidarity, the recognition of the other as an equal and an equally producer of knowledge. The form of regulation that has come to prevail makes solidarity unthinkable, unnecessary or even dangerous. After all, docile bodies don't need solidarity and strangers don't deserve it. *Horror vaccui* has been operative in this regard also: if there are no other types of relevant agents, solidarity, rather than being a missing link, has no place in scientific discourse.

In a period of self-reflexivity, one may ask if the insight into the epistemology of blindness is not in itself a blind insight. Not necessarily, is my answer. On the other hand, one has to admit that that might indeed be the case. The potential for an epistemology of seeing lies in the above-mentioned tension, intrinsic to modernity, between knowledge-as-regulation and knowledge-as-emancipation. The latter, as I said, has been totally marginalized by modern science but has not vanished as a virtual alternative. In point of fact, its presence as absence is what makes possible the epistemology of blindness.

An epistemology of seeing is one that inquires into the validity of a form of knowledge whose point of ignorance is colonialism and whose point of knowing is solidarity. While in the hegemonic form of knowledge we know by creating order, the epistemology of seeing poses the question of whether it is possible to know by creating solidarity. Solidarity as a form of knowledge is the recognition of the other both as an equal whenever difference makes her or him inferior and as different whenever equality jeopardizes his or her identity. Having been oversocialized by a form of knowledge that knows by creating order in nature as well as in society, we cannot easily practice or even imagine a form of knowledge that knows by creating solidarity both in nature and in society. To overcome the difficulties I propose, as a prolegomena to this new form of knowledge, three epistemological démarches: the epistemology of absent knowledges; the epistemology of absent agents; revisiting representation and its limits.

The Epistemology of Absent Knowledges

While analyzing the limits of interpretation and evaluation in modern science above, I stressed that the sociology of absences is a crucial démarche to identifying the blinders that limit interpretation and evaluation. However, such sociology is not possible if not grounded on an epistemology of absences. In order to identify what is missing and why it is missing we must rely on a form of knowledge that does not reduce reality to what exists. I mean a form of knowledge that aspires to an expanded conception of realism that includes suppressed, silenced or marginalized realities, as well as emergent and imagined realities. Once again, in a self-reflexive turn, we may ask if the knowledge that identifies the absences is not the same that legitimated the conditions that suppressed the possibility of alternative realities now being identified as absences. My answer is twofold. First, we will not know it until the consequences of this knowledge are evaluated in terms of the solidarity capital they are able to create. Second, there will be always absences that will not be noted. These constitute the void, which, rather than being stigmatized by our *horror vaccui*, should be contemplated by our *captatio benevolentiae*.

The epistemology of absent knowledges starts from the premise that social practices are knowledge practices. The nonscience-based practices, rather than being ignorant practices, are practices of alternative, rival knowledges. To favor one form of knowledge against another must be grounded in criteria, which, as I said in chapter 1,must emerge out of argumentative, context-dependent, pragmatic and democratically generated audiences. Moreover, none of them in isolation can guarantee the emergence and flourishing of solidarity. The objective will be rather the formation of constellations of knowledges geared to create surplus solidarity. I call the formation of such constellations the ecology of knowledges[11]. The ecology of knowledges confronts the monoculture of scientific knowledge and rigor with the identification of other knowledges and criteria of rigor that operate credibly in social practices. Such contextual credibility must be deemed a sufficient condition for the knowledge in question to have enough legitimacy to participate in epistemological debates with other knowledges, namely with scientific knowledge.

The central idea of the epistemology of absent knowledges is that there is no ignorance or knowledge in general. All ignorance is ignorant of a certain knowledge, and all knowledge is the overcoming of a particular ignorance (Santos 1995, 25). This principle of incompleteness of all knowledges is the condition of the possibility of epistemological dialogue and debate among the different knowledges. What each knowledge contributes to such a dialogue is the way in which it leads a certain practice to overcome a certain ignorance. Confrontation and dialogue among knowledges is confrontation and dialogue among the different processes through which practices that are ignorant in different ways turn into practices that are knowledgeable in different ways.

This we may call a new common sense, the outline of which I presented in chapter 1. The epistemology of absent knowledges tries to rehabilitate common sense, for it recognizes in this form of knowledge some capacity to enrich our relationship with the world. As I showed in chapter 1, modern science built itself against common sense, which it deemed superficial, illusory and false. Common sense was the name given to all forms of knowledge that did not meet the epistemological criteria that modern science established for itself. The distinction between science and common sense was made possible by what I called the first epistemological break. It distinguishes between two forms of knowledge: truthful knowledge and false knowledge. However opposed, these two epistemic entities entail each other, since one does not exist without the other. They are indeed part of the same cultural constellation that in our time gives signs of closure and exhaustion. In sum, common sense is as modern as modern science itself. The distinction between science and common sense is thus made both by science and by common sense, but it has different meanings in each case. When made by science, it signifies the distinction between objective knowledge and mere opinion or prejudice. When made by common sense, it signifies the distinction between an incomprehensible and prodigious knowledge and an obvious and obviously useful knowledge. It is then far from being a symmetrical distinction. Besides, when made from the point of view of science, the distinction has a power that is excessive in relation to the knowledge that makes it possible. Like all specialized and institutionalized knowledge, science has the power to define situations beyond what it knows about them. That is why science can impose, as an absence of prejudice, the prejudice of pretending to have no prejudices.

I proposed in chapter 1 the concept of a *double epistemological break* as a way out of this stalemate. The objective of the double epistemological break is to further the production of knowledge-as-emancipation as an emancipatory common sense itself: beyond the conservative prejudice and the incomprehensible prodigy, I proposed a prudent knowledge for a decent life. The emancipatory common sense is a discriminating common sense (or unequally common, if you like), constructed so as to be appropriated in a privileged way by the oppressed, marginalized or excluded social groups, and actually strengthened by their emancipatory practice. This leads me to the second *démarche* toward an epistemology of seeing.

The Epistemology of the Absent Agents

As we saw above, mainstream social sciences and specially mainstream economics have reduced the variety and wealth of social agency to two types of individuals—docile bodies and strangers—neither of which is fit to sustain a social practice based on knowledge-as-emancipation. The monopoly of subjectivity that they have conquered explains why at the beginning of the twenty-first century the crisis of social regulation, rather than prompting the opportunity for a new surge of emancipatory ideas, forces and energies, feeds on the symmetrical crisis of social emancipation. Thus, social regulation does not have to be effective in order to flourish; it flourishes simply because individual subject is unable both to know and to desire how to know and to desire beyond regulation.

As a result, the invention of a new emancipatory common sense based on a constellation of knowledges oriented towards solidarity must be complemented by the invention of individual and collective subjectivities capable of basing their social practice on that constellation of knowledges and willing to do so. Otherwise, no matter how neatly elaborated, emancipatory knowledges will turn gradually and insidiously, like Escher's drawings, into regulatory knowledges.

The epistemology of absent agents is thus a quest for destabilizing subjectivities, subjectivities that rebel against conformist, routinized, repetitive social practices and are energized by experimenting with liminality, that is, with eccentric or marginal forms of sociability. Against a political economy of representation that proliferates residues, the epistemology of seeing proliferates emergent qualities grounded in different knowledge practices and let them compete in the social fields, thus converting them into fields of social experimentation. The epistemology of blindness has promoted a construction of social practice based on the distinction between structure and agency. The apparent equality between the two terms of the distinction has been used to transform structure into a more or less iron-cage determination of agency. The result is the mediocrity of either docile bodies or strangers. The epistemology of seeing, on the contrary, will promote a construction of social practice based on the distinction between conformist action and rebellious action, or better, *action-with-clinamen* with an explicit preference for the latter.[12]

The decentering of conformism and thus of docile bodies through rebellious action must be complemented by the decentering of indifference and the strangers it breeds. Though this may be controversial because it evokes Carl Schmitt's political theory, I think that against indifference, which is the hallmark of political liberalism, it is necessary to revive the dichotomy friend-foe. Probably the most dilemmatic difficulty confronting critical theory today lies in the blurring of the distinction between friend and enemy. Critical theory has always presupposed a question—which side are we on?—and the answer to it. It is not surprising that assorted kinds of neopositivists have managed to delegitimate this question by trashing the normative claims that underlie it. More surprising, however, is the condition of those, mostly among the youngest generation of social scientists, who, though they would like to answer the question and take sides, see, sometimes with anguish, the seemingly increasing difficulty in identifying alternative positions in relation to which it would be imperative to take sides. This difficulty can be sociologically explained by the increasing opaqueness of the enemy. Without enemies there is no need for friends. If there are no friends there is no purpose in exercising solidarity[13]. There is nothing authoritarian or antidemocratic in the dichotomy of friends and enemies, as long as the dichotomy is established by nonauthoritarian democratic means.

Revisiting the Limits of Representation

The limits of representation, which, as we saw, are particularly drastic in mainstream economics, derive their credibility from the scientific actions they make possible. As the epistemology of absences confronts these actions with their human consequences by appealing to alternative knowledges and agencies that might generate other actions and produce alternative consequences, the limits of representation of mainstream science lose their monopoly of representation and are forced into a discursive competition with other knowledges and alternative forms of representation. Whenever this competition breaks out, the convincing power of the arguments cannot be derived from logical principles but rather from pragmatic considerations, from the "last things" called for by W. James, that is, from the human consequences of alternative courses of action. Such competition, however, is not a competition about consequences. It is rather a competition about the linkages between consequences and the political economy of analytical procedures that may sustain them in real life. In this chapter I refer to the analytical procedures themselves but I am aware that their viability and credibility lie on the political economy of their linkages with consequences in social life.

The epistemology of absence, both of absent knowledges and absent agents, enables us to revisit the limits of representation in mainstream social sciences: the limits of the representation of relevance, identification, duration, and interpretation/evaluation. Seen from the perspective of the constellation of emancipatory knowledges propounded here, they lose their dilematic nature. I will limit myself to indicate, in a brief note, some of the possible ways they can be overcome.

Concerning the limits of relevance, I propose two démarches: the *transcale* and the *curious perspective*. Since different knowledges privilege different scales of phenomena, the constellation of knowledges I am proposing here suggests that we learn how to translate among different scales. The limits of a representation on a given scale become more visible when we compare that representation with a representation on a different scale. Transcaling is thus a démarche that permits us to contrast limits of representation with the purpose of elucidating what is at stake in the choice among alternative criteria of relevance.

Transcaling presupposes a certain unlearning of current criteria of relevance determination. It invites us to consult social reality through different cognitive maps operating at different scales. The learning process consists in raising the consciousness of the limits—contrasting representation with orientation, position with movement—without getting paralyzed. A higher consciousness of limits is at the core of the kind of prudent knowledge I am proposing here, a form of knowledge that teaches us how to keep consequences under the control and within the sight of the actions that cause them.

Curious perspective is the search for a different angle from which the proportions and hierarchies established by normal perspective are destabilized, and their claim of a natural, orderly and faithful representation of reality accordingly subverted. In the seventeenth century, the artists and art teachers began to criticize Alberti's proper perspective for being fully manifest and comprehensible. They then began to explore "how rules of perspective can magnify or diminish, multiply or distort the image" (Gilman 1978, 34). Their idea was that the illusion of reality was something not to take too seriously, rather to take as play and play with. According to Gilman, "the world implied in the writings of later perspectivists is shifting, multifaceted, and ambiguous" (1978, 34).

In my view, this curious perspective, both playful and unsettling, must be brought in in the determination of degrees of scientific relevance. The criteria of relevance based on a mathematical rigidly established perspective tend to be reified by the recurrent and unproblematic use. Reification means, in this context, the conversion of the illusion of reality into a compressed, faithful reproduction of reality. On the contrary, the curious perspective reconstitutes the creative processes at the core of modern sciences, a production of illusions that rather than imitating society reinvent it.

Concerning the limits of identification, the epistemology of seeing invites us to shift our priorities: from an excessive focus on what we already know too well, that is, methods-based detection, to a focus on what we know less and indeed are knowing less and less, that is theory-based recognition. Since this discrepancy is exclusive of modern science, the recourse to alternative knowledges will unsettle the resolution levels to which we are used. It is necessary to raise our demands to an ever finer-grain resolution only possible in the context of constellations or ecologies of knowledges.

In remote sensing photography resolution is highly dependent upon target contrast. "A high-contrast target is one in which there is a large density difference between bright and dark areas" (Avery and Berlin 1992, 37). The improvement of resolution level with which we analyze society may require the invention of highly contrasted social practices, even when the surface of such practices, as the earth in itself, is deceptively low contrast. The generation of high-contrast and of multi-contrasted resolution is made possible by the transcaling and curious perspectives, which are characteristic of cognitive processes inside constellations of knowledges.

Concerning the limits of the representation of duration, the procedures already indicated will help to see that social reality is a more or less sedimented terrain, a geological construct made of different regulations composing different layers, all of them in force together but never in a uniform fashion, all of them in the same moment but always as a momentary convergence of different temporal projections. Koselleck's conception of "the contemporaneity of the noncontemporaneous" (1985), which is derived from Heidegger and Gadamer, may be useful to capture the complexity and unevenness of social, political, legal, or epistemological copresence. In the context of an ecology of knowledges, the analytical potential of this conception is maximized, because it is made self-reflexive, complex, uneven, and open to sociological variation itself. Although, in general, all social sciences bring together in a given time-space different temporalities and spatialities, some social sciences—which we may call performative—emphasize the flat contemporaneity, that is to say, the uniqueness of the encounter, while others—which we may call self-reflexive—emphasize the complex contemporaneity, that is, how the roots of the different practices or entities brought together at given point in time signal the presence of different ways of being contemporaneous. Of all social sciences, mainstream economics is the most performative. It reproduces the forms of power and knowledge that best suit its horizons of expectations. Whatever is brought into the analytical field (issue, social groups, cognitive maps, normative orderings) is somehow pulled by the roots, so as to become coeval with whatever else is brought together into analysis. The momentary and pragmatic suspension of noncontemporaneity apparently favors the elimination of hierarchies among social temporalities, thereby enhancing the possibility of one temporality absorbing other competing temporalities.

Like transcale, curious perspective, and multi-contrasted resolution, *intertemporality*, made visible by the complex contemporaneity of the noncontemporaneous, turns the question of duration into one of the most complex ones. Probably for this reason it is the question that mainstream economics has most caricatured through the kind of the compression of time and the flattening out of sequences in which it excels.

Finally, concerning the limits of interpretation and evaluation, the epistemology of both absent knowledges and absent agents provides the key to transcend the limits by raising the consciousness of their number and resilience. The richer the parameters that define authorship, intelligibility, and purposefulness, the greater the need to submit narrowly defined technological applications of knowledge to political and ethical contestations. In the process we will move from a paradigm of technical application of science to a paradigm of edifying application of prudent knowledges, knowledges that transform research objects into solidary subjects and urge knowledge-based actions to navigate prudently within the sight of consequences.

Enlightened by both the epistemology of blindness and the epistemology of seeing, it is possible to envisage the emergence of a prudent knowledge for a decent life, a knowledge which, by knowing from colonialism to solidarity, recognizes the order that bounds experiences and expectations, actions and consequences, except when order is itself a form of colonialism. The ultimate aspiration is all too human, an aspiration, which I call *advanced normality*.

NOTES

1. I develop in detail this characterization of Western modernity in Santos (1995).

2. A detailed analysis of these cartographic mechanisms can be read in Santos (1995, 456–73; 2002, 417–38).

3. On the roles and limits of maps, see, among others, Monmonier1991,1993; Campbell 1993; MacEachren 1994.

4. According to Monmonier, "Perhaps the most enigmatic problem in cartography is the generalization to a much smaller scale of thematic data, such as land use, mapped at a larger scale" (1985,111).

5. John Ruskin (n.d., 328): "perspective can, therefore, only be quite right by being calculated for one fixed position of the eye of the observer; nor will it ever appear *deceptively* right unless seen precisely from the point it is calculated for."

6. On the use of remote sensing in cartography, see Monmonier(1985, 89–100). As happens with scale and perspective the determination of the type and level of resolution is both a technical and a political problem. Concerning the latter and just as an example, high-resolution remote sensing systems can collect sensitive environmental data that polluters would prefer be kept from an alert and apprehensive public (Monmonier 1985, 185).

7. On this topic, see Deagan (1989).

8. On the problem raised by the Rapid Rural Appraisals, see Chambers (1992), Richards (1995), and Sapsford and Singer (1998).

9. For a recent treatment of these methodological issues, see Meneses (1999).

10. I develop the idea of the sociology of absences in Santos 2004.

11. By ecology I mean the practice of assembling diversity by way of identifying and promoting sustainable interactions among heterogeneous partial entities.

12. On the concept of action-with-clinamen, see Santos (1998). In Democritus and Epicurus clinamen is the capacity of the atoms to swerve or deviate from predetermined trajectories thus invalidating deterministic conceptions of reality.

13. At its deepest roots, the crisis of the Welfare State lies much less on a largely manipulated fiscal crisis, than on the ideological inculcation of the vanishing friends and their replacement by a sea of strangers, at best indifferent, at worst potentially dangerous.

REFERENCES

Andrews, L. 1995. *Story and Space in Renaissance Art: The Rebirth of Continuous Narrative*. Cambridge: Cambridge University Press.

Avery, T. A., and G. L. Berlin. 1992. *Fundamentals of Remote Sensing and Airphoto Interpretation*. New York: Macmillan.

Bachelard, Gaston. 1972. *La formation de l'esprit scientifique*. Paris: J. Vrin.

Binford, L. R. 1981. "Behavioral Archaeology and the 'Pompeii Premise.'" *Journal of Anthropological Research* 37 (3):195–208.

Borges, Jorge Luis. 1974. *Obras completas*. Buenos Aires: Emecé.

Bowles, Samuel. 1998. "Endogenous Preferences: The Cultural Consequences of Markets and other Economic Institutions." *Journal of Economic Literature* 36:75–111.

Campbell, John. 1993. *Map Use and Analysis*, 2nd ed. Dubuque, IA: Wm. C. Brown Publishers.

Chambers, R. 1992. "Rural Appraisal. Rapid, Relaxed and Participatory." *IDS Discussion Papers* 311:1–90.

Deagan, Kathleen. 1989. "Tracing the Waste Makers." Special issue, *Archeology* 42 (1):56–61.

Dunnel, Robert. 1989. "Hope for an Endangered Science." Special Issue, *Archeology* 42 (1):63–66.

ERDAS. 1997. *ERDAS Field Guide*. Atlanta: Erdas International.

Gamble, C. 1989. *The Paleolithic Settlement of Europe*. Cambridge: Cambridge University Press.

Gilman, E. B. 1978. *The Curious Perspective: Literary and Pictorial Wit in the Seventeenth Century*. New York: Yale University Press.

Keates, J. S. 1982. *Understanding Maps*. London: Longman.

Koselleck, Reinhart. 1985. *Futures Past. On the Semantics of Historical Time*. Cambridge, MA: MIT Press.

Lapouge. M. G. 1897. "The Fundamental Laws of Anthropo-sociology." *Journal of Political Economy*.

MacEachren, Alan. 1994. *Some Truth with Maps: A Primer on Symbolization and Design*. Washington, DC: Association of American Geographers.

Meneses, Maria Paula G. 1999. "New Methodological Approaches to the Study of the Acheulean From Southern Mozambique." PhD Diss., Rutgers University.

Monmonier, Mark. 1981. *Maps. Distortion and Meaning*. Washington, DC: Association of American Geographers.

———. 1985. *Technological Transition in Cartography*. Madison: The University of Wisconsin Press.

———. 1991. *How to Lie with Maps*. Chicago: Chicago University Press.

———. 1993. *Mapping It Out: Expository Cartography for the Humanities and Social Sciences*. Chicago: Chicago University Press.

Muehrcke, P. C. 1986. *Map Use*, 2nd ed. Madison, WI: J. P. Publications.

Nietzsche, Friedrich. 1971. "Rhétorique et langage." *Poétique II*, 5:99–144.

Racine, J. B., et al. 1982. "Escala e acção. Contribuições para uma interpretação do mecanismo de escala na prática da Geografia." *Revista Brasileira de Geografia* 45.

Richards, P. 1995. "Participatory Rural Appraisal: A Quick and Dirt Critique." *PLA Notes* 24:13–16.

Rifkin, Jeremy. 1987. *Time Wars: The Primary Conflict in Human History*. New York: Simon & Schuster, 1987.

Ruskin, John. n.d. *St. Mark's Rest. Lectures on Art, Opening the Crystal Palace, Elements of Perspective*. New York.

Santos, Boaventura de Sousa. 1995. *Toward a New Common Sense: Law, Science, and Politics in the Paradigmatic Transition*. New York: Routledge.

———. 1998. "The Fall of the *Angelus Novus:* Beyond the Modern Game of Roots and Options." *Current Sociology* 46 (2):81–118.

———. 2002. *Toward a New Legal Common Sense: Law, Globalization and Emancipation*. London: Butterworths.

———. 2004. "*A Critique of the Lazy Reason: Against the Waste of Experience.*" In Wallerstein, Immanuel, ed. 2004 The modern World-System in the kong buree. New York: Paradigm Publishers.

Sapsford, D., and H. Singer. 1998. "The IMF, the World Bank and Commodity Prices: A Case of Shifting Sands?" *World Development* 26 (9):1653–60.

Sharer, R. J., and W. Ashmore. 1987. *Archaeology: Discovering our Past*. Palo Alto, CA: Mayfield Publishing.

Stone, G. D. 1981. "The Interpretation of Negative Evidence in Archaeology". *Atlal (University of Arizona, Department of Anthropology), Occasional Papers* 2:41–53.

Veblen, T. 1898. "Why Is Economics Not an Evolutionary Science?" *The Quarterly Journal of Economics* 12:56–81.

Wallerstein, Immanuel, ed. 2004. *The Modern World-System in the* Long Durée. New York: Paradigm Publishers.

Wittgenstein, L. 1973. *Tractatus Logico-Philosphicus*. Frankfurt: Suhrkamp.

About the Contributors

Orlando Fals Borda is professor emeritus at the National University of Colombia, where he was also dean of the Faculty of Sociology for many years. He was elected deputy of the Colombian National Constituent Assembly, and served also a vice-minister of agriculture in Colombia. He is the founder of the Foundation of Participatory Research in Bogotá, in 1970. He is a recipient of Guggenheim, Kreisky, and Hoffman Awards.

Orlando Fals Borda is widely recognized as one of the leading theorists and practitioners of participatory action research. Among his publications, (coeditor with Muhammad Anisur Rahman) *Action and Knowledge: Breaking the Monopoly with PAR* (1991); (editor) *People's Participation: Challenges Ahead* (1998); *The Origins and Challenges of Participatory Action Research* (1999); *Acción y espacio: autonomías en la nueva república* (2000); *Por qué el socialismo ahora?: Retos para la Izquierda Democrática* (2003); and *Ante la crisis del pais: ideas-acción para el cambio* (2003).

Arturo Escobar is professor of anthropology at the University of North Carolina at Chapel Hill, where he is director of the Institute of Latin American Studies. He is also a research associate of the Instituto Colombiano de Antropología e Historia, Bogotá.

He has published widely on subjects related to anthropology, Latin American studies, political ecology, development studies, cultural studies of science and technology, political economy, theories of complexity, and the North-South relations. Among his many publications are: *Encountering Development: The Making and Unmaking of the Third World* (1995); (coeditor with Gustavo Lins Ribeiro) *World Anthropologies: Disciplinary Transformations within Systems of Power* (2006); (coeditor with Jai Sen, Anita Anand, and

Peter Waterman) *The World Social Forum: Challenging Empires* (2004); (coeditor with Sonia Alvarez and Evelina Dagnino) *Cultures of Politics/Politics of Cultures: Revisioning Latin American Social Movements* (1998).

Olival Freire Jr. is professor of history of physics at the Federal University of Bahia, Brazil. His research focuses on the history of physics in the twentieth century, by exploring the changes in the quantum controversy and the consensus and dissension in physics. In 2004 he was awarded a Dibner Senior Fellowship in History of Science.

Olival Freire Jr. has published in journals such as *Historical Studies in the Physical and Biological Sciences* and *Studies in History and Philosophy of Modern Physics*. Among his publications are: *David Bohm e a Controvérsia dos Quanta* (1999); *Science and Exile: David Bohm, the Cold War, and a New Interpretation of Quantum Mechanics* (2005).

Joan Fujimura is professor of sociology and former director of the Holtz Center for Science and Technology Studies at the University of Wisconsin–Madison. She has written on developments in genetics, molecular biology, biotechnology, biomedicine, HIV-AIDS research, and transnational bioscience in Japan and the United States. Her current research includes studies of race, medicine, and genetics, and of systems biology research. Joan Fujimura's recent publications comprise "'Sex Genes': A Critical Socio-Material Approach to the Politics and Molecular Genetics of Sex Determination" (*Signs*, 2006) and "Postgenomic Futures: Translations across the Machine-Nature Border in Systems Biology" (*New Genetics and Society*, 2005). She is also author of *Crafting Science: A Sociohistory of the Quest for the Genetics of Cancer* (1996).

Leonard Kaplan is professor of law at the University of Wisconsin–Madison. His research interests include several areas of jurisprudence, legal process, law, theology and state, civil procedure, criminal law, criminal procedure and evidence, as well as religious studies. In 2002, Professor Kaplan received the Academy's Michael Zeegers Lifetime Achievement Award for distinction in the pursuit of Scholarship, Pedagogy, and Human Rights Initiatives in the field of Law and Mental Health. Leonard Kaplan has coedited six books and written a large number of articles. He was a co-founder and coeditor-in-chief of a journal, *Graven Images: Studies in Culture, Law and the Sacred*. Also, and together with Beverly Moran he edited *Aftermath: The Clinton Impeachment and the Presidency in the Age of Political Spectacle* (2001).

Hugh Lacey is emeritus professor at Swarthmore College, Department of Philosophy. Previously, he also taught at the University of São Paulo

(Brazil), University of Sydney, and University of Melbourne (Australia). His research interests include history and philosophy of science, philosophy of psychology, philosophy of social science, philosophy and biotechnology. In his research, Lacey has sought to explore the role of values in scientific inquiry by challenging the predominant view from the perspectives of postmodernists, feminists, radical-ecologists, religious fundamentalists, etc. Hugh Lacey has published many articles on science education; his recent books include: *Is Science Value Free? Values and Scientific Understanding* (1999); *Psicologia Experimental e Natureza Humana: Ensaios de Filosofia de Psicologia* (2001); and *Values and Objectivity in Science: The Current Controversy about Transgenic Crops* (2005).

Richard Lee is professor of sociology at Binghamton University and Director of the Fernand Braudel Center. He supervises a research-working group on the "Cultural Forms of the World-System." Richard Lee's research agenda focuses on the long-term intellectual and disciplinary structures of knowledge formation in writings that range across the sciences, social sciences, and humanities. His most recent publications include: *Life and Times of Cultural Studies: The Politics and Transformation of the Structures of Knowledge* (2003); and the collections *World-Systems Analysis: Contemporary Research and Directions* (edited with Gerhard Preyer, 2004) and *Overcoming the Two Cultures: Science versus the Humanities in the Modern World-System* (coedited with Immanuel Wallerstein, 2004).

Francisco Louçã is professor of economics at the Higher Institute of Economics and Management of the Technical University of Lisbon. He is also a member of the Portuguese Parliament. In 1999 he was awarded the prize of the History of Economics Association for the best article published in an international scientific journal. Francisco Louçã has published widely, both in subjects related to the contemporary political situation of Portugal and Europe, as well as on economics. His research focuses on how the economy evolves through time, in order to offer a critical analysis of global capitalist development and design possible post-capitalist alternatives. Among his publications are: *Turbulence in Economics* (1997); *Is Economics an Evolutionary Science?* (with Mark Perlman, 2000); *As Time Goes By: From the Industrial Revolutions to the Information Revolution* (with Chris Freeman, 2001); and *Ensaio geral—passado e futuro do 25 de Abril* (edited with Fernando Rosas, 2004).

Maria Paula Meneses is a senior researcher at CES and also teaches at the Faculty of Economics of Coimbra University. Until 2002 she was an assistant professor of anthropology at Eduardo Mondlane University, Mozambique. Currently her research interests include identity processes, the production and reproduction of knowledges, postcolonial theory, epistemology, articulations between religion and politics, alternative mechanism of conflict resolution. As a scholar working on issues related to knowledge production, Maria Paula Meneses has been involved in research as well as in science policy making, all of them with a strong emphasis on Africa. Among her publications, coeditor with Boaventura de Sousa Santos and João Carlos Trindade are: *Law and Justice in a Multicultural Society: The Case of Mozambique* (2006); and "'When there are no problems, we are healthy, no bad luck, nothing: Towards an emancipatory understanding of health and medicine" (in Boaventura de Sousa Santos, editor, *Another Knowledge Is Possible: Beyond Northern Epistemologies*, 2007).

Walter Mignolo is professor of literature and romance studies at Duke University. He is also the academic director of Duke in the Andes, an interdisciplinary program in Latin American and Andean Studies in Quito, Ecuador at Pontificia Universidad Católica del Ecuador and the Universidad Politécnica Salesiana. Since 2000, he has directed the Center for Global Studies and the Humanities, a research unit within the John Hope Franklin Center for International and Interdisciplinary Studies at Duke University. Walter Mignolo has published widely on textual and literary theories, global coloniality, critical cosmopolitanism, and modern/colonial world system. His most recent books include: *The Darker Side of the Renaissance: Literacy, Territoriality and Colonization* (1995); *Local Histories/Global Designs: Coloniality, Subaltern Knowledges and Border Thinking* (2000); and *The Idea of Latin America* (2006).

Luis E. Mora-Osejo was, until his death, professor emeritus of biology and director of the Institute of Natural Studies at the National University of Colombia. He also served as rector of the University of Nariño, and as president of the Colombian Academy of Exact, Physical, and Natural Sciences. Luis Mora-Osejo was a recipient of the Guggenheim Award. Luis Mora-Osejo had many scientific publications mostly dedicated to his research field, botany.

João Arriscado Nunes is associate professor of sociology at the School of Economics of the University of Coimbra, and a researcher at the Centre for Social Studies of the same university. He served on the Council of the European Association for the Study of Science and Technology (EASST). His main research interests are in science and technology studies, political sociology, and social and cultural theory. João Arriscado Nunes has published extensively on biomedical research in Portugal; the globalization of science; the regulation of human reproduction and genetics; science, technology and culture; the politics of public participation in science and technology. He has coedited *Enteados de Galileu: a semiperiferia no sistema mundial da* *ciência* (2001); *Reinventing Democracy: Grassroots Movements in Portugal* (2006), and has just completed a book on oncobiological research in Portugal (forthcoming).

Anna Carolina Regner is professor of philosophy at the Federal University of Rio Grande do Sul, Brazil. Currently she is a visiting professor at Stanford, in the Program in History and Philosophy of Science. Anna Carolina Regner's research has been mainly focused on the epistemological, metaphysical, and methodological aspects of the Darwinian theory of Natural Selection as these aspects are approached from a broader contextual and historical point of view. She has written several books on Darwin in Brazil as well as a number of other articles on related topics. In the book *Charles Darwin—Notas de viagem: a tessitura social no pensamento de um naturalista* (1988), she analyzes the contribution of the passages on social and political phenomena in Darwin's diary on the Beagle for the later theory of natural selection.

João Ramalho-Santos is a reproductive biologist at the Center for Neuroscience and Cell Biology of the University of Coimbra (Portugal), where he also holds an associate professorship in biochemistry, cell, and developmental biology. He has been a visiting scientist at the Oregon Health & Science University (Portland, OR) and the University of Pittsburgh (PA), and has published over fifty original manuscripts in specialized journals, including *Nature, Nature Medicine, Developmental Biology, Human Reproduction and Biology of Reproduction.* Recently he held an associate editor position in *Human Reproduction* and currently performs the same duties for *Human Reproduction Update.* In the spare time he doesn't have, he studies sequential art and graphic novels.

Boaventura de Sousa Santos is professor of sociology at the School of Economics, University of Coimbra (Portugal) and Distinguished Legal Scholar at the University of Wisconsin–Madison Law School. Boaventura de Sousa Santos has published widely on globalization, sociology of law and the state, epistemology, democracy, and human rights. Among his most recent books are: *Towards a New Legal Common Sense: Law, Globalization, and Emancipation* (2002); *The Rise of the Global Left: The World Social Forum and Beyond* (2006); *A gramática do tempo: para uma nova cultura política* (2006); (editor) *Democratizing Democracy: Beyond the Liberal Canon* (2005); (editor) *Another Production Is Possible: Beyond the Capitalist Canon* (2006); and (editor) *Another Knowledge Is Possible: Beyond Northern Epistemologies* (2007).

Isabelle Stengers is professor of the philosophy of science at the Free University of Brussels whose work is recognized as one of the most important in the field of philosophy of science today. In 1993 she received the grand prize for philosophy from the French Academy. Isabelle Stengers has written extensively on philosophy of science, more specifically on the argument of authority used by modern science. Among her most important works are: *Power and Invention: Situating Science* (1997); *Cosmopolitiques* (in seven volumes); *La guerre des sciences aura-t-elle lieu?* (2001); *La sorcellerie capitaliste* (with Philippe Pignarre, 2005); and *La Vierge et le neutrino. Quel avenir pour les sciences?* (2006).

Frank Summers is a diplomat in clinical psychology of the American Board of Professional Psychology, and associate professor of clinical psychiatry and the behavioral sciences at Northwestern Medical School. He is also training and supervising analyst at the Chicago Institute for Psychoanalysis. Frank Summers has published extensively on object relations theories and their application to the process of psychoanalytic therapy. His books and papers in professional journals elucidate his theory that psychoanalytic therapy is a process of self creation in which the therapist plays the dual role of understanding current patterns and facilitating the creation of new ways of being and relating. Among his many publications are: *Object Relations Theories and Psychopathology: A Comprehensive Text* (1994); and two clinical monographs explicating his theory of psychoanalytic therapy: *Transcending the Self: An Object Relations Model of Psychoanalytic Therapy* (1999) and *Self Creation: Psychoanalytic Therapy and the Art of the Possible* (2005).

Peter Taylor, biologist by training, is professor in the Program in Critical and Creative Thinking, at the University of Massachusetts, where he is chair of the Department of Curriculum and Instruction of the Graduate College of Education. His research interests deal with the interconnections between science, technology, and society; more specifically, his focus is on critical thinking and reflective practice, especially about social analysis of ecological change and concepts and methods used in analyses of biosocial development. Peter Taylor has published widely, both books and articles in specialized journals such as *Biological Theory, International Journal of Epidemiology, Science as Culture*, and *Biology and Philosophy*. Among his publications are: *Unruly Complexity: Ecology, Interpretation, Engagement* (2005); and *Changing Life: Genomes, Ecologies, Bodies, Commodities* (edited with Saul E. Halfon and Paul N. Edwards, 1997).

Stephen Toulmin is professor at the Center for Multiethnic and Transnational Studies of the University of Southern California School of International Relations. He was educated at Cambridge University where he received his Ph.D. in philosophy ("An examination of the Place of Reason in Ethics"). Stephen Toulmin has held teaching positions in many countries and served as a fellow for the American Academy of Arts and Sciences, Honorable Doctor of Technology at the Royal Institute of Technology in Stockholm, Sweden, as well as an appointed position with the Government of Austria. Stephen Toulmin is famous for his research on the practices of reasoning, by linking the theory of knowledge to the historical development of concepts and practices. His most influential publications include: *The Return to Cosmology: Postmodern Science and the Theology of Nature* (1985); *Cosmopolis: The Hidden Agenda of Modernity* (1990); and *Return to Reason* (2001).

Shiv Visvanathan is professor at the Dhirubhai Ambani Institute of Information and Communication Technology in Ahmedabad, India. Currently, he is a visiting scholar at the Consortium for Science, Policy, and Outcomes, University of Arizona. His research interests center on cultures of knowledge, sociology and philosophy of science, history of technology, social movements, globalization, culture, and the politics of environmentalism. He is presently working on history of biotechnology in India. Shiv Visvanathan has a large number of publications, and he is a regular contributor to the national newspapers in India. His books include: *Organizing for Science: The Making of an Industrial Research Laboratory* (1985); *A Carnival for Science: Essays on Science, Technology and Development* (1997); and (editor) *Foul Play: Chronicles of Corruption 1947–1997* (1998).

Peter Wagner is professor of social and political theory at the European University Institute and professor of sociology at the University of Warwick. Peter Wagner has carried out extensive research on the sociology of modernity with an emphasis on Europe. His most recent publications on social and political philosophy and historical and political sociology include: *Theorizing Modernity: Inescapability and Attainability in Social Theory* (2001); *A History and Theory of the Social Sciences* (2001); *The Economy as Polity: The Political Constitution of Contemporary Capitalism* (2005); and (editor) *The Languages of Civil Society* (2006).

Immanuel Wallerstein is senior research scholar at Yale Sociology Department. Until 2005 he was the director of the Fernand Braudel Center for the Study of Economies, Historical Systems, and Civilization at Binghamton University. Immanuel Wallerstein's scholarly career can be divided into three overlapping subjects: Africa and colonialism/nationalism; the modern world-system; social science and the structures of knowledge. Indeed, his research on the characteristics of modern capitalism aimed at understanding the history and development of the modern world gave rise to the well-known theory of world-system analysis. He has written many books and articles on geopolitics and sociology. His most recent works include:

The End of the World as We Know It: Social Science for the Twenty-First Century (2001); *The Decline of American Power: The U.S. in a Chaotic World* (2003); *The Uncertainties of Knowledge* (2004); *World-System Analysis: An Introduction* (2004); and *European Universalism: The Rhetoric of Power* (2006).